W0107241

SYMBOLIC COMPUTATION

Artificial Intelligence

Springer Series
SYMBOLIC COMPUTATION - *Artificial Intelligence*

N. J. Nilsson: Principles of Artificial Intelligence. XV, 476 pages, 139 figs., 1982

J. H. Siekmann, G. Wrightson (Eds.): Automation of Reasoning 1. Classical Papers on Computational Logic 1957-1966. XXII, 525 pages, 1983.

J. H. Siekmann, G. Wrightson (Eds.): Automation of Reasoning 2. Classical Papers on Computational Logic 1967-1970. XXII, 638 pages, 1983.

L. Bolc (Ed.): The Design of Interpreters, Compilers, and Editors for Augmented Transition Networks. XI, 214 pages, 72 figs., 1983.

R. S. Michalski, J. G. Carbonell, T. M. Mitchell (Eds.): Machine Learning. An Artificial Intelligence Approach. XI, 572 pages, 1984.

L. Bolc (Ed.): Natural Language Communication with Pictorial Information Systems. VII, 327 pages, 67 figs., 1984.

J. W. Lloyd: Foundations of Logic Programming. X, 124 pages, 1984.

A. Bundy (Ed.): Catalogue of Artificial Intelligence Tools. XXV, 150 pages, 1984. Second, revised edition, IV, 168 pages, 1986.

M. M. Botvinnik: Computers in Chess. Solving Inexact Search Problems. With contributions by A. I. Reznitsky, B. M. Stilman, M. A. Tsfasman, A. D. Yudin. Translated from the Russian by A. A. Brown. XIV, 158 pages, 48 figs., 1984.

C. Blume, W. Jakob: Programming Languages for Industrial Robots. XIII, 376, 145 figs., 1986.

Christian Blume Wilfried Jakob

Programming Languages for Industrial Robots

Springer-Verlag
Berlin Heidelberg NewYork
London Paris Tokyo

Authors

Christian Blume
Gerhart-Hauptmann-Straße 5
D-7505 Ettlingen

Wilfried Jakob
Taunusstraße 20
D-1000 Berlin 41

Translator

Klaus Selke
Department of Electronic Engineering
University of Hull, Cottingham Road
Hull HU6 7RX, United Kingdom

Originally published under the title "Programmiersprachen für Industrieroboter" by Vogel-Verlag, Würzburg (Federal Republic of Germany), © 1983 by Vogel-Verlag, Würzburg
Translation, completely revised and extended, by Springer-Verlag 1986

ISBN-13:978-3-642-82747-1 e-ISBN-13:978-3-642-82745-7
DOI: 10.1007/978-3-642-82745-7

Library of Congress Cataloging-in-Publication Data. Blume, Christian. Programming languages for industrial robots. (Symbolic computation. Artifical intelligence) Rev. translation of: Programmiersprachen für Industrieroboter. Bibliography: p. Includes index. 1. Robots, Industrial. 2. Programming languages (Electronic computers) I. Jakob, Wilfried. II. Title. III. Series. TS 191.8.B5613 1986 670.42'7 86-26000
ISBN-13:978-3-642-82747-1 (U.S.)

2145/3140-543210

Preface

Previous works on industrial robots dealt with "programming" and "programming languages" only in passing; no comparison was made between characteristics of the individual programming languages.

This book, therefore, gives a detailed account of industrial robot programming and its environment. After introducing basic concepts special attention is paid to the language constructs relevant to robot programming. The features of various elements of the languages examined are compared. The languages are based on the following concepts:

SRL	-	high-level programming language based on AL with PASCAL elements (University of Karlsruhe, F. R. G.)
PASRO	-	integrated into PASCAL, based on the geometrical data types of SRL (I. I.-BIOMATIC Informatics Institute, Freiburg, F. R. G.)
AL	-	derived from the high-level programming language ALGOL (Stanford University, U. S. A., and University of Karlsruhe, F. R. G.)
AML	-	high-level programming language, influenced by PL/1 (IBM, U. S. A.)
VAL	-	language specifically developed for robots (Unimation, U. S. A.)
HELP	-	mixture of high-level language elements and robot language elements and real-time processing (DEA, Italy)
SIGLA	-	a simple machine language (Olivetti, Italy)
ROBEX	-	based on NC programming (Technical College (RWTH), Aachen, F. R. G.)
RAIL	-	high-level programming language for industrial robots with elements for graphic processing (Automatix, U. S. A.)
IRDATA	-	general software interface between programming and robot controller (Association of German Engineers (VDI), F. R. G.)

Elements of individual languages are given as examples in the discussion on general concepts - where the relevant language relates to the problem discussed. Our knowhow ist based partly on own experience or implementations (SRL, PASRO, AL, VAL, IRDATA) and partly on company data. Since company data does not always give the necessary detailed information, it is possible that certain elements are not discussed although they are part of that language.

This book, however, is not intended as an aid in learning any of the languages introduced, but rather to give a better understanding of the special problems and techniques of robot programming and to instil in the programmer's mind the need for exact and methodical work. Designers and users of robot systems are usually familiar with the peculiarities of robot programming but not with the systematic construction and formulation of algorithms. On the other hand, the computer scientist is conversant with the latter but not with the particular problems and requirements of robot technology. The book is, therefore, meant as an interface between the various disciplines so that both - robot technologist and computer scientist - can communicate at the same level using the same terminology.

Since we are dealing with languages, the presentation of computer science problems takes up more space than the basic robot technology necessary for using industrial robots. However, we start out by explaining important robot-technology terms for the computer scientist. When discussing data in particular, the special requirements of movement programming are presented and new data types are derived from the geometrical presentation.

The German edition of this book was first published in 1983 and was really intended as a manuscript for a lecture on "Programming of Robots" by author Christian Blume. This edition has been completely revised and now also includes the languages SRL, PASRO, and AML as well as the IRDATA software interface and elements for implicit programming such as RODABAS robot database. The reason for this is that the systems discussed have either undergone further technological development or will have an impact on the market, or both.

We are grateful to the following companies and institutes for documentation and information given: To IBM in particular for the demonstration of AML/E in Munich. We thank the German office of Unimation for the detailed explanations of VAL. To DEA and Olivetti go our thanks for kindly letting us have descriptions of the HELP and SIGLA languages. We thank the Machine Tool Laboratory (WZL) of the Technical College Aachen for so readily answering all our intricate questions regarding the ROBEX language developed there. In connection with instructions to a visual system we introduce an extract of RAIL; we are grateful to Automation for sending us the relevant documentation.

We thank all colleagues of the IRDATA working group of the VDI for closely cooperating and allowing us to publish the results.

In connection with the Karlsruhe AL implementation we are grateful for the kind support provided by Stanford University and particularly by Dr. Tom Binford.

For the close cooperation during the PASRO implementation we thank I. I.-BIOMATIC Informatics Institute GmbH and their manager, G. R. Koch.

We are also greatly obliged to K. K. W. Selke, electronics engineer at the University of Hull, U. K., for his translation and for accommodating all our requests for alterations.

And last but not least, we are very much indebted to the Head of the Department of Process Computer Technique at the University of Karlsruhe, Prof. Dr.-Ing. U. Rembold, and PSI GmbH, Berlin, for their kind support of our work. Our thanks go to all colleagues and friends who helped us with their advice and their recommendations.

C. Blume
W. Jakob

Table of Contents

1 Introduction

By now industrial robots are used in many production and manufacturing processes such as handling, welding, or spray-painting. Some of these jobs will continue to be manageable in the future with simple programming methods, e.g., adjustment or teach-in methods. For assembly and more complex handling operations, however, more highly developed programming methods have become necessary. Therefore, work is in progress, on an international level, to develop new programming languages and systems for industrial robots. The emphasis is on integrating sensors into the software of the robot controller, on including data from a CAD system, and on making available programming tools which are fairly simple - in the eyes of the user - but nonetheless provide for flexible programming. This can only be achieved by using industrial robot programming languages which have - contrary to the teach-in methods - the advantages of auto-documentation, communicativeness, being open to corrections, and offering offline programming. The languages discussed here meet the requirements in various ways and more or less completely. The system layouts, their use and the implementation of the individual languages differ greatly and sometimes can hardly be compared.

SRL (Structured Robot Language), with the IRDATA interface (Industrial Robot DATA), was defined by the authors and implemented at the University of Karlsruhe. The procedure and hardware used are described in Chap. 8.

The teaching and demonstration system PASRO (PAScal for RObots) was designed by author Christian Blume for BIOMATIK and has been implemented on a number of computers and under various operating systems and for two demonstration robots; see Chap. 8 as well. PASRO includes PASCAL completely so that for PASCAL no separate comparison is made as in the first German edition. PASRO could be implemented with the aid of MODULA, ADA or (especially for DEC computers) with MicroPower/Pascal extending at the same time it by multi-tasking elements, i.e., the parallel running of programs or subprograms.

AL (Assembly Language) was already developed by Stanford University, in the 1970s and implemented on a computer and robot configuration not suitable for industrial applications. Under a technology transfer in the early 1980s a new implementation was undertaken by the University of Karlsruhe and a number of improvements were made. The basic Karlsruhe implementation was carried out in PASCAL on a PDP 11/34 and was used to program a PUMA 500 robot by Unimation. AL has served as basis for a number of high-level robot programming languages in the United States, Europe, and Japan.

As nearly as in the 1970s, IBM announced a very advanced implicit programming system, AUTOPASS (AUTOmated PArts Assembly System). However, it is not known to what extent this has also been implemented. In the 1980s IBM introduced their own robots, which can be programmed with AML (A Manufacturing

Language), a high-level programming language. AML has a few peculiarities not usual in a robot language and not included in any of the languages discussed in this book. AML is formula-oriented, i.e., each instruction produces a value which may be reused. AML instructions, therefore, may be combined with each other in almost any way (orthogonality), and although this provides for flexible programming it has inherent risks because it permits a vague programming style. AML may also be used interpretatively, which is somewhat unusual for a language with PASCAL-like block structure and strings. On the other hand, the syntax for movement instructions and other commands not used for program structuring are kept very simple, which is the reason why hardly any AML syntax diagrams have been included. AML comprises more than 140 functions, and the demarcation between specific applications subprograms and those of the "system core" are fairly flexible.

There is a simple version of AML available, AML/E (Entry), which has been implemented on an IBM PC. It permits a teach-in by using terminals and a simple graphic presentation of the programmed position in a two-dimensional display of the IBM SCARA robot's working space (see also Chap. 8).

HELP is used by Messrs. DEA (Digital Electronic Automation) to program the PRAGMA robot. On the one hand it comprises robot and specific system functions while, on the other, it also has elements of high level-languages and real-time programming. The PRAGMA robot moves in Cartesian coordination, which means the programmer's positioning instructions can be directly processed by the controller. It is, nevertheless, surprising that no provision is made for data to indicate position and orientation of the robot. A new version is currently under development, and the hardware has also been modified.

Stanford University's close proximity and personal relationships have influenced the development of Unimation's VAL (Variable Assembly Language), a language which initially was determined by the robot's functions. As a consequence the first version of this language had highly sophisticated concepts like frame variables but was severely handicapped by missing enquiry facilities for coordinate values. A new version, VAL-II, consists of VAL-I - with minor limitations - which has been described in the first edition of this book. The present edition includes references to VAL-II elements but no syntax diagrams.

Olivetti's SIGLA (SIGMA LAnguage) is for programming two-armed SIGMA portal robots. These robots are driven by stepping motors, the values for which must be explicitly stated by the programmer. Since the syntax of SIGLA consists only of the statement of a two-lettered keyword and the enumeration of parameters, it is counted among the simple programming languages. Nonetheless, SIGLA can do more than solve simple problems since it is capable of parallel processing of program parts (tasks) and sensor enquiries, although SIGLA programs are not very clear or easy to read.

Besides SRL, ROBEX (ROBot EXapt) of Aachen Technical College is also a German development. It is based on EXAPT (EXtended APT), which is a technological extension of APT (Automatically Programmed Tools). Originally, ROBEX was implemented with severe limitations as far as "standard" EDP is concerned, e.g., lack of variables and arithmetics.

The new version for microcomputers, ROBEX-M, has partly overcome these handicaps, although the restrictive syntax (e.g., 6 characters for a keyword) has been maintained and the actual NC parts for geometric path calculation are still missing. Current development trends bring ROBEX and SRL closer together since SRL, in turn, is being extended to an implicit system.

In connection with programming algorithms for picture processing on the control computer, Messrs. Automatix have developed RAIL. RAIL is a high-level programming language with structuring facilities similar to those of PASCAL; it is not covered in detail.

Although the IRDATA code (Industrial Robot Data) was designed as an interface between programming and controlling - and not as a new language - it is discussed here. The elements of the robot programming language are transferred into IRDATA code by the language converter and transmitted to the control unit. IRDATA is a logical interface, i.e., control functions and their parameters are stated but not the hardware interface or the transfer protocol. The main advantage of IRDATA is that in future the control unit of a robot may be linked with almost any programming system which can generate IRDATA code.

The comparison of the above languages includes detailed tables as far as this is at all possible with the variety of language elements. For quick reference we conclude with a summary of the most important features of the languages in Appendix L.

Although most of the languages introduced here have no rules for formatting a program text, for reasons of clarity the examples will be indented in accordance with the program structure and reserved words and predefined identifiers will be written in uppercase letters.

2 Fundamentals

In this chapter, basic terms and concepts of robot technology and computer sciencewill be explained. These also include terms already familiar to programmers, such as variable, stack and subroutine, so that engineers with backgrounds in automatic control technology or manufacturing may find an easier introduction to the subject. Also, an explanation of these fundamental concepts is very useful in helping to understand different robot languages.

2.1 Terminology

Since most of the terminology introduced here is fundamental, it will not be formally defined, but just described verbally and explained by simple examples.

Unfortunately, there is no exact definition for the most important term **industrial robot**. Even the name industrial robot, chosen to distinguish it from the robot of science fiction literature, is not universally valid. Terms like *robot, manipulator* or *universal transfer device* are used, and one author even calls it a *robot manipulator.* The VDI (Verein Deutscher Ingenieure, the association of German engineers) has defined an industrial robot as follows (VDI Entwurf 2860 Gründruck):

Industrial robots are universally applicable devices with several axes of motion, which may be freely programmed (i.e., altered without mechanical interference) with respect to their sequence of motion and angles or positions and may be guided by sensors - if applicable. They are equipped with grippers, tools or other means of production and are able to perform manipulation and/or production tasks.

Hence it is important to be able to program the device to perform movements in three-dimensional space with motors actuating axes of motion. Programming without mechanical interference is taken for granted saying in computer science, but there are some simple *pick-and-place devices* which are programmed by repositioning mechanical endstops. An industrial robot may be equipped with sensors (effectively as peripherals). **Sensors** are devices which are able to register physical signals like light waves, pressure or joint positions, and can convert these to electrical signals (analogue or digital) and pass them on to the computer controlling the robot. This sensory information may then be analyzed and the program flow modified accordingly. At the end of the robot, a *gripper, tool* or other *means of production* is attached; these are collectively called **effectors**. These effectors also have to be controlled and may be equipped with sensors themselves. Figure 2.1 shows three typical industrial robots.

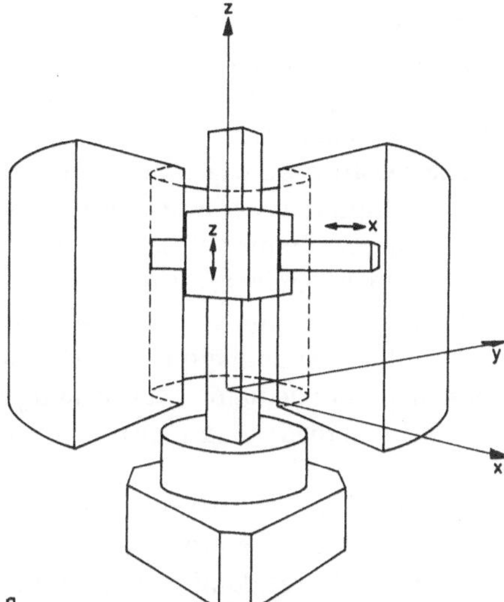

Fig. 2.1 a-c: Working envelopes of industrial robots.
a Working envelope of a robot with two prismatic and one revolute joint.

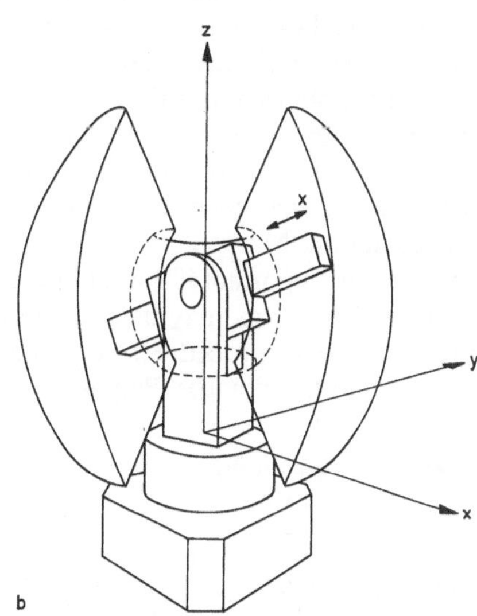

b Working envelope of a robot with one prismatic and two revolute joint.

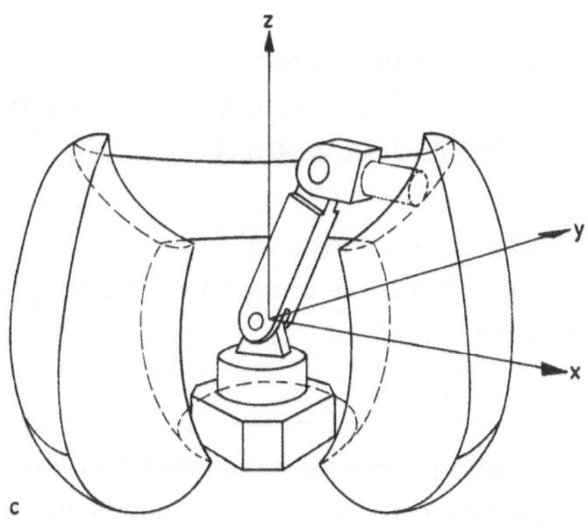

c Working envelope of a robot with three revolute joints.

A robot may perform a variety of activities, because of its mechanical construction and its actuators. These may include moves with or without sensory guidance, approaching grippers or tools with or without sensory monitoring, or modification of the program flow.

The construction which determines the exterior form of an industrial robot is extremely important as its *dimensions* and *geometry* determine the *work space* and hence the possibilities of moving to any position and orientating the gripper arbitrarily, which requires at least six degrees of freedom.

A **degree of freedom** is an independent axis of motion whose movement cannot be realized by a combination of other axes. An *axis of motion* is implemented by a combination of actuator, gearing and joint of the industrial robot.

Individual **robot joints** may execute translations (prismatic joints) or rotations (revolute joints). The move parameters are distances or angles, respectively. Direct programming of complex movements using these joint angles and/or translations is very cumbersome for a robot with several axes because the programmer cannot easily envisage the resultant motion of the effector.

Thus, motion is specified in **Cartesian coordinates**, which are much easier to comprehend for humans, and then translated into joint angles or distances. This is the object of **coordinate transformations**, as well as the calculation of Cartesian coordinates from values of position sensors in the joints. All problems concerning geometry and different kinds of motion are analyzed in the field of **kinematics**. Forces causing the motion and moving masses are neglected. Usually this is sufficient, but a *dynamic model* may be necessary which takes all forces acting upon the robot into consideration. Moments of inertia, gravitation, centrifugal and coriolis forces are particularly important if the robot is to follow a given trajectory precisely.

Industrial robots may be programmed either textually or without any explicit program text as procedures containing only **playback sequences** do not need any programming languages. Playback sequences are programmed as follows:

Manual programming
(arranging endstops)

Setting procedures
(robot is moved to desired position and orientation without its own actuators)

Sequence programming
(robot is moved by its own actuators, but guided by the operator who "shows" the robot the path of motion)

Master-slave programming
(robot is remotely controlled by an operator guiding a smaller master. Its motions are then transferred to the slave).

Here, the robot is programmed by being driven to the individual points, and the generated code usually consists of coordinates of points on the path. Only the **teach-in procedure** enables specifications beyond the pure motion programming, e.g.,

- speed of the move
- duration of motion
- programmed delays
- simple program loops
- simple program branching on interrupts (sensory signals)
- special functions (like type of control, single step mode).

The programmer needs to activate appropriate keys to initiate desired functions. In order to specify numerical parameters like duration, he enters a sequence of numerical digits for the constant. Teach-in already contains a minimal amount of text. Each function may be entered by a special key, so that there is no principal difference in teach-in and textual programming, in which operations and data are entered symbolically in the form of character strings. Symbols are the **vocabulary** of a programming language; the rules of grammar specifying the allowed combination of the vocabulary are contained in the **syntax**. Such a syntax for programming languages is restricted and will be defined formally (Sect. 2.3). The meaning of the symbols or their combination, i.e., the **semantics**, is neglected. A further point to consider is the practical organization in a language, i.e., the **pragmatics**. This may be illustrated by an example. Let the vocabulary of a simple language consist of the following words

V = (car, rose, is, large, red, fast)

and the syntax of the rules

1. Three elements of the vocabulary form a sentence.
2. The first element of a sentence is either "car" or "rose".
3. The second element is "is".
4. The third element of a sentence is one of the symbols "large", "red", or "fast".

The six possible sentences in this language are

Car is large
Car is red
Car is fast
Rose is large
Rose is red
Rose is fast.

The meaning of the sentences is intuitively obvious because the symbols are taken from the English language, and hence the semantics are known. Thus the sentence "rose is fast" is identified to be wrong, because no rose can be associated with the characteristic fast. So, the fourth rule of the grammar may be modified as:

4. The third element of a sentence is one of the symbols
   a) "large", "red", or "fast" if the first element is "car";
   b) "large" or "red" if the first element is "rose".

This example also illustrates that restrictions in the semantics may be expressed as syntax rules and thus the border between semantics and syntax is fluid. The modified rule 4 shows, in addition, that a sentence may not be chosen out of context, but that the choice of the third element depends on the context so far, or more precisely on the first element.

In the example above, the terms **character** and **symbol** were used, and although they are known, they shall be defined precisely here. Hence a character is understood to be (according to DIN 44300): "*An element of a defined finite set of (differing) elements. This set is called a* **character set**."

Examples for these sets are the set of school marks or the set of ASCII codes, which is very important in data processing. The sign "↷" may signify *turning clockwise* or *circular up and down moves in the air.* A character has a meaning; both together form a **symbol**. Symbols obviously depend on the language used. "↷" may just as well signify *to turn right.* A symbol may consist of several characters; it is then called a **word symbol**. There are many word symbols in programming languages, e.g., BEGIN, GOTO or MOVE. On the other hand, one character may have several meanings and hence represent several symbols. For example, the character "*" in the language SRL may have the following significance:

- scalar multiplication:	FACTOR * LENGTH
- vector stretching	5 * V
- vector cross product	V1 * V2
- vector rotation	R * V
- vector transformation	F * V
- chaining of two rotations	R1 * R2
- frame transformation	F * F

The programmer has to be aware of this, since programming systems recognize meanings of symbols in the context they are written, i.e., the type of operands. The operands belong to the **data objects** of a program. If they are defined with relationships among them and they are not just isolated, a **data structure** has been formed. A list of a linear sequence of equal objects is a simple example for a data structure. The relation consists of the pointer to the next element (except for the last one), which is realized as an address in a memory location.

A further basic term of mathematics and computer science is **algorithm**. This usually signifies a general method or procedure for solving a class of problems. Algorithms may be described in different ways and in great detail. Computer scientists describe algorithms in programming languages, in which language constructs represent a (finite) set of processing steps. Applications of algorithms then correspond to the execution of programs by computers. The following restrictions have to be observed:

1. The program consists of a finite number of programming steps which are limited by the storage capacity and which describe the algorithm completely.
2. The effect of each processing step is defined uniquely, as is the execution of the next step after the current one.

3. Each processing step lasts for a finite amount of time. Unfortunately, this does not mean that complete algorithms or programs always come to an end. For example, a (usually unintentional) endless loop may occur. As well as that there are some commands in which the execution time is infinite, like waiting for a signal which will never be set.

The term **task** is also of central importance. It contains a program section (or a complete program) which executes a relatively limited function and which is regarded as the smallest unit by the supervisory flow control.

Task examples are user programs or input/output routines. Strictly speaking, one should refer to them as *sequential* tasks because a linear time sequence of processing has been tacitly assumed. However, several processes may be run in *parallel* if more than one processor is available, or *quasi parallel* if only one processor is available. The kernel (*operating system*) divides the processor time into arbitrary units and assigns them to different tasks. Apart from sequencing of input/output, further functions of the *operating system* include assignment of hardware (peripherals, files, memory) to processes as well as file organization and reactions to special conditions. Last but not least, interrupts have to be considered. **Interrupts** cause a hardware break in the program flow, store the current status of the program, and execute a previously defined routine. Interrupts may be initiated from any device, sensor, or from safety switches on a robot. The operating system offers certain functions for the run-time system, which may also contain an interpreter, robot controller and an interactive part for the teach-in. This is implemented on the robot control computer. After specification and analysis of a problem, implementations contain the design of algorithms for solutions, programming of instruction successions describing that algorithm, and testing of the executable code on a computer. Apart from the run-time system, which executes user programs, a **programming system** is necessary, usually running on a background computer. Programming and run-time systems are dealt with in greater detail in Chap. 8.

2.2 Concepts in Computer Science

In connection with the development of programming languages, various concepts have evolved which form the basis for many high-level languages. Most of all, these concepts concern themselves with the organization and administration of program data as well as constructs for complex program flow. Hence structured programming is supported, making the program transparent to the programmer. Although only the robot languages SRL, AL, PASRO - and with restrictions also AML and HELP - realize concepts introduced here, these standards continue to influence further developments in programming systems for robots. For example, the language ROBEX has been developed from an NC machine background and thus does not contain variables. All machine moves are calculated during compilation and passed on as constants. A modification is developed, however, so that basic concepts of computer sciences can be incorporated.

2.2.1 Variables

A basic introduction of the **variable concept** follows, since many mechanical and automatic-control engineers are not familiar with it. They are considered from a programmer's viewpoint and with respect to the organization of memory during program execution.

2.2.1.1 Treatment of Variables from a Programmer's Viewpoint

During programming of industrial robots, variables are necessary, among other things, for *processing sensory information.* They permit storage of sensor values and their evaluation at various points in the program. Additionally, such *arithmetic operations* may be carried out as may be necessary in calculating new positions. A variable is represented in the program by a **name**, where the choice of the name is arbitrary, like ALFA or BIG. In general, a type specification is associated with a variable. The type has been identified in the **declaration** by the programmer. Afterwards, the programmer can refer to the variable at any place in the program where a reserved location is needed to which a value can be assigned or whose current value can be processed.

```
VAR switch   : BOOLEAN;
    distance : vector;
    avalue   ,
    spec     : INTEGER;
      .
      .
      .
avalue := 5;
READ (spec);
avalue := avalue + spec;
```

Program section 2.1: Declaration examples in PASRO

Program section 2.1 declares the variable `switch` to be of type BOOLEAN and the variable `distance` to be of type VECTOR. After the variables `avalue` and `spec` have been declared to be of type INTEGER, `avalue` is assigned the value 5. The content of variable `spec` is read from a peripheral (e.g., terminal) and `avalue` is recalculated. Its old value of 5 is incremented by the value of `spec`, which could only be known at execution time.

2.2.1.2 Treatment of Variables During Program Execution

A variable is realized during the execution of a program by a *memory address* (location) and *memory contents* (value). The assignment

```
avalue := 5;
```

causes the value 5 to be written into the memory location with the address of avalue. (The assignment "avalue‹-›address" is dealt with later.) The instruction

```
READ(spec);
```

reads a value from a terminal and stores it in the memory location of the variable spec. The type specification INTEGER informs the system to convert the number without a decimal point into its binary representation.

As an example of this: Let the above program section be executed and let the user enter 12 in the READ instruction. Thus the memory has the following contents:

variable avalue	with address 1020	5
variable spec	with address 1080	12

before the execution of the instruction

```
avalue := spec + avalue;
```

afterwards, the memory changes to:

variable avalue	with address 1020	17
variable spec	with address 1080	12.

2.2.1.3 Evaluation of Variable Adresses

The compiler evaluates *relative addresses* of variables during the translation of user programs. For this purpose it generates a *symbol table*, in which it enters names and relative addresses of variables. The relative address is evaluated by a counter which is cleared at the start and incremented after each entry into the symbol table by the amount of storage needed for that variable.

Example:

```
VAR high,p1:    INTEGER;
amount,interest: REAL;
```

The compiler evaluates the relative addresses as

name	relative address
high	0
p1	2
amount	4
interest	8

The assumption is made that an INTEGER variable needs two bytes and a REAL four bytes; also, the addressing is done in bytes.

Of course, the compiler could also generate *absolute addresses*, if so desired. However, this may lead to difficulties if several programs are present or if memory is allocated dynamically. Thus, absolute addresses are evaluated only at *run-time* by adding the relative address to a start address of the data section.

2.2.1.4 Representation of Variable Adresses

As already described, the compiler calculates (usually relative) addresses with the help of symbol tables and declarations. Apart from these addresses, specification of the *nesting depth of blocks* is also necessary in **block-orientated languages**, so that a complete address specification consists of the pair

‹block nesting depth›,‹relative address› .

The nesting depth gives information about the number of *blocks* in which the variable is embedded. This is essential for dynamic storage allocation (Sect. 2.2.3).

An example:

```
MOVE (55.3, 27, 10.5, 0, 70.8, 15)
```

The significance of the real values is not of interest here, but it is explained in Sect. 2.3.1 and 4.2.2.

The code generated has the structure

5000	positioning
1200	MOVE command
0	constant identification
1	number of points
55.3	
27.0	values of
10.5	motion
0.0	point
70.8	
15.0	

If P1 is a variable of type FRAME, which consists of exactly those six values to describe the orientation of a target point, then the statement from above may be specified as

```
MOVE P1
```

and the code generated is

5000	positioning
1200	MOVE command
1	variable identification
1	number of points
0	block nesting depth
56	relative address of variable P1

The interpreter can now find the values of the frame variable P1 under the relative address 56. Without using variables, the code generated in the first case uses 16 bytes, whereas the use of a variable reduces it to six.

2.2.2 Principle of Stack Operation

The use of stacks has proven itself in many applications, such as passing parameters, reserving memory space for variables or calculating arithmetic or logical expressions. A **stack** is an area of memory into which values are written by the **last-in-first-out** (LIFO) procedure. This is described in the following:

First, a pointer is set at the start of the stack area. If some data are to be stored on the stack, the stack pointer is decremented first and the data entered as the "lowest" element. Access can only be gained to the "lowest" element by reading the data and then incrementing the stack pointer by one. It now refers to the previous element on the stack. Data entry is called *push* and data removal *pop.* A stack usually "grows" from a high to a low address. There is no reason, in principle, why it could not go grow into the other direction, i.e., incrementing the stack on data entry.

2.2.3 Block Structuring

Many programming languages which belong to the ALGOL family in the widest sense contain **block structuring** capabilities. The programmer divides his program into relatively self-contained units, which may be marked by BEGIN and END keywords. Each block may contain data *declarations* and *instructions* as well as other blocks (block nesting).

2.2.3.1 Scope and Life Span of Variables

The segmenting of programs into blocks has no direct influence on their performance, except in procedures and functions. However, variables and constants declared in a block are only accessible within this block and blocks contained in it, but not outside the one they are declared in. This is called the **scope** of a variable.

In Fig. 2.2, the variable b is a *local variable,* while the variables a and v are *global* within block 2. If the names of any variables inside a block are the same as the ones in the outer block, the inner variables override the outer ones, which are no longer accessible inside. The variable a declared in block 1 has a **life span** lasting throughout block 1, but its scope it interrupted by block 2 (see Fig. 2.3). The life span of v,

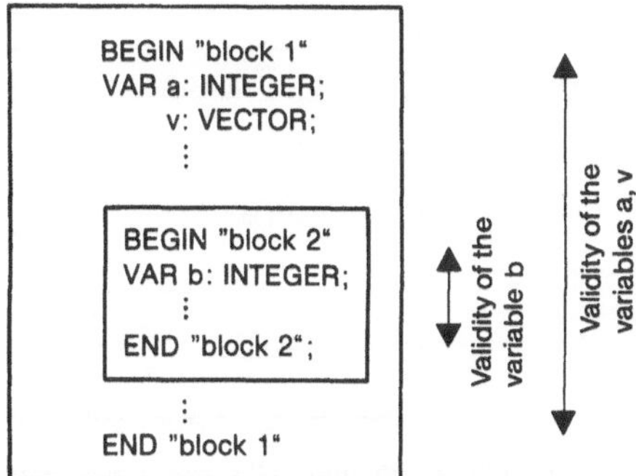

Fig. 2.2: Scope of variables

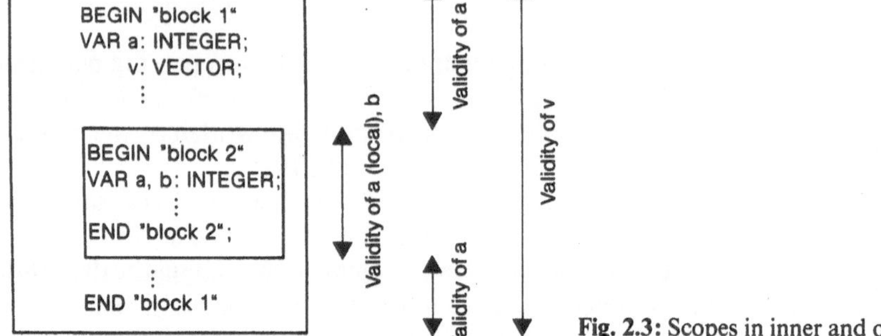

Fig. 2.3: Scopes in inner and outer blocks

however, covers the whole of block 1, whereas variables **a** (local) and **b** in block 2 cover only block 2.

Apart from the symbols BEGIN and END, blocks may also be defined by procedure declarations, functions or task definitions.

2.2.3.2 Block Structures and Memory Allocation

Not only is the program itself better structured by using blocks, but utilization of *memory space* for program data is also improved. The life span of a variable is limited to the block in which it was declared, and hence it only needs a memory location when the particular block is executed. Thus the necessary memory space is allocated during a program run before the execution of any instructions in the block and freed on block exit.

Dynamic storage allocation during the execution of program section 2.2 is illustrated in Fig. 2.4.

It is quite plain that locations 1006 to 1010 are used for the variables **interest** and **bank** (during execution of block 1) as well as for the variables **oven** and **expo** (during execution of block 2).

2.2.3.3 Memory Organization on the Stack

Memory for variables can be allocated on the stack because the block most recently executed is released first in a nested block structure. An area of memory is reserved

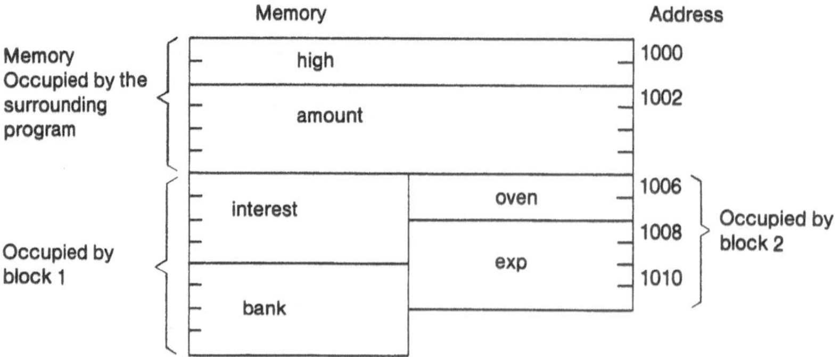

Fig. 2.4: Dynamic memory allocation for variables

```
PROGRAM resolution;

  VAR
    high:   INTEGER;
    amount: REAL;

   PROCEDURE block1;

     VAR
       interest: REAL;
       bank:     REAL;

     BEGIN (* body of block 1 *)
                .
                .
                .
           (* instructions *)
                .
                .
                .
     END (* body of block 1 *) ;

   PROCEDURE block2;

     VAR
       oven: REAL;
       expo: REAL;

     BEGIN (* body of block 2 *)
                .
                .
                .
           (* instructions *)
                .
                .
                .
     END (* body of block 2 *) ;

 BEGIN (* body of resolution *)
                .
                .
                .
           (* instructions *)
                .
                .
                .
 END (* body of resolution *) ;
```

Program section 2.2: Example for dynamic memory allocation in PASRO

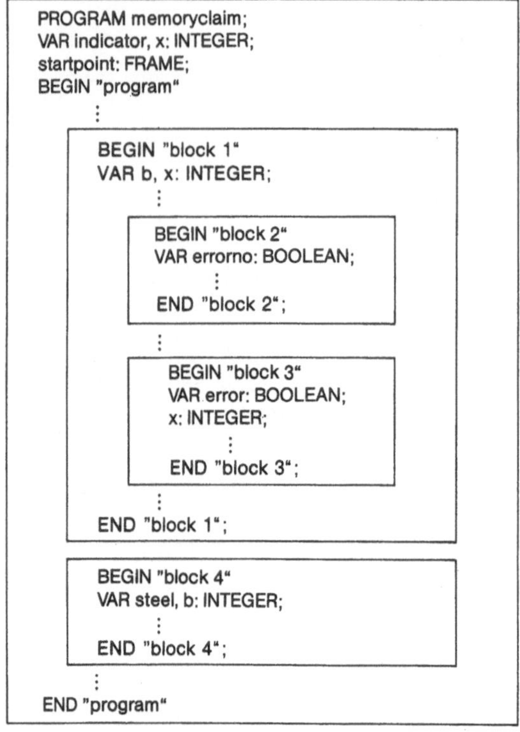
PROGRAM memoryclaim;
VAR indicator, x: INTEGER;
startpoint: FRAME;
BEGIN "program"
BEGIN "block 1"
VAR b, x: INTEGER;
BEGIN "block 2"
VAR errorno: BOOLEAN;
END "block 2";
BEGIN "block 3"
VAR error: BOOLEAN;
x: INTEGER;
END "block 3";
END "block 1";
BEGIN "block 4"
VAR steel, b: INTEGER;
END "block 4";
END "program"

Fig. 2.5: Memory allocation, scope and life span for variables during program execution

on the stack for variables on block entry. Any further inner blocks append new data spaces at the bottom of the stack. As soon as the processing of that block is finished, the data area is released, i.e., always the last one generated. Memory allocation, life span and scope of variables during program execution in the example program are illustrated in Fig. 2.5.

If no *recursive* procedures (see Sect. 2.2.5) and no dynamic variables are permitted in the language, total memory allocation and variable addresses can be evaluated during compilation.

2.2.4 Subroutines, Procedures, Functions and Macros

Subroutines, procedures and functions were introduced to reduce typing effort during programming as well as program length, if similar and frequently repeating program sections occur. Macros, however, only perform a *text substitution* and are processed before a program run. They do not shorten the length of a program. One effect of macros and particularly of procedures is the possibility of reusing algorithms which have been written before. This should not be underestimated, especially in larger programs, and procedures also aid program structuring. Such structuring can then lead to hierarchical constructions, so that individual procedures may process a partial task completely and deliver results to calling procedures.

errorno

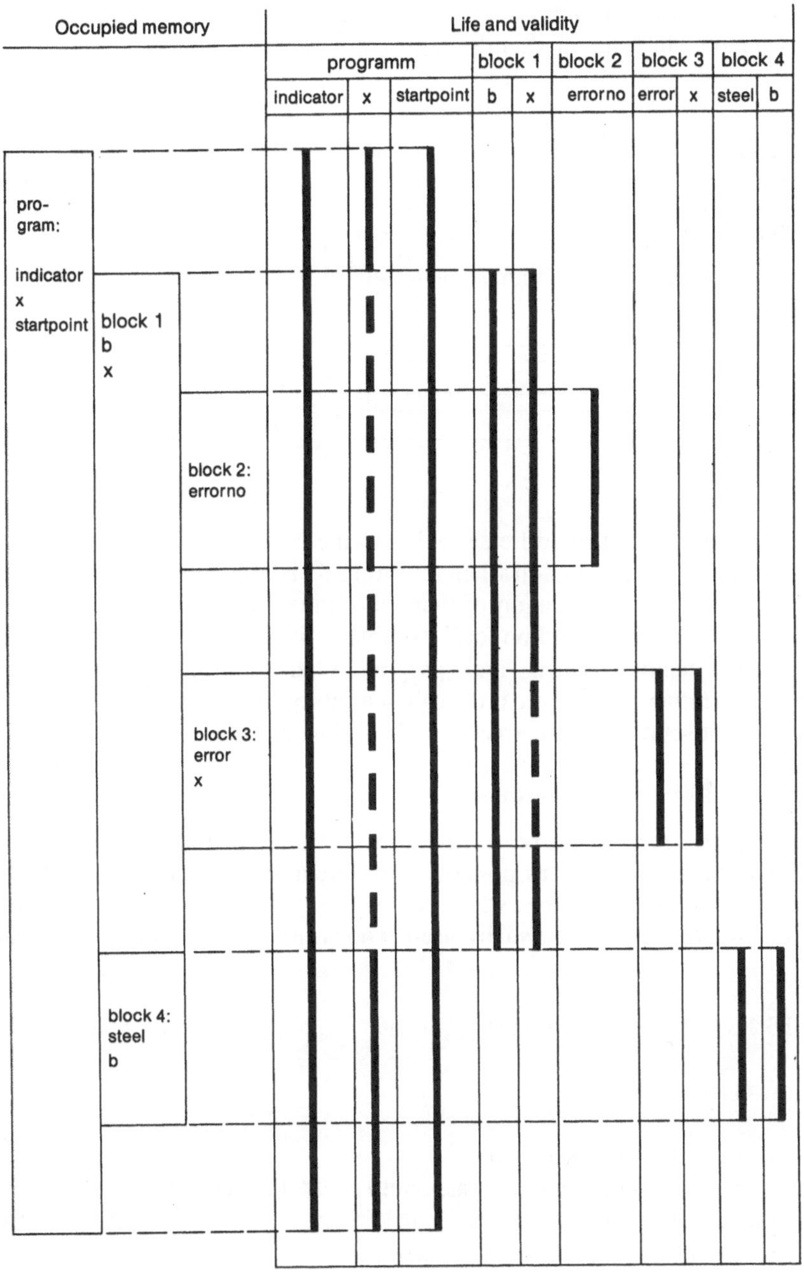

Occupied memory	Life and validity									
	programm			block 1		block 2	block 3		block 4	
	indicator	x	startpoint	b	x	errorno	error	x	steel	b

pro-
gram:

indicator
x
startpoint

block 1
b
x

block 2:
errorno

block 3:
error
x

block 4:
steel
b

2.2.4.1 Subroutines

A subroutine is a separate program section which may be called from any point in the program. After the subroutine has been processed, the return jump goes to the next sequential instruction after the subroutine call (see Fig. 2.6).

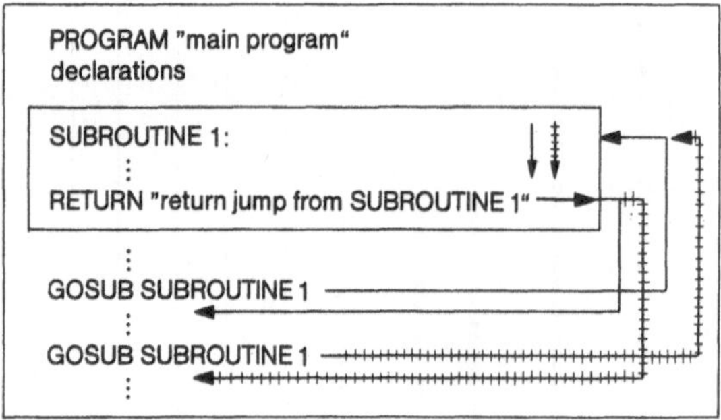

Fig. 2.6: Subroutine calls

Organization of return addresses is left to the **run-time system**, which stores the address of the next instruction when a subroutine is called in order to continue the main program after completion of that subroutine. The simplest form of subroutines has no parameters and no local variable definitions. In addition, routines are processed sequentially, i.e., the main program is halted during the execution of subroutines. A further special case is the call of a subroutine from itself (**recursion**). Various forms of subroutines are treated in the following sections.

2.2.4.2 Procedures and Functions

In order to use subroutines repeatedly under differing conditions, they should adjust themselves as flexibly as possible to different applications. Thus subroutines are equipped with parameters, which can be changed on every call. The terms subroutine and procedure are usually used synonymously, but languages containing

```
PROCEDURE framedistance (targetframe: FRAME; distance: SCALAR);

VAR armframe : FRAME;

BEGIN_PROCEDURE "framedistance"
  armframe := robotframe (puma_robot);  (* Current robot position *)
  IF length (targetrame.TRANSL - armframe.TRANSL) < distance THEN
    SMOVE puma_robot TO targetframe
  ELSE
    errorflag := TRUE;
END_PROCEDURE "framedistance";
```

Program section 2.3: Procedure declaration in SRL

only subroutines without parameters or local data cannot claim to contain the procedure concept. Thus procedures and functions are introduced here as "better" subroutines.

The use of parameters is illustrated in the following example. A procedure is to calculate the distance between the current robot position and any other point of motion (i.e., frame; see also Sect. 2.3.1) and to compare it with a given value which may vary with each procedure call. Should the distance exceed a specified value, a (global) error flag is set, otherwise the robot move is to continue to the specified point. Hence, a procedure `framedistance` is defined with two parameters:

1. a frame, to which the robot is to be moved,
2. a specified distance.

Program section 2.3 shows the implementation of such a program segment in SRL. The procedure would be called as follows:

```
framedistance (hook, 2);
```

or even:

```
framedistance (FRAMEC(xrot,VECTORC(5,10,7)), base+3);
```

The first call of the procedure determines the position between the robot position and the frame `hook`. If it is less than 2 cm, the robot is moved to the position `hook`, or the flag `errorflag` is set. In the second call, the robot is to be moved to the explictly defined frame with a distance specified by the vector (5,10,7). The distance with which it has to be compared has yet to be calculated; it is the sum of the variable `basis` and the constant 3.

The *procedure declaration* consists of the specification of the procedure name `framedistance`, the parameter list with the entries `targetframe` and `distance`, a variable declaration part with the local variable `armframe`, and the procedure body containing the actual implementation of the algorithm. The parameters in the list and in the procedure body are called *formal* parameters because they only reserve space for the *actual* parameters, which are specified at each procedure call. There are several different **parameter passing mechanisms** to substitute formal parameters by actual ones. However, the general rule holds that the sequence of parameters, i.e., their position, determines the assignment of actual to formal parameters and

```
PROCEDURE coordaxisvalue (p: INTEGER);
BEGIN
  axis := YAXIS;          (* Global variable axis is changed     *)
  vector1.XAXIS := 10;    (* Global variable vector1 is changed  *)
  x := p;                 (* Parameter p assigned to the local   *)
                          (* variable x                          *)
END;
```

Program section 2.4: Procedure for different parameter passing mechanisms in a hypothetical language

that the types of formal and actual parameters have to agree. Hence formal parameter types are specified in the procedure header.

The most important parameter passing mechanisms will be explained by examples. A procedure has been written according to program section 2.4. The procedure changes the values of both global variables **axis** and **vector1** which are of type VECTOR. The formal parameter p of type INTEGER is assigned to the variable x. The following sequence of instructions delivers different values for x according to the passing mechanism (see program section 2.5).

```
axis := XAXIS;
vector1 := VECTOR (20, 7, 13);
coordaxisvalue ( vector.axis );
```

Program section 2.5: Call of the procedure declared in program section 2.4 in a hypothetical language

a) **Call-by-value**
The actual parameter is evaluated at the time of the procedure call, and the resultant value is substituted in all places where it is used in the procedure body. Thus,

```
vector.axis = VECTOR (20,7,13).XAXIS
```

is evaluated in the example, i.e., the x-coordinate of the vector is selected and computed. Consequently, the variable x contains the value 20 after the procedure call.

b) **Call-by-reference**
A pointer to the actual parameter is passed with the procedure call, i.e., the formal parameter now represents the variable passed as an actual parameter. Consequently, a change in value of the variable also changes the actual parameter. So the call in the example is equipped with a pointer to the x-coordinate of the vector. Thus, the statement

```
vector1.XAXIS := 10;
```

assigns the value of 10 to the x-coordinate of the variable **vector1**. The variable x contains this value after the procedure exit.

c) **Call-by-name**
This parameter passing mechanism is rarely realized because its implementation is extravagant and it uses a comparatively large amount of time during program execution. This is a result of *textually substituting* formal parameters by actual ones. Values are only determined when the procedure processing arrives at the point where the parameters are used. They are evaluated each time they appear, and each time new values are possible. This also leads to a parameter passing which is not clear. In the given example, the actual parameter **vector1.axis** is evaluated just before the parameter p is assigned to the global variable x. Now the y-coordinate of **vector1** is selected, because **axis** has been assigned the value **yaxis** in the procedure body. The y value is written into x.

Some **side effects** may occur in assignments to parameters or generally global variables, particularly with the call-by-reference or call-by-name mechanisms. They are mainly used for passing results "out" of a procedure, but if they are not used carefully variables may be changed which are regarded as constants in the main program.

Thus, functions are provided apart from procedures, which should avoid side effects in any case and return results to the main procedure in a different manner. For that purpose, the name of the function is regarded as a variable of a specified type to which the result of the function is assigned. This value may then be used in instructions or expressions in the main program because the function call is no longer a separate unit but is given by its name and parameter list. For example, the following PASRO function evaluates the sum of all the components of a vector:

```
FUNCTION vectorsum (v: VECTOR) : REAL;
BEGIN
  vectorsum := v.X + v.Y + v.Z
END;
```

This function may then be called in the assignment to a `REAL` variable `basevalue`:

```
basevalue := 5 + vectorsum (targetvector);
```

Standard functions are special because the programmer need not define them; they are already offered by the system. They are usually used in the evaluation of basic mathematical functions like square root, sine, tangent, or similar. An example is:

```
distance := SIN (rotation) * vectorlength;
```

The sine of the angle `rotation` is evaluated, this value is then multiplied by the `vectorlength`, and the result stored in the variable `distance`.

2.2.4.3 Macros

Contrary to subroutines, macros are processed not during program execution but at compile time, so that they are not recognizable in the executable program. Macros *substitute text* only, ranging from simple string substitution up to generating text by using parameters, internal macro definitions and nested macro calls (see COLE [2.1]). A *macro definition* is the assignment of a macro name to a macro body. A *macro name* may consist of any character string, just like a variable name. A *macro body* is defined by a sequence of characters and/or macro calls. A *macro call* is initiated by specifying the macro name (with possible parameters), and the *macro generator* then substitutes the name by the character sequence in the macro body.

Consider the example (AL):

```
DEFINE inch = ≠ * 2.54*cm ≠;
```

This macro definition assignes the macro name `inch` to the macro body `* 2.54*cm`. The character "≠" serves as a delimiter, which will not be used in the "surrounding" programming language. The macro call contained in the instruction

```
distance := base + 35 inch;
```

is changed after the processing by the macro generator to

```
distance := base + 35 * 2.54*cm;
```

Definitions of macros with parameters are carried out correspondingly:

```
DEFINE movement (targetframe) =
# MOVE arm TO targetframe
WITH DURATION = 5*sec; #
```

and the call within the instruction sequence

```
x := 200;
movement (deposit)
OPEN hand;
movement (magazine)
```

generates the following text

```
x := 200;
MOVE arm TO deposit
WITH DURATION = 5*sec;
OPEN hand;
MOVE arm TO magazine
WITH DURATION = 5*sec;
```

According to the type of generator, macro definitions may be entered from external libraries at the beginning or distributed throughout the program. In any case, they have to be known to the macro generator before they can be called.

Advantages of macro usage are:

- Little typing effort.
- Important sequences of commands, which are used repeatedly, are programmed by experienced programmers as macros and are then available for general usage.
- Consistent usage of macro calls improves program legibility and standardization.

Disadvantages are contained in the larger memory requirements and longer compilations. Thus, advantages and disadvantages of using macros have to be weighed carefully in each application.

2.2.5 Recursion

Some high-level languages permit the call of a procedure from the procedure itself or from another procedure. The first case is termed **direct recursion**, and the second may lead to **indirect recursion**, if two procedures call each other (see Fig. 2.7). So far, the general opinion has been that such abstract games are unnecessary in robot programming. However, an example will demonstrate the use of recursive procedures, and it will be shown that the same problem can be solved by other methods, but only with difficulty.

A small demonstration program (program 2.6) realizes the control of a robot between arbitrary start and endpoints. The motion is to be executed with constant acceleration as specified in the program, until maximum velocity is reached or braking has to begin. The braking is carried out by a constant deceleration, which has

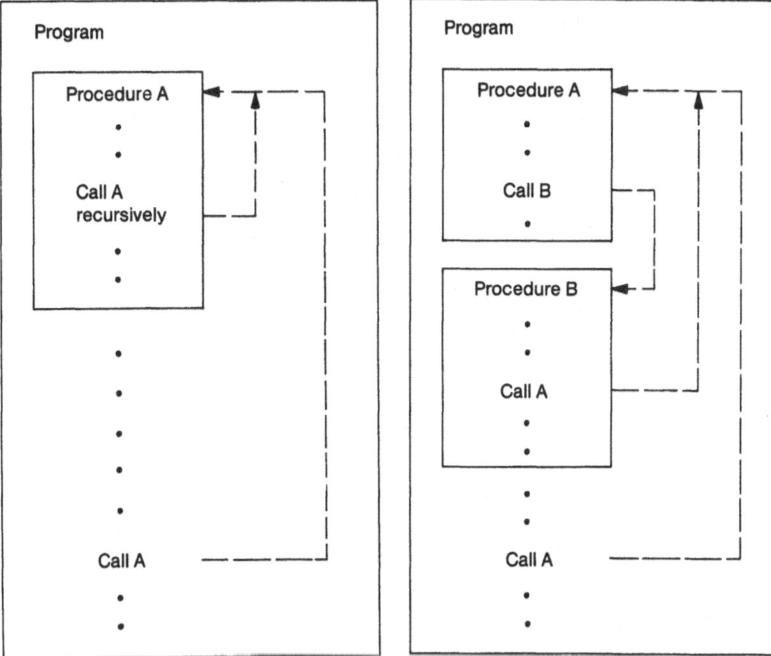

Fig. 2.7: Recursive procedure calls

the same (negative) value as the acceleration. The robot controller itself does not contain the possibility to execute an instruction like

```
< move robot arm from A to B
with constant acceleration/deceleration C >
```

It does have, however, a command `VIAMOVE` which drives the robot to an intermediate point on the motion path. When it arrives at this point, a certain velocity is reached, which leads on to the next point. The problem is the generation of a sequence of `VIAMOVE` instructions such that a constant acceleration or deceleration is enforced on passing through intermediate points. The following relations hold for acceleration, velocity, distance and time:

$$\Delta s = \frac{a}{2}\Delta t^2 + v_1 \cdot \Delta t$$

$$\Delta v = a \cdot \Delta t = v_2 - v_1$$

where

Δs = distance between two intermediate points
Δv = velocity increase between intermediate points
Δt = specified time interval dt
a = chosen acceleration between a and b

```
PROGRAM motionharmony;

  CONST
    vmax = 80;                 (* Max. robot velocity            *)
    dt = 0.25;                 (* Time interval of 250 ms for    *)
                               (* control                        *)
  VAR
    startp       ,             (* Start position                 *)
    targetp      : FRAME;      (* End position                   *)
    halfdistance,              (* half way point                 *)
    b            : REAL;       (* acceleration/deceleration      *)


  PROCEDURE movement (distance, v: REAL);

    VAR
      a : REAL;                (* constant or variable           *)
                               (* acceleration                   *)

    BEGIN_PROCEDURE "movement"
      a := b;
      distance := ABS(distance) + 0.5*SQR(dt) + v*dt;
      IF halfdistance < 0 THEN
        distance := - distance;
      v := v + a*dt;
      IF (ABS(distance) < ABS(halfdistance)) AND (v <= vmax) THEN
      BEGIN
        VIAMOVE puma TO startp.TRANSL.XAXIS + distance
          WITH VELOCITY = v;
        movement (distance, v);
        VIAMOVE puma TO targetp.TRANSL.XAXIS - distance
          WITH VELOCITY = v;
      END;
    END_PROCEDURE "movement" ;

  BEGIN_PROGRAM
    startp.TRANSL.XAXIS  :=  50;    (* X coordinate value   50 cm   *)
    targetp.TRANSL.XAXIS := 150;    (* X coordinate value  150 cm   *)
    halfdisance := (targetp.TRANSL.XAXIS - startp.TRANSL.XAXIS) / 2;
    b := 40;                        (* Constant accln/decln of      *)
                                    (* 40 cm/(sec*sec)              *)
    movement (0,0);
    MOVE puma TO targetp;
  END_PROGRAM.
```

Program 2.6: Program with recursive procedure calls for motions with constant acceleration in SRL

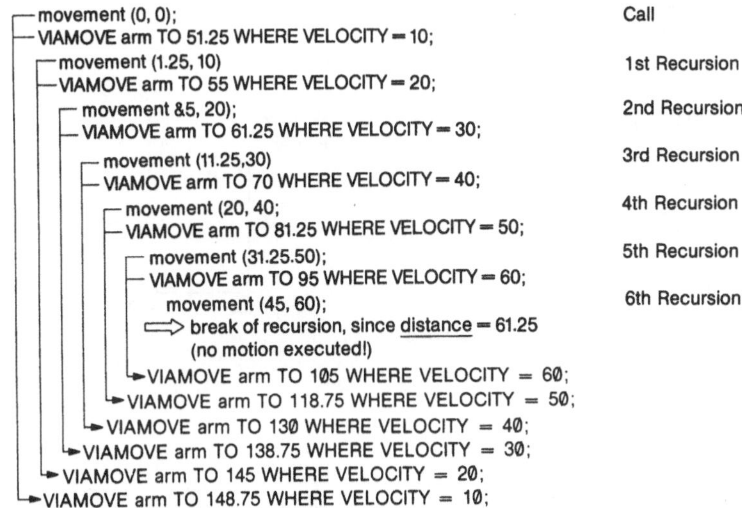

movement (0, 0);
VIAMOVE arm TO 51.25 WHERE VELOCITY = 10;
movement (1.25, 10)
VIAMOVE arm TO 55 WHERE VELOCITY = 20;
movement &5, 20);
VIAMOVE arm TO 61.25 WHERE VELOCITY = 30;
movement (11.25,30)
VIAMOVE arm TO 70 WHERE VELOCITY = 40;
movement (20, 40;
VIAMOVE arm TO 81.25 WHERE VELOCITY = 50;
movement (31.25.50);
VIAMOVE arm TO 95 WHERE VELOCITY = 60;
movement (45, 60);
break of recursion, since distance = 61.25
(no motion executed!)
VIAMOVE arm TO 105 WHERE VELOCITY = 60;
VIAMOVE arm TO 118.75 WHERE VELOCITY = 50;
VIAMOVE arm TO 130 WHERE VELOCITY = 40;
VIAMOVE arm TO 138.75 WHERE VELOCITY = 30;
VIAMOVE arm TO 145 WHERE VELOCITY = 20;
VIAMOVE arm TO 148.75 WHERE VELOCITY = 10;
Call
1st Recursion
2nd Recursion
3rd Recursion
4th Recursion
5th Recursion
6th Recursion

Fig. 2.8: Processing of a recursion sequence

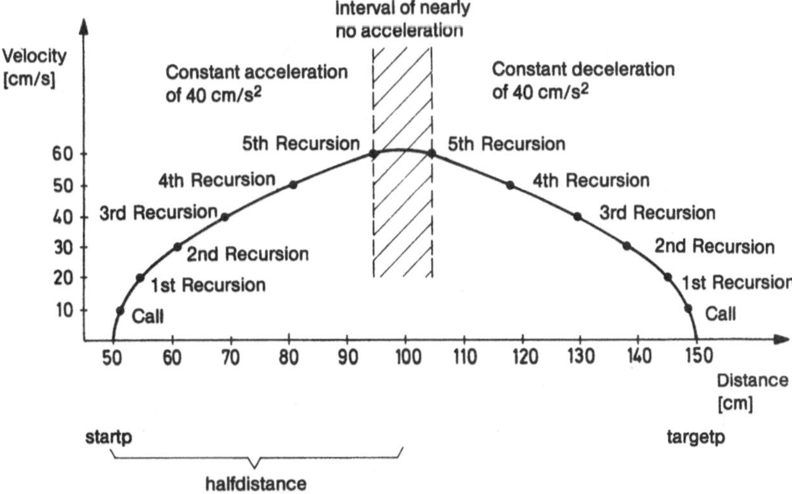

Fig. 2.9: Velocity-distance diagram of the VIAMOVE instructions of the recursive procedure movement

v_1 = velocity at first intermediate point
v_2 = velocity at second intermediate point

Let `startp` be an arbitrarily chosen start point and `targetp` be the endpoint of the movement. Also, let the motion coincide exactly with the x-axis for simplicity, and hence only the x-coordinate values of `startp.TRANSL.XAXIS` and `targetp.TRANSL.XAXIS` are considered. Additionally, the mathematical functions SQR (for squaring) and ABS (absolute value) are needed for the small demonstration program.

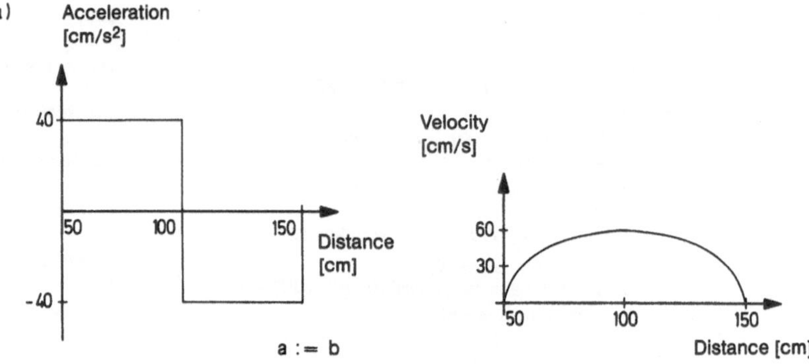

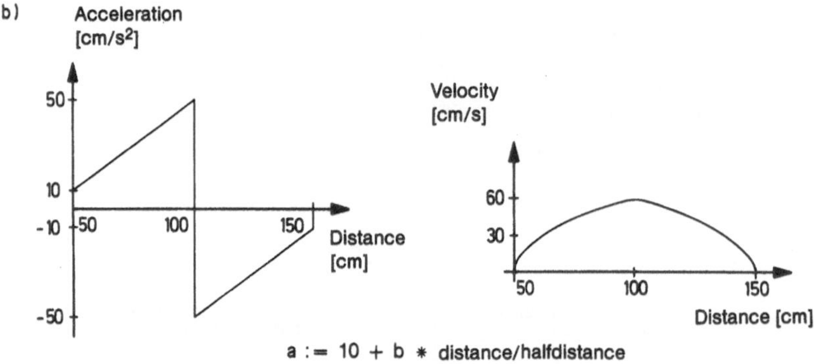

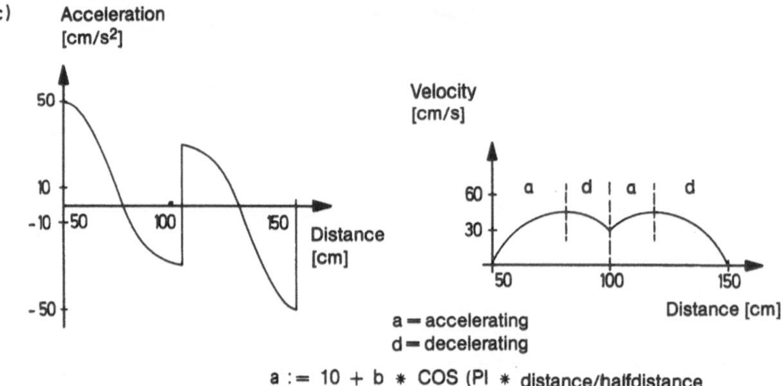

Fig. 2.10 a-c: Motion patterns with
a constant acceleration/deceleration,
b linearly rising/falling acceleration/deceleration,
c cosine-shaped acceleration/deceleration

Program execution results in **recursive calls** with the parameters described in Fig. 2.8, and the main program finishes the move by the command

```
MOVE puma TO 150;
```

so that the robot stops.

Acceleration and deceleration are symmetrical; the value may be specified arbitrarily (within device specific limits) (see Fig. 2.9). The attentive reader will have noticed that line 13 is not necessary, and that the expression

```
0.5 * a * SQR(dt)
```

in line 14 is constant, as well as the expression `a * dt` in line 17. These expressions for constant acceleration are only evaluated once and not each time a recursion is entered. However, if the acceleration within a move is to be variable, for example small at the beginning and end but large in the middle, a new value for the acceleration is evaluated in line 13, such as

```
a := 10 + b + distance/halfdistance;
```

Thus this small program can generate freely any harmonic motion, such as for spray painting or grinding (see Fig. 2.10).

The end of the recursion is determined by the termination condition in line 18:

```
(ABS (distance) < ABS (halfdistance)) AND (v <= vmax)
```

The first comparison tests whether the distance covered still lies within the first half of the total distance. This is no longer true in the sixth recursion call, and no further recursion call follows. The second comparison only tests for an exceeding of maximum velocity. If this is so, a recursion does not follow either and the robot is moved with nearly maximum velocity until braking begins.

The term **incarnation** is useful for a general understanding of procedure calls and recursions. An incarnation is the execution of a procedure called by a procedure. Thus, the first incarnation of the procedure `movement` in the example is invoked by a call in the main program in line 30:

```
movement(0,0);
```

However, before this incarnation is finished, the second incarnation of the same procedure is started by the first recursive call within the procedure itself in line 21:

```
movement(distance,0);
```

Hence, the values of the parameters `distance` and v are stored as *local variables* (as well as the locally defined variable a). The second incarnation generates a new data set. According to this scheme, the execution of each further recursive procedure call is interrupted until the termination criterion is fulfilled. Then the last incarnation of the procedure is finished and the next-to-last one continued, i.e., the `VIAMOVE` instruction in line 22 is executed with the data `targetp.TRANSL.XAXIS - distance = 106.25` and `v = 60`. Thereafter, all further incarnations are executed with their corresponding data and instructions and the incarnation terminated. The main program then finishes the motion off with the `MOVE` instruction (line 31).

The main advantage of the recursion is the fact that any number of points may be calculated without previously establishing a data array (see Sect. 3.1.5.1). Data for the deceleration during braking is needed so that the motion may be symmetrical and the exact start of braking may be calculated. However, since the exact number of points is not known in advance or cannot be evaluated, an array with unnecessarily large dimensions would have to be generated, thus wasting memory space. Dynamic memory reservation would also need large amounts of storage and in addition consume much computation time. Thus, recursion is not only an elegant solution but also problem-orientated.

2.2.6 Processes, Tasks and Coroutines

The terms **process** and **task** introduced in Sect. 2.1 can now be described in greater detail. A process is a program section which executes an incarnation with each call, just like a procedure. An incarnation may be started not only by the program but also by external events (such as interrupts). It may be executed in parallel to other processes or programs. This may also include the parallel existence of several incarnations of the same process or the marking of a succession of incarnations which are processed sequentially. A process may also wait for a specific event such as the termination of another process. The introduction of process states has proved very useful in this context. Thus a process may be in the *dormant* state before it starts. If

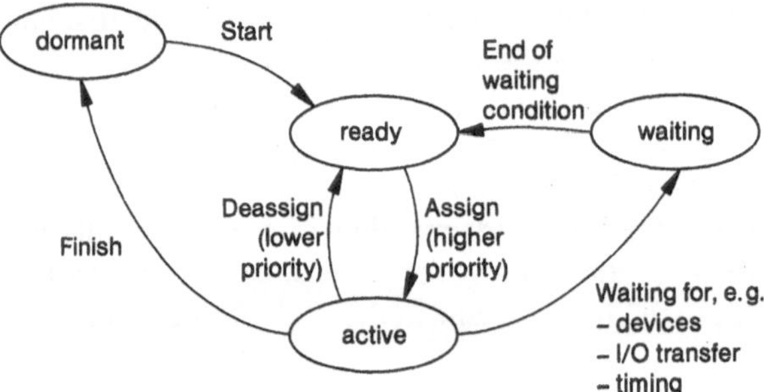

Fig. 2.11: Examples for process states and state transitions

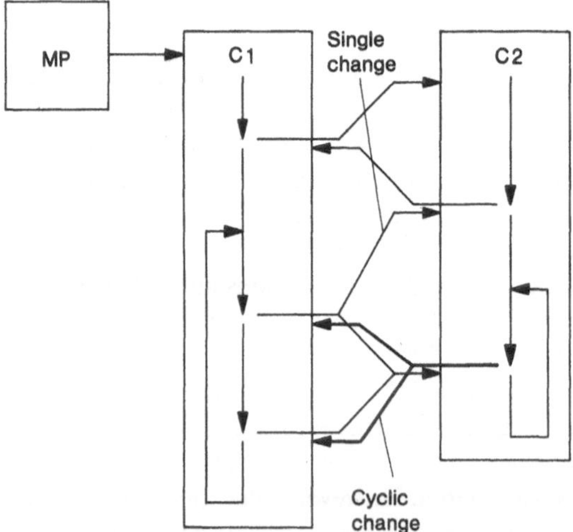

Fig. 2.12: Execution of two coroutines C1 and C2 with cyclic changes of return addresses

it begins, it may have to stay in a *ready* state because of other processes until it becomes *active*. A *wait* state indicates the waiting for other processes to finish, like the input or output to peripheral devices. Fig. 2.11 shows an example for a state diagram with these four process states (others are also possible).

A change of a process state from one to another is carried out by the **scheduler** of the operating system. Just as in procedures, incarnations of processes may be started and finished at different places in the some programming languages. A generalization of this concept leads to **coroutines**, which are interrupted by a special command after their incarnation and then may be continued at the same place (see also ROHLFING [2.2]). Thus an interplay of coroutines with equal priorities can take place, in contrast to a hierarchical flow structure of processes. An example is shown in Fig. 2.12. After the main program "MP" has been started, the coroutine "C1" is run cyclically. The first run also starts the coroutine "C2", which is also executed cyclically. This will be terminated by a system command or by switching off the computer.

2.2.7 Synchronization

Some of the previously described sequential processes, which run in parallel, occasionally have to be *synchronized*. Synchronization is particularly necessary if one process can only be executed after another has finished (Fig. 2.13).

Such synchronization may be realized by **synchronization variables** which are accessible to both processes (Fig. 2.14). This is not an optimum solution, however, since process 1 permanently tests the state of the synchronization variable "s" and thus wastes processor time (see also WETTSTEIN [2.3]).

If process 1 has a higher priority than process 2, it may even be wrong, because then the continuing interrogation inhibits the setting of the synchronization variable. Thus, it is far better to leave the synchronization to an **operating system**, especially if several processes are to be organized. **Semaphores** (Greek: sign carriers) are employed for this, and these are only accessible to the programmer by function calls to the operating system. If the semaphores are binary, the P-operation - usually represented by a WAIT instruction - initiates the operating system or interpreter to interrogate an internal semaphore. If it is zero, the process is halted by the WAIT instruction, otherwise it is continued. The V-operation - usually represented by a SIGNAL instruction - continues a process which has previously been halted by a WAIT instruction (Fig. 2.15). Thus a waiting process is deactivated after a WAIT and does not use any processor time. It only continues to claim processor time after it has been activated by a SIGNAL instruction.

The general form of each semaphore has a waiting list associated with it, so that several processes can wait for a release (WETTSTEIN [2.3]). Now the internal synchronization variable is implemented as a counter, which is incremented and decremented. The effects of the WAIT and SIGNAL instructions are as follows:

```
WAIT instruction:
s := s - 1;
IF s >= 0 THEN the process continues execution
ELSE the process is put into a wait state and entered into the waiting list of
     the semaphore s;
```

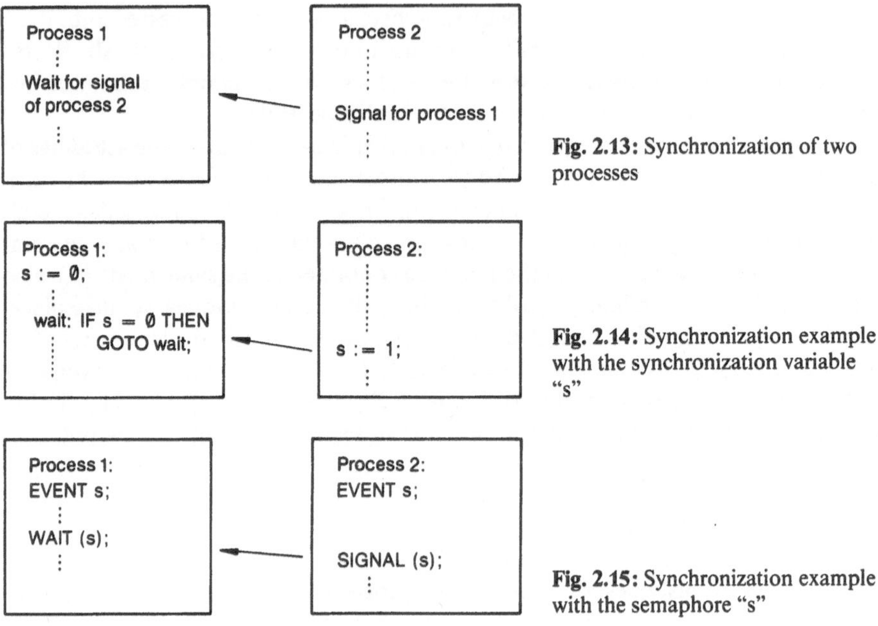

Fig. 2.13: Synchronization of two processes

Fig. 2.14: Synchronization example with the synchronization variable "s"

Fig. 2.15: Synchronization example with the semaphore "s"

```
SIGNAL instruction:
s := s + 1;
IF s <= 0 THEN a process on the waiting list is activated;
The process continues its execution;
```

Semaphores are usually initialized with a 1. Fig. 2.16 shows three processes, all accessing a magazine cyclically. However, only one robot can have access at one time, because otherwise collisions might occur. The process section in which the magazine is accessed, is termed an exclusive or **critical section**. If the second process arrives at a critical section, its WAIT instruction only causes s = 0. If process 3 arrives at the same critical section, the WAIT instruction deactivates process 3 and enters it into the waiting list of the semaphore s. The semaphore is set to -1. When process 2 exits the critical section, the SIGNAL instruction increments s to null and also acti-

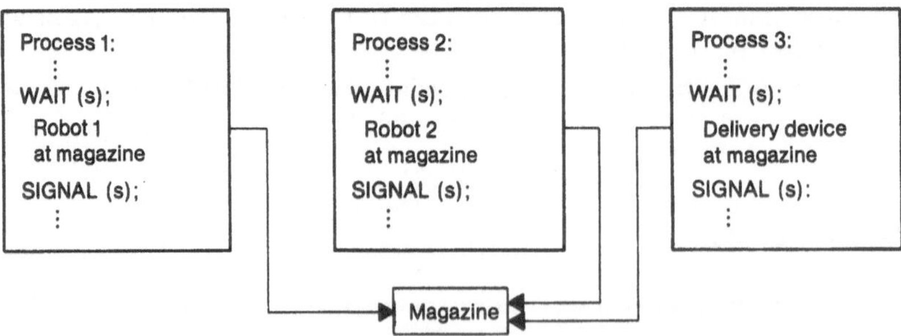

Fig. 2.16: Synchronized access to a magazine

vates process 3 from the waiting list. If, in the meantime, process 1 arrived at the critical section, it would also have been deactivated. The semaphore now has the value -1, and only after the SIGNAL instruction of process 3, can process 1 enter the critical section.

2.3 Concepts for Robot Languages

The new area of applications, *robotics*, demands new and special techniques, which are determined by the robot itself. This is, of course, reflected in the design of programming languages for industrial robots, which may be biased towards certain applications. In addition, these approaches attempt to reach three aims:

1. Improvement of operating and programming comfort for users.
2. Following on from current standards.
3. Reduction of complexity during system development.

The first point needs further explanation. Of course, any program system usually aims to improve operating and programming comfort. However, robot programming needs to solve the special problem of describing *spatial movements* and whole movement sequences by linguistic means. From the implementation point of view, the simplest way to do this is a succession of robot coordinates which are passed on to the robot controller. However, specification of robot moves in angles for revolute joints or distances for prismatic joints is very cumbersome and highly time-consuming for programmers.

One improvement is the representation in Cartesian coordinates, so that the programming system has to transform Cartesian into robot coordinates and vice versa. Nevertheless, all points on a path have to be specified, i.e., the startpoint, possible intermediate points, at which directions and/or speeds may be changed, as well as the endpoint. This may be very demanding on the geometric imagination of humans, particularly in more complex cases. Thus any programming system strives to realize a higher level, on which programmers express trajectories as desired functions and not by individual points for paths or moves. These high-level functions may include "grip an object" or "mount object A on top of object B" and are orientated toward the imagination and intellectual capacity of humans. Hence, a system has to be capable of looking up missing or implicitly given data from a data base on its own and to generate corresponding motions.

2.3.1 The Concept of Frames

In one form or another, the concept of frames is used in nearly all programming systems as an attempt to describe orientation and position of robots geometrically. The endpoint of a tool or the gripping point between jaws (**Tool Center Point, TCP**) is usually referred to as *robot position. Orientation* specifies the direction (from above,

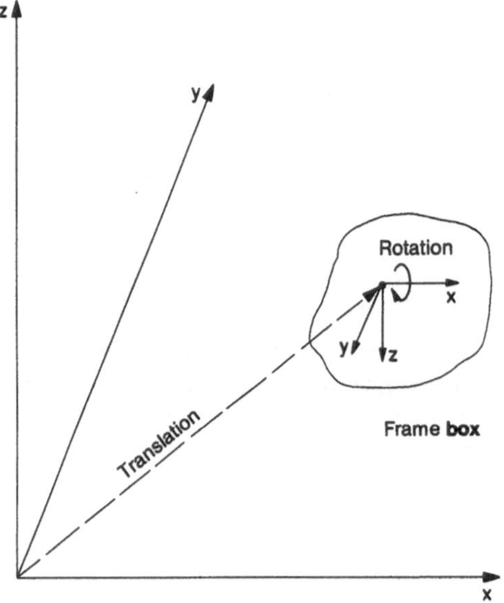

Fig. 2.17: Geometric representation of the frame box

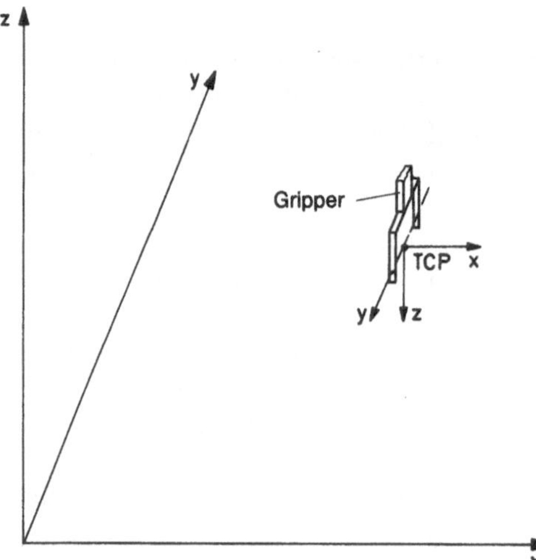

Fig. 2.18: Connections between frames and position and orientation of grippers

below, sideways, etc.) and twist of tools or grippers with which a position is approached. In contrast to programming in robot coordinates (i.e., joint angles and distances), *frames* describe the position of a motion point in Cartesian coordinates and orientation by rotation specifications around the robot or gripper axes. A reference system for these details has been defined at system generation, and such a *base coordinate system* may have its origin at the corner of the work table, with the z-axis pointing straight upward, the x-axis pointing forward, and the y-axis to the side. It is also possible to visualize a frame as a copy of the base coordinate system, where the origin is defined by a translation vector. The orientation of this frame in space may be changed by rotation demands around the main axes (Fig. 2.17). For example, a programmer may specify a motion point (and its associated orientation of gripper/tool) in SRL by declaring an object of data type FRAME. He may then assign the explicit values of a three dimensional vector to one (or more) rotation(s).

Here is an example:

```
box: FRAME       (* frame declaration *)
.
.
.
box := FRAMEC (ROTC(YAXIS,90), VECTORC(60,80,20));
```

This frame now represents a coordinate system having its origin at X = 60 cm, Y = 80 cm and Z = 20 relative to the base coordinate system. In addition, the frame has been rotated by 90 degrees around the y-axis, such that its z-axis points along the x-direction of the base and its x-axis points vertically downwards (Fig. 2.17).

If an industrial robot is moved to the position and orientation of a frame, the following actions result in the AL programming system:

- The endpoint of the translation vector is situated in between the jaw or at the TCP.
- The gripper points along the z-axis of the frame coordinate system.
- The y-axis passes through both gripping points of the gripper (Fig. 2.18).

This also allows frame definitions *relative* to other frames, in which the translation vector refers to the origin of the frame referred to and the rotations to its axes. The geometric representation is shown in Fig. 2.19.

Such relative frame definitions are of particular interest if changes in values of one frame automatically cause changes in relatively defined frames according to relations specified. For example, this situation arises if robots are translated or rotated or if an object posesses several gripping points and all their values change during repositioning. Programming in robot coordinates would have to program each individual point. Cartesian coordinate systems additionally represent *common reference systems* for robots and sensors.

A frame is represented internally by a 3 x 3 rotation matrix R and a 1 x 3 vector v. The rotation matrix may represent its own data type ROTATION and may be defined

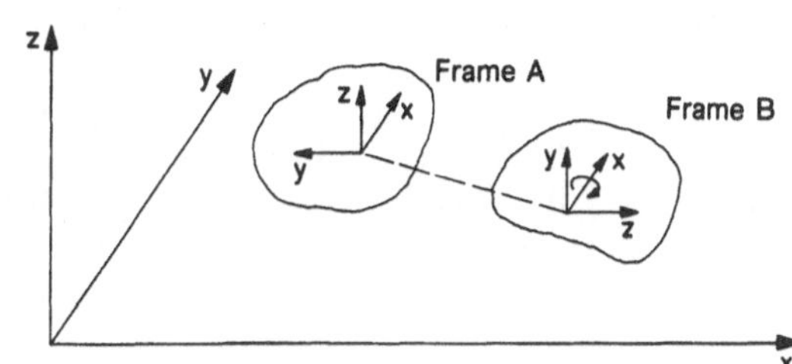

Fig. 2.19: Frame relations

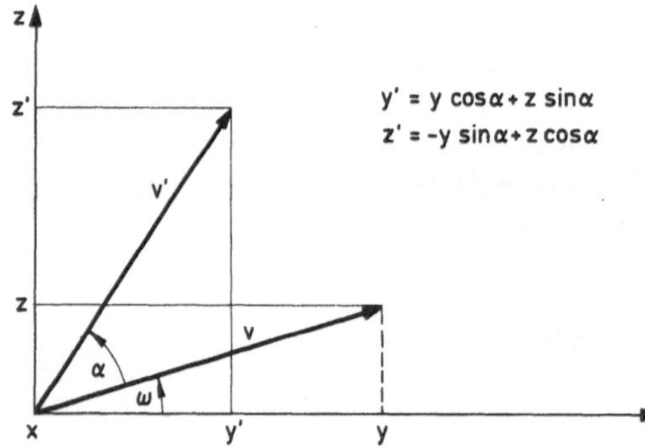

Fig. 2.20: Derivation of rotation around x-axis

in such manner that multiplication by a vector v results in a vector v', which contains the result of a rotation of v around an axis specified by the rotation matrix and angle:

$$v' = R \cdot v$$

A *rotation* around the x-axis by an angle alpha is described by

$$R_x = \begin{bmatrix} 1 & 0 & 0 \\ 0 & \cos\alpha & -\sin\alpha \\ 0 & \sin\alpha & \cos\alpha \end{bmatrix}$$

corresponding to the geometric derivation in Fig. 2.20. A rotation around the y-axis is represented by

$$R_y = \begin{bmatrix} \cos\alpha & 0 & \sin\alpha \\ 0 & 1 & 0 \\ -\sin\alpha & 0 & \cos\alpha \end{bmatrix}$$

A rotation around the z-axis is of the form

$$R_z = \begin{bmatrix} \cos\alpha & -\sin\alpha & 0 \\ \sin\alpha & \cos\alpha & 0 \\ 0 & 0 & 1 \end{bmatrix}$$

Example:
The vector $v^T = (15, 100, 20)$ is to be rotated around the x-axis by 30 degrees. The corresponding rotation matrix R_x then holds the values

$$R_x = \begin{bmatrix} 1 & 0 & 0 \\ 0 & \cos\alpha & -\sin\alpha \\ 0 & \sin\alpha & \cos\alpha \end{bmatrix} = \begin{bmatrix} 1 & 0 & 0 \\ 0 & 0.87 & -0.5 \\ 0 & 0.5 & 0.87 \end{bmatrix}$$

After the rotation, the vector v' has the following values

$$v' = R_x \cdot v = \begin{bmatrix} 1 & 0 & 0 \\ 0 & 0.87 & -0.5 \\ 0 & 0.5 & 0.87 \end{bmatrix} * \begin{bmatrix} 15 \\ 100 \\ 20 \end{bmatrix} = \begin{bmatrix} 15 \\ 77 \\ 67.4 \end{bmatrix}$$

It is sometimes more convenient to combine rotations and translations and complete it to a 4 x 4 matrix, the **Denavit-Hartenberg-matrix** (or DH-matrix) (see also DENAVIT [2.4]).

$$\text{DH-matrix} = \begin{bmatrix} [R] & [v] \\ 000 & 1 \end{bmatrix}$$

The DH-matrix describes a *homogeneous transformation* or *frame.* Its mathematical background shall not be considered any further here (compare with BLUME [2.5]). For example, let a point `target` be declared as:

```
target := FRAMEC (ROTC(YAXIS,90), VECTORC(55,-37,23));
```

The frame representing the point internally has the values:

$$DH_{target} = \begin{bmatrix} 0 & 0 & 1 & 55 \\ 0 & -1 & 0 & -37 \\ -1 & 0 & 0 & 23 \\ 0 & 0 & 0 & 1 \end{bmatrix}$$

It is obvious from the columns of the rotation matrix that the x-axis of the frame points into the opposite direction of the base z-axis and the z-axis of the frame points along the base x-axis.

Representing frames in DH format, sometimes called *transformation matrices,* eases calculation problems of rotation and translation considerably. Rotation and translation matrices are extended to 4 x 4 DH-matrices and multiplied by the frame matrix. However, the position vector is not allowed to change during the multiplication by a rotation.

Example:
Let the frame `target` rotate around the z-axis by 90 degrees and translate by the vector X = -55 cm, Y = 37 cm and Z = 0 cm. The resulting frame is called `newframe`:

$$\text{newframe} = \begin{bmatrix} 1 & 0 & 0 & -55 \\ 0 & 1 & 0 & 37 \\ 0 & 0 & 1 & 0 \\ 0 & 0 & 0 & 1 \end{bmatrix} * \begin{bmatrix} 0 & -1 & 0 & 0 \\ 1 & 0 & 0 & 0 \\ 0 & 0 & 1 & 0 \\ 0 & 0 & 0 & 1 \end{bmatrix} * \begin{bmatrix} 0 & 0 & 1 & 55 \\ 0 & -1 & 0 & -37 \\ -1 & 0 & 0 & 23 \\ 0 & 0 & 0 & 1 \end{bmatrix}$$

$$\text{newframe} = \begin{bmatrix} 1 & 0 & 0 & -55 \\ 0 & 1 & 0 & 37 \\ 0 & 0 & 1 & 0 \\ 0 & 0 & 0 & 1 \end{bmatrix} * \begin{bmatrix} 0 & -1 & 0 & 55 \\ 0 & 0 & 1 & -37 \\ -1 & 0 & 0 & 23 \\ 0 & 0 & 0 & 1 \end{bmatrix}$$

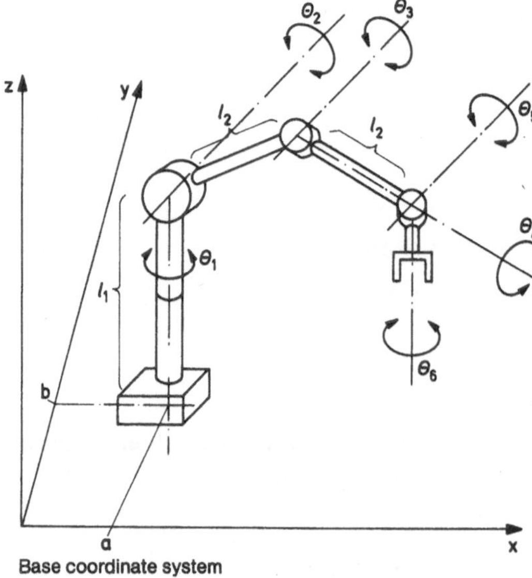

Fig. 2.21: Geometry of a robot with six joints and a gripper

$$\text{newframe} = \begin{bmatrix} 0 & -1 & 0 & 0 \\ 0 & 0 & 1 & 0 \\ -1 & 0 & 0 & 23 \\ 0 & 0 & 0 & 1 \end{bmatrix}$$

Now `newframe` has an orientation such that its x-axis still points in the opposite direction of the base z-axis, its x-axis points in the negative x-direction, and its z-axis points along the base y-vector. Its origin is at X = 0, Y = 0 and Z = 23.

2.3.2 Coordinate Transformations and Trajectory Planning

The frame concept allows the user to specify position and orientation in world coordinates, like Cartesian or cylindrical coordinates. However, they can be used directly only in exceptional circumstances because the robot is controlled in its own coordinates. Usually, programming or run-time systems contain modules which carry out *coordinate transformations.* A transformation from *world* to *robot* coordinates is necessary to reach frames which have been defined by explicit values. On the other hand, if a programmer wants to express a point taught by the teach-in procedure as a frame, the inverse transformation from *robot* into *world coordinates* is needed.

The following section will explain both transformations by examples (PAUL [2.6]). **Trajectory planning**, which enables a controlled movement between specified frames, is also briefly discussed.

The derivation of *transformation equations* is demonstrated by the example of a hypothetical robot with six joints which has a structure similar to familiar industrial robots like the PUMA 600. Figure 2.21 shows the geometry of that robot and its

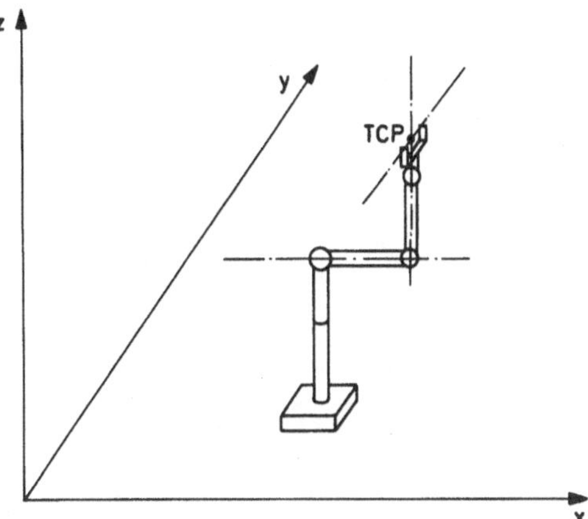

Fig. 2.22: Robot in base configuration

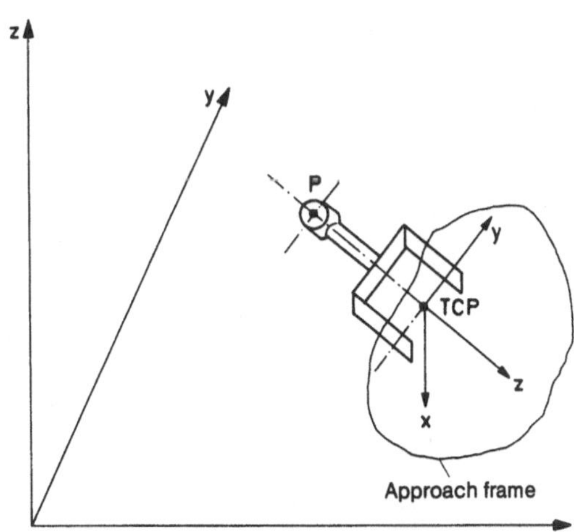

Fig. 2.23: Frame, gripping point and point P to be headed for

joint nomenclature. A base configuration (Fig. 2.22), i.e., when all joint angles are at zero, is defined as:

- $theta_1$ is zero if the arm points along the x-axis of the base coordinate system;
- $theta_2$ is zero if the upper arm of the robot is parallel to the xy plane;
- $theta_3$ is zero if the lower arm points vertically up;
- $theta_4$ is zero if a straight line passing through both gripping points on the jaws is parallel to the y-axis;
- $theta_5$ is zero if the gripper points along the direction of the last robot joint;
- $theta_6$ is zero if the same condition as for $theta_4$ is fulfilled.

It is also possible to subdivide the approach to a position and orientation where $theta_1$, $theta_2$ and $theta_3$ determine the position and $theta_4$, $theta_5$ and $theta_6$ finalize orientation. The point laid down by $theta_1$, $theta_2$ and $theta_3$ is no longer the actual gripping point (or Tool Center Point, TCP), but it is translated along the negative di-

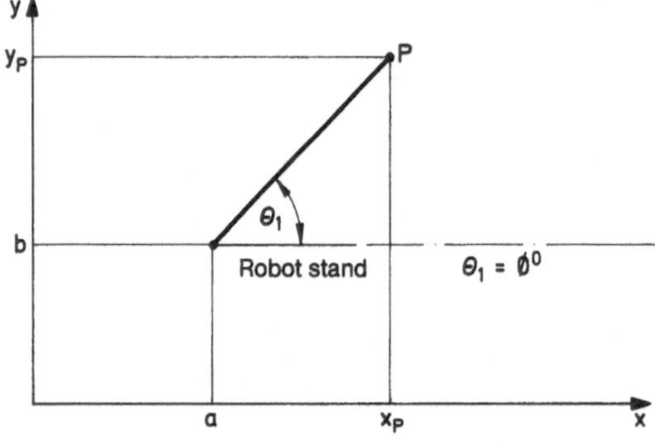

Fig. 2.24: Calculation of the angle $theta_1$

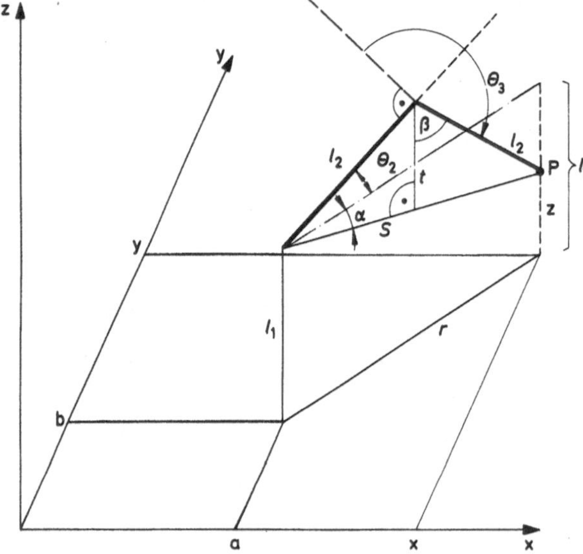

Fig. 2.25: Derivation of angles $theta_2$ and $theta_3$

rection of the gripping direction by the length of the tool/gripper. Fig. 2.23 shows this point P.

Let the desired frame F be programmed *offline* by a programmer:

```
F:=FRAMEC(ROTC(YAXIS,90)*ROTC(XAXIS,30),VECTORC(24,18,10));
```

Thus the gripping point is at x = 24 cm, y = 18 cm and z = 10 cm, and the frame coordinate system has been turned by 30 degrees around the x-axis and then by 90 degrees around the y-axis. The frame F is represented internally by:

$$F = \begin{bmatrix} 1 & 0 & 0 & 24 \\ 0 & 1 & 0 & 18 \\ 0 & 0 & 1 & 10 \\ 0 & 0 & 0 & 1 \end{bmatrix} \begin{bmatrix} 0 & 0 & 1 & 0 \\ 0 & 1 & 0 & 0 \\ -1 & 0 & 0 & 0 \\ 0 & 0 & 0 & 1 \end{bmatrix} \begin{bmatrix} 1 & 0 & 0 & 0 \\ 0 & 0.87 & -0.5 & 0 \\ 0 & 0.5 & 0.87 & 0 \\ 0 & 0 & 0 & 1 \end{bmatrix}$$

$$F = \begin{bmatrix} 0 & 0.5 & 0.87 & 24 \\ 0 & 0.87 & -0.5 & 18 \\ -1 & 0 & 0 & 10 \\ 0 & 0 & 0 & 1 \end{bmatrix}$$

The third column of the matrix is a unit vector along the z-direction of the frame coordinate system, which is expressed in coordinate values of the base coordinate system:

$$(0,0,1)^T_{frame} = (0.87, -0.5, 0)^T_{basis}$$

Let the gripper length GL be 6 cm away from the revolute joint $theta_5$ to the actual gripping point. The point to arrive at is then calculated as

$$P = P_{frame} - (0.87,\ -0.5,\ 0)^T * GL$$

$$P = (24, 18, 10)^T - (0.87,\ -0.5,\ 0)^T * 6$$

$$P = (18.8, 15, 10)^T$$

The angle $theta_1$ then drives the robot to the point P on the xy-plane. Fig. 2.24 shows a simplified derivation of the angle $theta_1$.

Let the center point of the robot stand be at x = a and y = b (relative to the base coordinate system), and let the point P have the coordinate values $x_p = 24$ and $y_p = 21$. The angle $theta_1$ is then given by

$$\tan theta_1 = \frac{y_p - b}{x_p - a}$$

$$theta_1 = \arctan((y_p - b)/(x_p - a))$$

Let

$$a = 14$$
$$b = 11$$

and thus $theta_1$ has the value

$$theta_1 = 45°$$

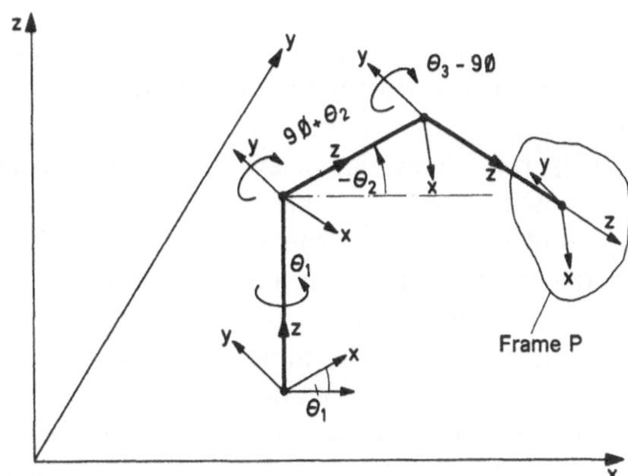

Fig. 2.26: Evaluation of orientation of frame P

The following numbers are needed for the evaluations of $theta_2$ and $theta_3$:

l_1 = 40 cm = distance from the xy-plane to the second revolute joint,
l_2 = 20 cm = lengths of upper and lower arm of the robot.

The schematic drawing in Fig. 2.25 shows the geometry for the derivation of angles $theta_2$ and $theta_3$.
Thus:

$$r = \sqrt{(x_p-a)^2+(y_p-b)^2}$$

$$s = \sqrt{(x_p-a)^2+(y_p-b)^2+(z_p-l_1)^2}$$

$$t = 0{,}5\ \sqrt{4 * l_2^2 - s^2}$$

Since the distances s, l_2 and l_2 form an isosceles triangle, it follows that

$$\tan(alpha - theta_2) = 2 * t/s$$

$$theta_2 = alpha - \arctan((\sqrt{4 * l_2^2 - s^2})/s)$$

$$\tan alpha = (l_1 - z_p)/r$$

The angle alpha is found by

$$alpha = \arctan((l_1 - z_p)/\sqrt{(x_p-a)^2+(y_p-b)^2})$$

and $theta_2$ may then be evaluated as

$$theta_2 = 38.24°$$

and $theta_3$ as

$$theta_3 = 270 - 2 * beta$$
$$theta_3 = 90 + 2 * alpha - 2 * theta_2$$

The angles alpha and $theta_2$ have already been evaluated and thus give

$$theta_3 = 170°$$

The determination of the last three angles for orientation is carried out in a different way. (This is possible because of the special robot geometry in which all three axes intersect at one point.) This idea is prompted by the fact that the orientation of the last robot joint may be evaluated simply with transformations and the already calculated angles $theta_1$, $theta_2$ and $theta_3$. The orientation of the gripper at the pick-up point is also known from the frame F, and it is possible to calculate the transformation matrix to arrive at that frame from the last robot joint. Only orientation is considered here, and all transformations are set to zero. Fig. 2.26 illustrates the rotations necessary in order to advance from the base coordinate system to frame P, which specifies the robot orientation by its z-axis.

First, joint 1 is rotated by an angle $theta_1$ around the z-axis (of the base coordinate system), so that its x-axis points in the direction of the second joint. Then, the new y-axis is rotated by $90+\text{theta}_2$ degrees, so that the resulting z-axis points in the direction of the second joint. Finally, the y-axis is turned by theta_3-90 degrees, so that the new z-axis points in the direction of the third robot joint, which has the gripper attached to it. Consequently, the frame P has the following orientation matrix

$$P=ROT(z,\text{theta}_1)*ROT(y,90+\text{theta}_2)*ROT(y,\text{theta}_3-90)$$

Please note that the following abreviations will be used:

$\sin \text{theta}_i = si$
$\cos \text{theta}_i = ci$
$\sin(\text{theta}_i+\text{theta}_k) = si*ck+ci*sk = sik$
$\cos(\text{theta}_i+\text{theta}_k) = ci*ck-si*sk = cik$

Thus P contains

$$P=\begin{bmatrix} c1 & -s1 & 0 & 0 \\ s1 & c1 & 0 & 0 \\ 0 & 0 & 1 & 0 \\ 0 & 0 & 0 & 1 \end{bmatrix} \begin{bmatrix} -s2 & 0 & c2 & 0 \\ 0 & 1 & 0 & 0 \\ -c2 & 0 & -s2 & 0 \\ 0 & 0 & 0 & 1 \end{bmatrix} \begin{bmatrix} s3 & 0 & -c3 & 0 \\ 0 & 1 & 0 & 0 \\ c3 & 0 & s3 & 0 \\ 0 & 0 & 0 & 1 \end{bmatrix}$$

$$P=\begin{bmatrix} -c1*s2 & -s1 & c1*c2 & 0 \\ -s1*s2 & c1 & s1*c2 & 0 \\ -c2 & 0 & -s2 & 0 \\ 0 & 0 & 0 & 1 \end{bmatrix} \begin{bmatrix} s3 & 0 & -c3 & 0 \\ 0 & 1 & 0 & 0 \\ c3 & 0 & s3 & 0 \\ 0 & 0 & 0 & 1 \end{bmatrix}$$

$$P=\begin{bmatrix} -c1*s2*s3+c1*c2*c3 & -s1 & c1*s2*c3+c1*c2*s3 & 0 \\ -s1*s2*s3+s1*c2*c3 & c1 & s1*s2*c3+s1*c2*s3 & 0 \\ -c2*s3-s2*c3 & 0 & c2*c3-s2*s3 & 0 \\ 0 & 0 & 0 & 1 \end{bmatrix}$$

$$P=\begin{bmatrix} c1*c23 & -s1 & c1*s23 & 0 \\ s1*c23 & c1 & s1*s23 & 0 \\ -\ s23 & 0 & c23 & 0 \\ 0 & 0 & 0 & 1 \end{bmatrix}$$

Now the frame P has to be rotated around its z-axis by an angle theta_4, then around its new y-axis by an angle theta_5, and finally around the new z-axis by theta_6 to make it coincident with frame F. (Only rotation is considered here.)

$$F=P*ROT(z,\text{theta}_4)*ROT(y,\text{theta}_5)*ROT(z,\text{theta}_6)$$

Pre-multiplying by the inverse matrix P^{-1} results in the equation

$$P^{-1}F=ROT(z,\text{theta}_4)*ROT(y,\text{theta}_5)*ROT(z,\text{theta}_6)$$

$$P^{-1}F=\begin{bmatrix} c4 & -s4 & 0 & 0 \\ s4 & c4 & 0 & 0 \\ 0 & 0 & 1 & 0 \\ 0 & 0 & 0 & 1 \end{bmatrix} \begin{bmatrix} c5 & 0 & s5 & 0 \\ 0 & 1 & 0 & 0 \\ -s5 & 0 & c5 & 0 \\ 0 & 0 & 0 & 1 \end{bmatrix} \begin{bmatrix} c6 & -s6 & 0 & 0 \\ s6 & c6 & 0 & 0 \\ 0 & 0 & 1 & 0 \\ 0 & 0 & 0 & 1 \end{bmatrix}$$

$$P^{-1}F=\begin{bmatrix} c4c5c6-s4s6 & -c4c5s6-s4c6 & c4s5 & 0 \\ s4c5c6+c4s6 & -s4c5s6+c4c6 & s4s5 & 0 \\ -s5c6 & s5s6 & c5 & 0 \\ 0 & 0 & 0 & 1 \end{bmatrix}$$

If the matrix P is inverted and if F is substituted into the general representation of rotations, it follows

$$P^{-1}F=\begin{bmatrix} c1c23 & s1c23 & -s23 & 0 \\ -s1 & c1 & 0 & 0 \\ c1s23 & s1s23 & c23 & 0 \\ 0 & 0 & 0 & 1 \end{bmatrix} \begin{bmatrix} tx & ox & ax & 0 \\ ty & oy & ay & 0 \\ tz & oz & az & 0 \\ 0 & 0 & 0 & 1 \end{bmatrix}$$

$$P^{-1}F=\begin{bmatrix} c1c23tx+s1c23ty-s23tz & c1c23ox+s1c23oy-s23oz & c1c23ax+s1c23ay-s23az & 0 \\ -s1tx+c1ty & -s1ox+c1oy & -s1ax+c1ay & 0 \\ c1s23tx+s1s23ty+c23tz & c1s23ox+s1s23oy+c23oz & c1s23ax+s1s23ay+c23az & 0 \\ 0 & 0 & 0 & 1 \end{bmatrix}$$

Equating the corresponding elements of both matrices for the expression $P^{-1}F$ delivers the following equations:

$$c4s5=c1c23ax+s1c23ay-s23az$$
$$s4s5=-s1ax+c1ay$$

Thus, $theta_4$ may be calculated as

$$theta_4=\arctan\left(\frac{-s1ax+c1ay}{c1c23ax+s1c23ay-s23az}\right)$$

$Theta_6$ may now be evaluated with the help of the equations

$$-s5c6=c1s23tx+s1s23ty+c23tz$$
$$s5s6=c1s23ox+s1s23oy+c23oz$$

This leads to the expression for $theta_6$

$$theta_6=\arctan\left(\frac{c1s23ox+s1s23oy+c23oz}{-c1s23tx-s1s23ty-c23tz}\right)$$

Finally, the value of $theta_5$ is computed from

$$s4s5=-s1ax+c1ay$$
$$c5\;=c1s23ax+s1s23ay+c23az$$

$$theta_5=\arctan\left(\frac{-s1ax+c1ay}{s4(c1s23ax+s1s23ay+c23az)}\right)$$

This algorithm may certainly be improved and optimized for a practical implementation. Thus the determination of $theta_5$ should distinguish the cases when $theta_4$ is 0 or 180 degrees because a different formula has to be chosen in order to calculate $theta_5$, so that a division by zero is avoided, or the missing information about the sign of $theta_4$ can be added.

The equations

$$c4s5 = c1c23ax + s1c23ay - s23az$$
$$c5 \quad = c1s23ax + s1s23ay + c23az$$

may be a starting point, and the angle $theta_5$ is then

$$theta_5 = \arctan\left(\frac{c1c23ax + s1c23ay - s23az}{c4(c1s23ax + s1s23ay + c23az)}\right)$$

Substitution of numerical values delivers the results

$$theta_4 = -\ 29.3°$$
$$theta_5 = \ 64.9°$$
$$theta_6 = \ 25.0°$$

Thus the joints have to be driven to the following angles

$$theta_1 = \ 39.8°$$
$$theta_2 = \ 38.24°$$
$$theta_3 = \ 170°$$
$$theta_4 = -\ 29.3°$$
$$theta_5 = \ 64.9°$$
$$theta_6 = \ 25°$$

by the controller in order to move the robot to the position and orientation programmed in frame F by

```
F := FRAMEC(ROTC(y,90)*ROTC(x,30),
VECTORC(24,18,10));
.
.
.
SMOVE arm TO F;
```

Such angle demands are absolute. Thus the controller has to evaluate the incremental angle relative to the current robot position A by

$$\Delta\Theta_i = \Theta_i - \Theta_{Ai}$$

However, there are several ways in which the controller can reach the new joint angles $theta_i$ (Sect. 2.3.3). In high-level languages, a programmer specifies a *trajectory* of a robot not only by start and end frames, but he may also give intermediate points which are passed with required velocities. This necessitates the analysis of the whole movement in a module for *trajectory planning*. It segments the motion path accordingly and evaluates a table of intermediate points which are approached in turn after specific time intervals (see also PAUL [2.7]).

In the final part, the transformation from robot into world coordinates is explained. This becomes necessary if a programmer has defined a point of motion by the teach-in procedure and not by explicit frame values. Thus the robot is driven to

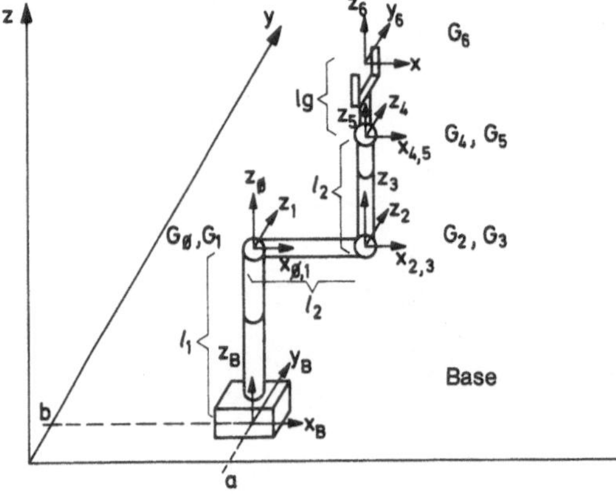

Fig. 2.27: Joint frames G_i of a six degree of freedom robot in base configuration

the desired position and orientation, the joint angles are interrogated, and the transformation matrix of the frame is calculated. Required coordinate transformations are obtained as follows:

To begin with, the origin of the base coordinate system is set to coincide with the base of the robot (see Fig. 2.27). Now each robot joint is assigned a frame, say G_1 to G_6. This has to be done by observing the following rules for the orientation of two successive frames:

1. The axis z_{i-1} lies along the axis of motion of the i-th robot joint.
2. The axis x_i is perpendicular to the axis z_{i-1} and points away from it.

The operations listed below arrive at frame G_i from frame G_{i-1}:

1. Rotate around z_{i-1} by an angle theta$_i$ so that x_{i-1} is parallel to x_i.
2. Translate a distance d_i along z_{i-1}.
3. Translate a distance a_i along x_i (i.e., the rotated x_{i-1}).
4. Rotate around x_i by an angle alpha$_i$ so that the z-axes correspond again.

Alpha$_i$, d_i and a_i have the values listed in Table 2.1 if the robot is in its base configuration with all theta$_i$ equal to zero (Fig. 2.27).

Table 2.1: Parameter values of the frame transformation of the robot

Θ	α	d	a
Θ_1	−90	0	0
Θ_2	0	0	l_2
Θ_3	90	0	0
Θ_4	−90	l_1	0
Θ_5	90	0	0
Θ_6	0	l_g	0

Transformations from frame G_{i-1} to frame G_i are carried out by multiplying the following four matrices representing their four respective operations:

$$T_{i-1,i} = ROT(z_{i-1}, \Theta_i) * TRANSL(0,0,d_i) * TRANSL(a_i,0,0) * ROT(x_i, \alpha_i)$$

$$T_{i-1,i} = \begin{bmatrix} \cos\Theta_i & -\sin\Theta_i & 0 & 0 \\ \sin\Theta_i & \cos\Theta_i & 0 & 0 \\ 0 & 0 & 1 & 0 \\ 0 & 0 & 0 & 1 \end{bmatrix} \begin{bmatrix} 1 & 0 & 0 & a_i \\ 0 & 1 & 0 & 0 \\ 0 & 0 & 1 & d_i \\ 0 & 0 & 0 & 1 \end{bmatrix} \begin{bmatrix} 1 & 0 & 0 & 0 \\ 0 & \cos\alpha_i & -\sin\alpha_i & 0 \\ 0 & \sin\alpha_i & \cos\alpha_i & 0 \\ 0 & 0 & 0 & 1 \end{bmatrix}$$

$$T_{i-1,i} = \begin{bmatrix} \cos\Theta_i & -\sin\Theta_i\cos\alpha_i & \sin\Theta_i\sin\alpha_i & a_i\cos\Theta_i \\ \sin\Theta_i & \cos\Theta_i\cos\alpha_i & -\cos\Theta_i\sin\alpha_i & a_i\sin\Theta_i \\ 0 & \sin\alpha_i & \cos\alpha_i & d_i \\ 0 & 0 & 0 & 1 \end{bmatrix}$$

Hence, starting at frame G_0, which only contains a translation relative to the base coordinate system, i.e.,

$$G_0 = \begin{bmatrix} 1 & 0 & 0 & a \\ 0 & 1 & 0 & b \\ 0 & 0 & 1 & l_1 \\ 0 & 0 & 0 & 1 \end{bmatrix}$$

the next frame G_i is obtained by successive post-multiplications (because the transformations are relative to the frame coordinate system) with the transformation $T_{i-1,i}$ until G_6 is reached, which delivers position and orientation of the gripper and thus the frame F at which the robot is currently positioned. Hence,

$$F = G_0 * T_{0,1} * T_{1,2} * T_{2,3} * T_{3,4} * T_{4,5} * T_{5,6}$$

The substitution of values for $alpha_i$, d_i and a_i into the transformation matrices $T_{i-1,i}$ results in

$$T_{0,1} = \begin{bmatrix} c1 & 0 & -s1 & 0 \\ s1 & 0 & c1 & 0 \\ 0 & -1 & 0 & 0 \\ 0 & 0 & 0 & 1 \end{bmatrix}$$

$$T_{1,2} = \begin{bmatrix} c2 & -s2 & 0 & l_2c2 \\ s2 & c2 & 0 & l_2s2 \\ 0 & 0 & 1 & 0 \\ 0 & 0 & 0 & 1 \end{bmatrix}$$

$$T_{2,3} = \begin{bmatrix} c3 & 0 & s3 & 0 \\ s3 & 0 & -c3 & 0 \\ 0 & 1 & 0 & 0 \\ 0 & 0 & 0 & 1 \end{bmatrix}$$

$$T_{3,4} = \begin{bmatrix} c4 & 0 & -s4 & 0 \\ s4 & 0 & c4 & 0 \\ 0 & -1 & 0 & l_2 \\ 0 & 0 & 0 & 1 \end{bmatrix}$$

$$T_{4,5}=\begin{bmatrix} c5 & 0 & s5 & 0 \\ s5 & 0 & -c5 & 0 \\ 0 & 1 & 0 & 0 \\ 0 & 0 & 0 & 1 \end{bmatrix}$$

$$T_{5,6}=\begin{bmatrix} c6 & -s6 & 0 & 0 \\ s6 & c6 & 0 & 0 \\ 0 & 0 & 1 & l_g \\ 0 & 0 & 0 & 1 \end{bmatrix}$$

These matrices are then multiplied out and the result is

$$T_{0,1} * T_{1,2} * T_{2,3} * T_{3,4} * T_{4,5} * T_{5,6} =$$

$$\left[\begin{matrix} c1c23c4c5c6-s1s4c5c6-c1s23s5c6-c1c23s4s6-s1c4s6 \\ s1c23c4c5c6-c1s4c5c6-s1s23s5c6-s1c23s4s6+c1c4s6 \\ -\;s23c4c5c6-\;c23s5c6\;+s23s4s6 \\ 0 \end{matrix}\right.$$

$$\begin{matrix} -c1c23c4c5s6+s1s4c5s6+c1s23s5s6-c1c23s4c6-s1c4c6 \\ -s1c23c4c5s6+c1s4c5s6+s1s23s5s6-s1c23s4c6+c1c4c6 \\ s23c4c5s6+c23s5s6+s23s4c6 \\ 0 \end{matrix}$$

$$\begin{matrix} c1c23c4s5-s1s4s5+c1s23c5 \\ s1c23c4s5+c1s4s5+s1s23c5 \\ -s23c4s5+c23c5 \\ 0 \end{matrix}$$

$$\left.\begin{matrix} (c1c23c4s5-s1s4s5+c1s23c5)l_g+(c1s23+c1c2)\,l_2 \\ (s1c23c4s5+c1s4s5+s1s23c5)l_g+(s1s23+s1c2)l_2 \\ (-s23c4s5+c23c5)\,l_g+(c23-s2)l_2 \\ 1 \end{matrix}\right]$$

Since

$$G_0=\begin{bmatrix} 1 & 0 & 0 & a \\ 0 & 1 & 0 & b \\ 0 & 0 & 1 & 0 \\ 0 & 0 & 0 & 1 \end{bmatrix}$$

the values of the frame matrix

$$F=G_0 * (T_{0,1} * T_{1,2} * T_{2,3} * T_{3,4} * T_{4,5} * T_{5,6})$$

$$F=\begin{bmatrix} tx & ox & ax & px \\ ty & oy & ay & py \\ tz & oz & az & pz \\ 0 & 0 & 0 & 1 \end{bmatrix}$$

are obtained by the matrix elements

$$tx=(c1c23c4-s1s4)c5c6-c1s23s5s6+(-c1c23s4-s1c4)s6$$
$$ty=(s1c23c4+c1s4)c5c6-s1s23s5c6+(-s1c23s4+c1c4)s6$$
$$tz=(-s23c4c5-c23s5)c6+s23s4s6$$

$$ox=(-c1c23c4+s1s4)c5s6+c1s23s5s6+(-c1c23s4-s1c4)c6$$
$$oy=(-s1c23c4-c1s4)c5s6+s1s23s5s6+(-s1c23s4+c1c4)c6$$
$$oz=(s23c4c5+c23s5)s6+s23s4c6$$

$$ax=(c1c23c4-s1s4)s5+c1s23c5$$
$$ay=(s1c23c4+c1s4)s5+s1s23c5$$
$$az=-s23c4s5+c23c5$$

$$px=((c1c23c4-s1s4)s5+c1s23c5)l_g+c1(s23+c2)l_2+a$$
$$py=((s1c23c4+c1s4)s5+s1s23c5)l_g+s1(s23+c2)l_2+b$$
$$pz=(-s23c4s5+c23c5)l_g+(c23-s2)l_2+l_1$$

In addition to trajectory planning, both kinds of coordinate transformations are needed in any convenient programming system, during teach-in as well as during program execution. This amount of calculation justifies a possible use of *multi-processor systems* (may include *arithmetic processors*) or *multi-computer systems.* Otherwise, the controlling processor will be occupied by these numerical calculations to such an extent that the control itself and the program flow may be influenced detrimentally.

The frame F is obtained again if the values $theta_1$ to $theta_6$ and l_1, l_2, and l_g of the example at the beginning of this section are substituted into the above expressions for the matrix elements of the frame. However, one should be aware of numerical errors arising from the complexity of the formula.

2.3.3 Types of Move Control for Industrial Robots

The compiler or interpreter represents a frame internally as a DH-matrix, after the desired position and orientation has been defined by the programmer. This matrix forms the base for the calculation of joint angles (or distances). It is left to the controller to drive the robot to the new set of joint angles $theta_i$. Only in the case of the simpler **PTP control** can this be done without intermediate values of individual joint angles or frames. Otherwise the controller has to execute an **interpolation**, where further intermediate values or frames are calculated between the start and end-point (see also BINDER [2.8]). It is called *distance-sliced interpolation* if it specifies individual increments along the path. Time intervals may vary in which the robot moves an incremental distance, and depend on the trajectory geometry and programmed velocity. The interpolation is called *time-sliced* if it calculates joint values at equal time intervals.

The controller is given start and end frames of motions in robot or Cartesian (world) coordinates, possibly modified by additional intermediate frames and velocities. Two types of control are distinguished in principle:

1. **Point to point** (PTP) control
2. **Continuous path** (CP, better termed "controlled path").

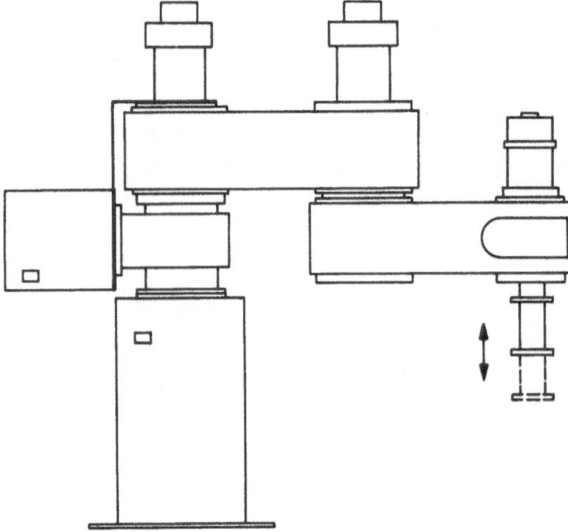

Fig. 2.28: Small Japanese robot with two vertical positioning axes (SCARA type)

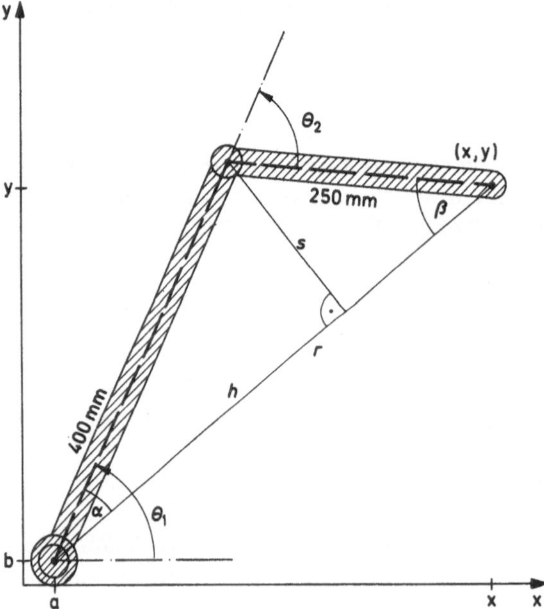

Fig. 2.29: Joint axes positions of the robot

In PTP control, all robot axes are moved with maximum velocity or acceleration, until individual joint positions are reached. Although moves of all axes start at the same time, they stop at different times because some joints are moved large angles and others only small ones. There is no functional interaction between individual axes, and the resulting motion of the gripper (or TCP) is unpredictable (for the programmer).

Functional interactions of individual joint moves are taken into account in the case of **continuous path** control. The simplest method merely delays all joints so that they all arrive at their endpoint at the same time. Besides **joint interpolation, linear**

interpolation in Cartesian coordinates is of great importance. In this case, the controller calculates the motion of the robot axes, such that the TCP moves on a straight line between start and end-point. **Circular** and **parabolic** interpolations have also been implemented, as well as other complex functions which are useful in surface following.

These three types of control are now introduced here, PTP, joint and linear Cartesian interpolation. Take a small robot with with four axes (Fig. 2.28), as it is used in the handling of lighter parts. We are only considering the two revolute joints theta_1 and theta_2. The coordinate transformations follow from Fig. 2.29:

a) robot to Cartesian coordinates:

$$x = 400 * \cos \Theta_1 + 250 * \cos (\Theta_1 + \Theta_2) + a$$
$$y = 400 * \sin \Theta_1 + 250 * \sin (\Theta_1 + \Theta_2) + b$$

b) Cartesian to robot coordinates:

$$r = \sqrt{(x-a)^2 + (y-b)^2}$$

$$h^2 + s^2 = 400^2$$

$$(r-h)^2 + s^2 = 250^2$$

$$h = \frac{48750}{r} + \frac{r}{2}$$

$$s = \sqrt{160000 - h^2}$$

$$\alpha = \arctan \left(\frac{s}{h}\right)$$

$$\beta = \arctan \left(\frac{s}{r-h}\right)$$

$$\Theta_1 = \arctan \left(\frac{y-b}{x-a}\right) + \alpha$$

$$\Theta_2 = -\alpha - \beta$$

The robot is to be driven from the position

$x = 435, y = 510$ bzw. $\Theta_1 = 60°$ $\Theta_2 = -25°$

to the position

$x = 252, y = 161$ bzw. $\Theta_1 = 70°$ $\Theta_2 = -140°$

PTP control now moves the axis theta_1 with maximum velocity by two degrees per time interval and theta_2 by ten degrees. Fig. 2.30 shows the resulting path of the end-point. The weaving motion is caused by the fact that theta_1 has already finished its move after a third of the total motion time. Fig. 2.34 shows the motions for both joints and all three types of control.

Joint interpolation reduces the moves of theta_1 to 0.87 degrees, so that both joints finish at the same time (Fig. 2.34). The resulting motion is circular (Fig. 2.31).

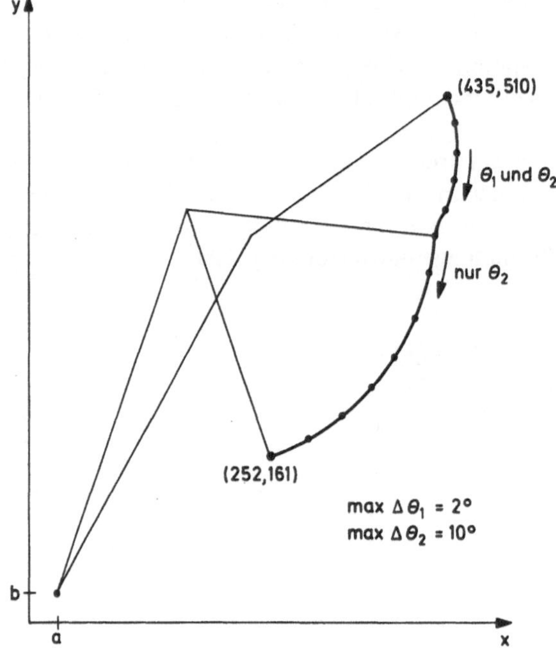

Fig. 2.30: PTP control

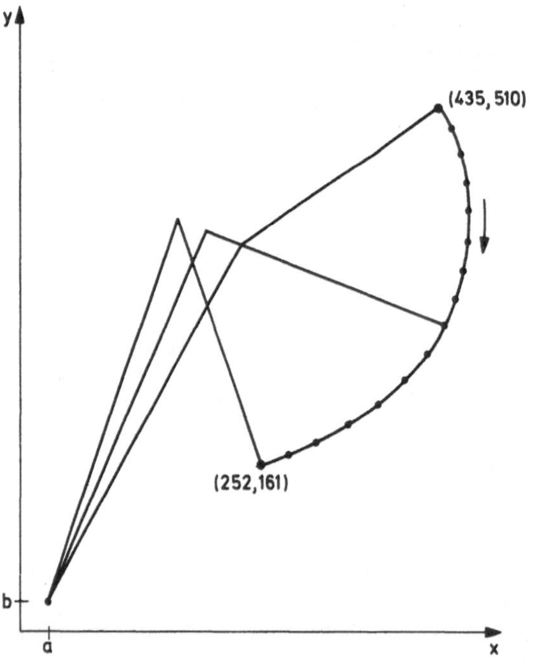

Fig. 2.31: Linear joint interpolation

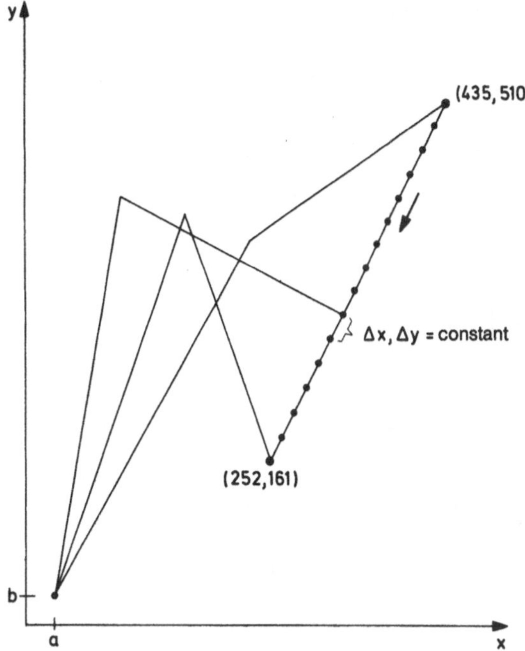

Fig. 2.32: Linear Cartesian Interpolation (continuous path control)

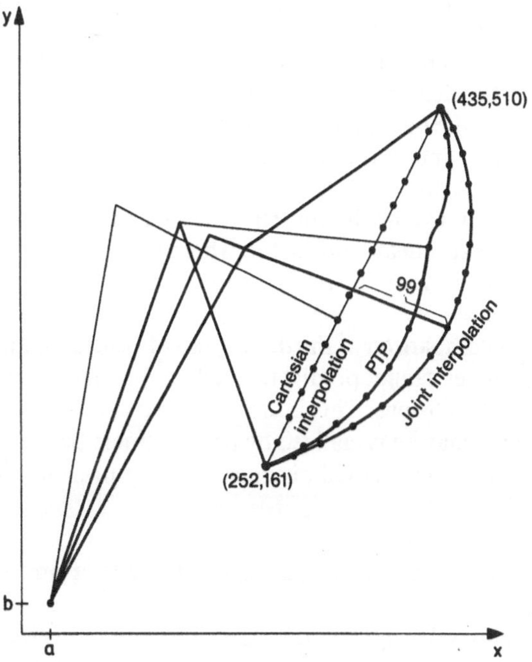

Fig. 2.33: Comparison of trajectories for various types of control

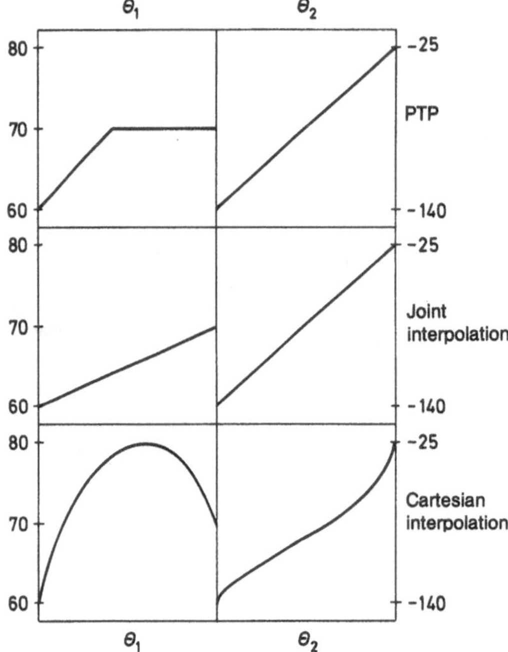

Fig. 2.34: Types of control for robots

Linear Cartesian interpolation does not calculate the intermediate points in robot coordinates, but in world coordinates. Thus these intermediate points are transformed into robot coordinates to which the robot is moved. This interpolation cannot be divided into such small parts as was possible by joint interpolation because of significantly higher computing efforts. Thus the resultant motion is not exactly a straight line, as is illustrated in an idealized way in Fig. 2.32, but it is a series of small concatenated arcs. The use of arithmetic processors keeps the intervals, and thus deviations, to a minimum. The resulting motion diagram shows nonlinear behavior for both axes $theta_1$ and $theta_2$, where $theta_1$ even exeeds the final value of 70 degrees and goes up to nearly 80 degrees before it is reduced back to the final seventy degrees.

A comparison of the three types of control in Figs. 2.33 and 2.34 shows that a programmer cannot predict the path and hence has to leave a wide corridor in order to avoid collisions. This is also true to a certain extent for joint interpolation motion, which may be visualized by programmers but deviates from a straight line by up to 99 mm - with a total distance travelled of nearly 400 mm.

2.3.4 Programming Languages on Mainframe Computers

High-level language developments for industrial robots are based on languages known from mainframe computers. Thus, all familiar concepts, such as variables, block structures, declarations, procedures and others, are transferred automatically into any new robot language. Similarly, instructions for program flow control, arithmetic and assignments to variables are adopted from the underlying languages like

ALGOL, FORTRAN or PASCAL, because they, too, are needed for robot programming.

Language constructs not needed (at present!) in the context of robotics are not adopted into the robot language, so that the language complexity is reduced. These may include text entry, string variables and access to files. However, current limits imposed upon implementations by memory space and speed of calculation will disappear even in the near future due to the steady price fall of memories and multiple processor systems. Thus the processing of character strings may well be implemented in a robot program when various user inputs have to be evaluated at different points in the program.

However the languages of mainframes may be restricted, robot languages have to be extended by instructions for the control of robots. Usually this is associated with the introduction of new data types such as VECTOR, ROTATION and FRAME (Sect. 3.1.4). Further commands may be designed in a new language in order to open/close grippers/tools, to set up and synchronize tasks (Sects. 2.2.6 and 7.2), to process sensory information, to input/output signals and to set up hardware-specific parameters. **Orthogonality** of the language has to be observed during language design stages, particularly in connection with arithmetic operations. For example, after the data type VECTOR has been introduced, addition should be extended to cover vectors. However, the programmer now has to be aware that the same symbol "+" represents two different operations (for numbers and vectors) and that a careless use may result in type conflicts, if it is attempted to add a vector and a real number. Another aspect of orthogonality is the use of simple data types during assignments to variables which are more complex. Unfortunately, this is not available in all languages, and program section 2.7 will illustrate this. A geometric variable of type VECTOR (Sect. 3.1.4) is defined by variables of the standard type REAL.

Although it should be taken for granted, some languages do not even allow access to individual components of complex data types, which is illustrated by the statement

```
IF vector[1] > 5 THEN PRINT ("Large reach for robot");
```

where `vector[1]` indexes the x-component of the vector.

There are several ways of deriving new languages for industrial robots from those on mainframes. In the case of SRL and AL, the mainframe languages PASCAL and ALGOL have provided the base for syntactic and semantic language definitions. However, new compilers have been written which check the new syntax and generate code.

A simpler way of implementation is to keep the reference language itself, as in the case of PASRO (PAScal for RObots), and to predefine new data structures and

```
vx, vy, vz: REAL;
vector1    : VECTOR;
vx := 5.1;
vy := 10;
vz := 17.5;
vector1 := VECTORC (vx, vy, vz);
```

Program section 2.7: Use of standard data types in declarations of geometric data types

Table 2.2: Programming languages and their derived robot languages

Programming language	Language for industrial robots
ALGOL (ALGOrithmic language)	AL (Assembly Language) and SRL (Structured Robot Language)
PASCAL	PASRO (PAScal for RObots) and SRL
PL/1 (Programming Language)	AUTOPASS (AUTOmated Parts ASsembly System) and AML (A Manufacturing Language)
FORTRAN (FORmula TRANslation)	NONAME (Stanford Research Institute)
BASIC (Beginner's All Purpose Instruction Code)	MAL (Multipurpose Assembly Language)
RTL/2 (Real Time Language)	INDA (INDustrial Automation)

procedures in that language. Thus, no new compiler is needed, but only extended system libraries into which new procedures and functions are entered, such as, for instance, robot moves.

In conclusion, the concept of using already present mechanisms and system program segments eases the implementation. On the other hand, the required knowledge in computer science may be a disadvantage during programming because programmers for robots in industry are educated mainly in production techniques.

As shown in Table 2.2, popular programming languages have already been used to derive robot languages. It is conspicuous, however, that no programming languages from process control have been used, although they already contain concepts for tasking, interaction with peripherals, interrupt and exception handlers (Sect. 4.9).

2.3.5 Robot-Specific Programming Languages

Manufacturers of industrial robots particularly have developed their programming systems by starting with the capabilities of their own robots. They usually sell a language of the simpler type which has been developed *around the function of the controller.* Thus motion commands may be highly differentiated, so that programmers may specify various types of control and interpolation. The syntax of such languages is usually kept simple so that compilers or interpreters can be implemented on systems with limited storage capacity. Possibilities for program flow control, subroutine techniques and arithmetic are drastically reduced, and thus structured programming is not properly supported in more complex programs.

2.3.6 NC Programming

Numerically controlled (NC) production systems have been developing programming systems since the 1950s (see also WECK [2.9]). The resulting programming languages do not only contain complex function specifications, such as demands for

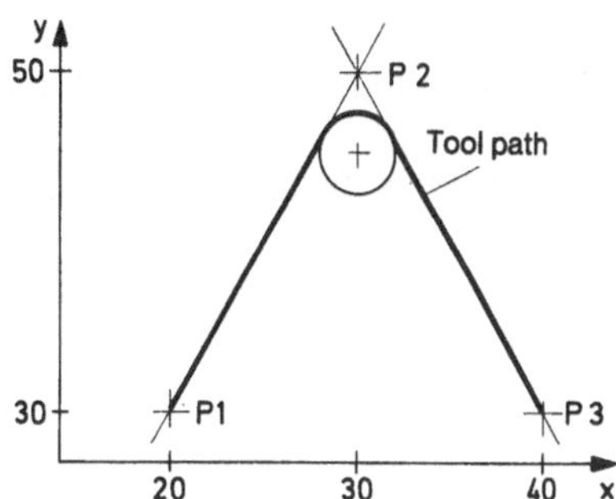

Fig. 2.35: Geometric data in NC programming

feed rates, hole drilling and control of milling machines, but also a descriptive part in which the geometry of work pieces is described. This uses *geometric conesections* (points, lines, circles and rectangles) to identify work-piece surfaces and positions. The data is evaluated at the time of program translation and substituted for the parameters of functions. This constitutes **implicit programming** because moves are not given explicitly but derived from geometric data. However, all data is often given as constants, so that no processing of variables is carried out at run-time and only very limited sensory interaction is possible.

The language APT (Automatically Programmed Tools) has achieved a special significance because it forms the base for a series of other languages. EXAPT (EXtended APT) has become known particularly in Germany as an adaptation of APT to special technologies. A program is structured into header data and the body. The first specifies tooling, geometric definitions, and technological definitions, which may include parameters for cutting threads, and the second contains instructions.

In NC programming, a programmer may specify a trajectory by geometric conesections, and the compiler (called *processor* in NC terminology) evaluates explicit values on the trajectory.

Example (Fig. 2.35):
- definition of point P1 by x = 20, y = 30
- definition of point P2 by x = 30, y = 50
- definition of point P3 by x = 40, y = 30
- definition of a straight line through P1, P2
- definition of a straight line through P2, P3
- definition of a circle of radius 5, touching both straight lines tangentially
- motion command to a milling machine to move from P1 to P3 along the straight lines and the arc connecting them.

In order to calculate the motion instructions explicitly, the NC processor needs a data base holding all necessary information about the robot and its environment. This world model is generally restricted to geometric and technological information.

The language ROBEX (ROBot EXapt), developed in WZL of the Technical University Aachen, is an extension of EXAPT for robot control. Program section 2.8 is useable for robots by specifying the machine tool as ROBOT1 and moving the ro-

```
MACHIN/MW1                specification of the nc machine
TRANS/100,200,0           position of the working piece on
                          the table
P1 = POINT/100,60,25      specification of point P1
                          x = 100, y = 60, z = 25
FROM/10,-10,0             start position of movement
GOTO/P1                   movement to P1
```

Program section 2.8: APT example

bot to point P1 with the GOTO instruction. ROBEX will be explained in greater detail later as the representative of NC concepts.

2.3.7 Production Schedules, Colloquial Language

Programming in high-level languages assumes adequate training, and thus existing specialists and fitters are all too often not usable as programmers after the introduction of industrial robots in their own production areas. Thus attempts are made to adjust robot languages to the production schedule of human labor, so that a programmer may find it much easier to familiarize himself with a language and implement programs in a shorter time. This requires *considerably larger programming and run-time systems.* A significant amount of information is missing in a program orientated toward production schedules, and indeed is not needed by humans, because their brains store a vast amount of information about the geometry of the workspace, tools, logical conditions, physical laws, jigs and many others. So, the instruction

```
< fit cover to base frame >
```

is perfectly adequate for humans to execute the fitting process. A program would have to perform the following:

→ Identify gripping points on the cover.
  → Find out that the cover is fixed by press studs.
  → Identify necessary orientations.
  → Generate explicit instructions:
    - Move robot to gripping point of cover.
    - Grip cover.
    - Move robot to fitting position.
    - Lower robot slowly and check that all press studs have engaged either by sudden force changes on the gripper or by vision systems.
    - Open gripper.
    - Depart robot.

A **planning module** in programming systems attempts to generate an instruction schedule according to the final state desired by the human (cover fitted to base

```
9  1. ASM SUPPORT BRACKET
10 P/U AND POSITION THE NUT IN THE NEST OF THE FIXTURE
11 1090037 NUT, CAR RET TAB          QTY 01
12 P/U, ORIENT AND POSITION THE BRACKET INTO THE
   FIXTURE WITH ITS TAB OVER THE NUT
13 1115191 BRKT ASM RAIL SUPPORT          QTY 01
14 P/U SCREW AND LOAD DRIVER
15 1107379 STUD, CR TAB INTLK          QTY 01
16 P/U, ORIENT AND POSITION THE INTERLOCK OVER
   THE BRACKET HOLE, WITH THE NOTCHED LUG UP
17 1117637 INTERLOCK, CR + TAB          QTY 01
18 P/U AIR DRIVER
19 DRIVE SCREW TIGHT
20 TORQUE 12.0 IN/LBS
21 ASIDE AIR GUN
```

1. OPERATE *nutfeeder* WITH *car-ret-tab-nut* AT *fixture.nest*
2. PLACE *bracket* IN *fixture* SUCH THAT *bracket.bottom*
CONTACTS *car-ret-tab-nut.top*
AND *bracket.hole* IS ALIGNED WITH *fixture.nest*
3. PLACE *interlock* ON *bracket* SUCH THAT
interlock.hole IS ALIGNED WITH *bracket.hole*
AND *interlock.base* CONTACTS *bracket.top*
4. DRIVE IN *car-ret-intlk-stud* INTO *car-ret-tab-nut*
AT *interlock.hole*
SUCH THAT TORQUE IS EQ 12.0 IN-LBS USING *air-driver*
ATTACHING *bracket* AND *interlock*
5. NAME *bracket interlock car-ret-intlk-stud car-ret-tab-nut*
ASSEMBLY *support-bracket*

AUTOPASS program for support bracket assembly.

Fig. 2.36: Production plan for support bracket assembly in AUTOPASS (LIEBERMAN [2.10])

frame). It starts with the (known) current state and needs an extensive **world model** (Sect. 2.3.8) which corresponds to the human knowledge of the workspace.

The AUTOPASS (AUTOmated Parts ASSembly) system of IBM makes an attempt at this schedule generation (see also LIEBERMAN [2.10]). Unfortunately, its not known how far the whole system has been implemented and tested. The compiler contains a **problem-solving part**, which checks user instructions for their validity and condition, performs necessary supplements and realizes a sequence of instructions in order to bring demands to a successful conclusion. If the generation fails, more information has to be entered by the operator.

The associated language contains instructions for assembly as well as for positioning and orientating components and operations to be performed with tools and fixtures. Additionally, characteristics of the robot may be declared, and spatial relationships between different objects specified. Figure 2.36 shows a production schedule for a brake assembly and the corresponding AUTOPASS program section.

One attempt even tries to express the task description in *quasi natural language* - it can only be quasi natural because only a *subset of the natural language* can be implemented for reasons of syntactic complexity and semantic analysis. Within the frame-work of an "Artificial Intelligence Project", the University of Milan developed the DONAU system (Domain Orientated NAtural language Understanding; see BERNORIO [2.11]; MAROY [2.12]). It permits the programming of a (simulated!) robot with a subset of Italian and produces a program in the robot language MICRO-PLANNER. However, since it was implemented in LISP on a UNIVAC 1108, the hardware requirements may be too severe for industrial implementation.

2.3.8 World Models

Some compilers for robot languages generate or use **world models**, which represent descriptions of robot environments (see TAYLOR [2.13]). **Data bases** contain information about:

- Positions of work pieces
- Relative positioning of work pieces
- Gripping points and orientations
- Physical characteristics of workpieces, like weight, surface condition, stiffness
- Changes executed so far in work piece positioning (sequence protocol of manipulation)
- Current robot position and orientation
- Fixed positions of feeders, conveyers, limits, storage pallets
- Manipulations already executed, e.g., assembly of two parts

A world model also contains a more general knowledge apart from the data valid or relevant for the appropriate robot programming, such as:

- Spatial geometry of working environment and robot
- Forbidden zones
- Logical connections or restrictions, e.g., that already assembled parts are heavier

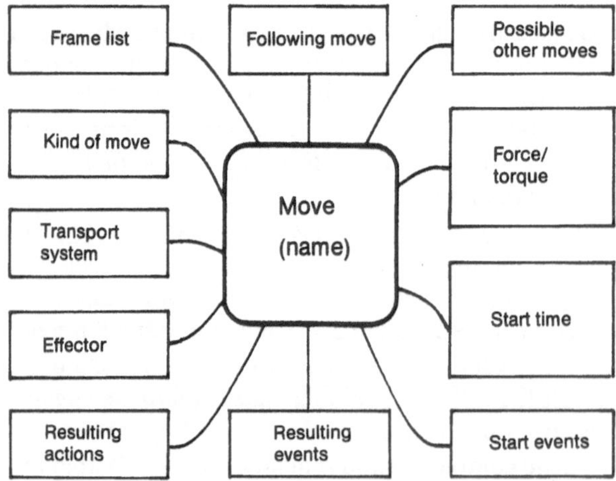

Fig. 2.37: Predefined description structure of a move in RODABAS

and may not longer be moved individually to different positions
- Boundary conditions, events

World models are used during compilation for implicit robot programming and also within limits in NC programming.

At the University of Karlsruhe, RODABAS (RObot DAta BAse) was developed as a relational data base (BLUME [2.14]). Its logical structure is derived from frame concepts which are used in the area of Artificial Intelligence (AI). Based on this relational data base, a "kernel" of a world model was specified for assembly operations. A scheme of descriptions was predefined and users may apply it to robots, sensors, move sequences and other features. Figure 2.37 shows the predefined structure of a move in RODABAS in which the user has to give only values and references. This structure of object descriptions should not be changed under normal circumstances, but if a change is necessary for special purposes, e.g., to add information about the temperature of an object, the structure can be easily modified.

3 Data Structures

This relatively abstract subject is dealt with here in some detail because the **power of a language** depends on potential data structures as well as on available commands. It could be assumed, particularly in robot languages, that the *provision of commands* is more important. It is, therefore, an intention of this book to show that the complexity of data structures is just as important in flexible robot applications with sensor integration.

Numerous programming examples are given in order to aid subject comprehension, and various solutions and representations are given in different robot languages.

Before we take a closer look at data objects, a peculiarity of SRL will be introduced. An SRL program may begin with a *system specification part*, in which the programmer can describe any hardware facilities used in his program, such as robots, effectors, sensors, interrupts, registers and input/output ports. Absolute addresses, data bases and time-out procedures can also be specified. The user assigns a symbolic name to the logical hardware address and thus determines the data structure for sensor input or input/output ports, for example. This considerably increases the readability and portability of SRL programs. The following example will illustrate this:

SRL:

```
SYSTEM_SPECIFICATION
  ROBOT:
    puma600 = ROBOT(0);
    jhrobot = ROBOT(1);
  EFFECTOR:
    ringgrip = GRIPPER(0) OF puma600;
    pargrip  = GRIPPER(1) OF jhrobot;
  SENSOR:
    visioninfo = CHANNEL(1);
    STRUCTURE visioninfo = RECORD
                             partno: INTEGER;
                             x,
                             y     : REAL;
                               .
                               .
                               .
                           END;
END_SYSTEM_SPECIFICATION;
```

During the execution of an SRL program, the sensor can now be activated and any sensor information read in by the following statements:

SRL:
```
INPUT (visioninfo);
IF visioninfo.part = 3 THEN
  SMOVE puma600 TO deposit;
```

3.1 Data Objects

Data objects are designated memory units of a particular **data type**. The concept of data types will be considered in more detail later. Memory units may consist of several memory cells (words, bytes, etc.). If a programmer assigns a **name** to such a data object, a **variable** is created. As explained in Sect. 2.1.1, variables can occupy one or more data cells into which data can be written or from which data can be read by the program, which knows the memory locations by their addresses. The use of addresses by the programmer is very involved and cumbersome and makes the program unintelligible, so a translator (*compiler* or *assembler*) allows symbolic calling of memory locations with **imaginary names**, arbitrarily chosen by the programmer.

Early programming languages allowed only a limited number of characters for the symbols. FORTRAN, for example, permits only six characters per name. Modern languages allow an unlimited number of symbols (within the limits of one line), but often only a certain number of characters from the left are regarded as significant. Additional characters are ignored and do not contribute to the differentiation of names, so that apparently different names are identical to the compiler (or assembler). Obviously, this can lead to troublesome errors. The first six to twelve characters are significant in ALGOL, from which AL is derived, and the first eight in PASCAL (although some compilers may allow more).

Some languages, like AL, SRL and PASRO, permit upper- and lower-case letters. However, no distinction is made between them. For example, `abc`, `ABC` and `aBc` are names for the same variable. This rule allows a clearer notation. Names may also contain digits, but they have to begin with a letter. Table 3.1 shows the particulars for the robot languages.

The significance of this convention may be best illustrated by the example of assigning an arithmetic evaluation:

FORTRAN:
```
STOCK = OLDSTK + INSTCK - OUTSTK
```

ALGOL:
```
STOCK = OLDSTOCK + INCOMMINGSTOCK
      - OUTGOINGSTOCK
```

PASRO:
```
Stock = Oldstock + Incommingstock
      - Outgoingstock
```

AL and SRL:
```
Stock = Old_Stock + In_comming_Stock
      - Out_going_Stock
```

AML:
```
STOCK = OLD_STOCK + IN_COMMING_STOCK
      - OUT_GOING_STOCK
```

VAL:

```
STOCK = OLDSTOCK + INCOMMINGSTOCK
STOCK = STOCK - OUTGOINGSTOCK
```

VAL-II:

```
STOCK = OLDSTOCK + INCOMMINGSTOCK
        - OUTGOINGSTOCK
```

HELP:

```
STOCK = OLDSTOCK + INCOMMINGSTOCK
        - OUTGOINGSTOCK
```

SIGLA: No expressions possible. There are only special counting registers.

ROBEX: There are no expressions in the present version.

The most legible solution is the possible use of the underline character "_" in AL, SRL and AML. Naturally, it is possible to use short names in all languages as in FORTRAN, but although they reduce typing effort, they preclude the possibility of *self-documenting nomenclature.* Self-explanatory names ease work considerably during later program debugging, changes or expansions and hence justify the increased tying effort during program writing.

An example should clarify the above-mentioned concept of data types: Assume the bit combination

0000000000110101

to be present in a 16-bit memory cell. How should this now be interpreted? Is it a

Table 3.1: Permissible and significant characters in identifiers. The significance of the special names of SIGLA is explained in further detail in Sect. 3.1.3.

Language	Significant	Permitted
SRL upper and lower case	unlimited	unlimited
AL upper and lower case	30	unlimited
PASRO upper and lower case	8 or more	unlimited
VAL upper case only	unlimited	unlimited
AML upper case*	unlimited*	unlimited*
HELP upper case only	6	unlimited
SIGLA	only I1...I16, P1...P16 and M1...M1023 allowed	
ROBEX upper case only	6	6

* Assumed, no contrary information in the manual

binary representation of the number "53" or the ASCII character "5", which has the same bit pattern? An indication of the data type is another important part of the information and is necessary for an unambiguous answer to this question. For example, the above case could be of type INTEGER or CHARACTER.

Apart from these technical grounds for the introduction of data types, there is another equally important reason from the viewpoint of the user: He does not want to become involved with the technicalities of computer sciences, but instead wants to work with data types that are well suited for his application. One such type, probably used in all applications, is INTEGER, representing a subset of the set of integers. It can only be a subset because no computer can represent the infinite set of integers, as no memory would be large enough. In contrast to these general, not problem-specific data types, there are problem-oriented types, for example the type VECTOR, which is particularly useful in robot applications. It consists of three elements of type REAL, and represents a subset of the set of all points in three dimensional space.

Obviously, the concept of data types is also useful in describing general and problem-oriented data structures. Different types are presented in detail in Sects. 3.1.3 to 3.1.6.

3.1.1 Declarations

A variable name is assigned to a data type by a declaration, through which the programmer attaches variables to certain data types, in SRL for example:

```
gripptranslation,boxedge,safedistance: VECTOR;
```

In this way, the compiler knows the three variables as well as the corresponding data types and can check during the use of the variables whether the intended operation is allowed for the data types involved. For example, vectors can be added, whereas a vector and a scalar cannot. Furthermore, the definition causes the compiler to allocate the necessary space for the variable.

Apart from this *explicit* form of declaration, there is the *implicit declaration* in certain instructions. For example, the assignment

```
newgripptranslation := gripptranslation + VECTORC(20,5,3);
```

is unambiguous, since the result on the right hand side delivers a vector and the variable `newgripptranslation` on the left hand side has to be of type VECTOR. This form of declaration seems to be more convenient for the programmer at first, since he can simply write down his program and declare a new variable whenever he needs one.

So, why is the explicit declaration still necessary, indeed, the only one permissible in many good and widely spread programming languages? For this see Fig. 3.1. The typing error in the variable `newgripptanslation` does not cause an error at the place where it occurs, but only in a later line, where something unknown and undefined has to be added to the vector `safedistance`. If it is kept in mind that there is usually more than one line between the two relevant ones, the effect on debugging

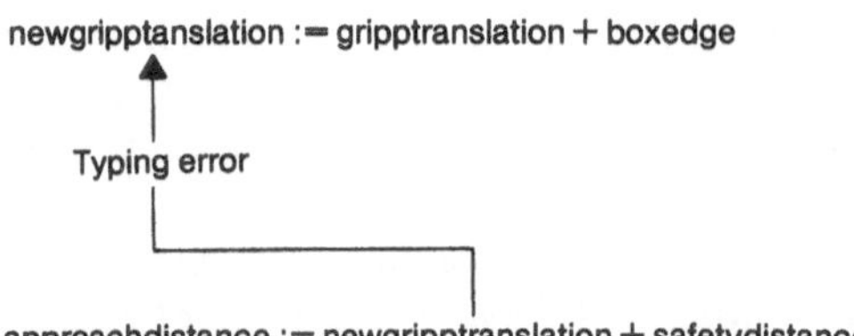

newgripptanslation := gripptranslation + boxedge
Typing error
approachdistance := newgripptranslation + safetydistance

Fig. 3.1: Implicit declarations and typing errors

Table 3.2: Forms of variable declarations

Form of declaration	PASRO	SRL	AL	VAL VAL-II	AML	HELP	SIGLA	ROBEX
Explicitly in program header	yes	yes	*	./.	*	*	./.	./.
Explicitly before use of variable	./.	./.	yes	./.	yes	yes	./.	./.
Implicitly	./.	./.	yes	yes	./.	yes	yes	yes

* Possible

can easily be estimated. It is even worse if `newgripptanslation` has been declared in a previous assignment. This time the typing error is not detectable until program execution.

The compiler will always notice typing errors with explicit variable declarations (unless the error results in a previously defined name), and the aforementioned effects cannot come into appearance. Therefore, explicit declarations increase **program security** and in addition encourage systematic software design, since the programmer is forced to identify the desired data objects before entering the program. A further argument for explicit declarations is improved program legibility, resulting from the declaration of all data in one program or program section (usually in the header).

Table 3.2 displays an overview of the permissible definitions in the different languages.

Although AML demands the explicit declaration of variables, it also permits a repeated declaration of the same identifier (in the same block level; see Sect. 2.2.3) in overwriting the old declaration. All other languages would generate a compiler error message. Such freedom in using declarations can have serious consequences in AML, for example, when a new variable is to be introduced during a later expansion of the program and the same identifier is used inadvertently. It is quite possible to overlook existing declarations as they do not have to be placed at the beginning. Thus it can be concluded that this feature is detrimental to program security.

Further objects can be defined apart from variables. The following list and Table 3.3 give a summary:

Table 3.3: Summary of possible declarations

	PASCAL	SRL	AL	VAL	AML	HELP	SIGLA	ROBEX
Constants	expl./ impl.	expl./ impl.	impl.	impl.	impl.	impl.	impl.	*
Dimensions	./.	./.	expl.	predef.	./.	predef.	./.	**
Data types	expl.	expl.	./.	./.	***	./.	./.	./.
Variables	expl.	expl.	****	impl.	expl.	expl./ impl.	impl.	*
Macros	./.	./.	expl.	./.	./.	./.	./.	expl.
Labels	expl.	./.	*****	impl.	impl.	impl.	impl.	impl.
Subroutines	expl.	expl.	expl.	impl.	expl.	impl.	impl.	./.
Procedures	expl.	expl.	expl.	./.	expl.	./.	impl.	./.
Functions	expl.	expl.	expl.	./.	expl.	./.	./.	./.
Interrupt routines	./.	expl.	expl.	expl.	expl.	./.	./.	./.
Tasks	./.	expl.	./.	./.	./.	******	./.	./.

* According to the publications to date, special constants are possible in ROBEX. This is the concern of the next section.

** Only the choice between the metric and imperial measuring system is possible.

*** Restricted to the implicit declaration of aggregates.

**** The American version contains only an explicit declaration. The Karlsruhe version also allows the implicit declaration because of the integration of the component POINTY into the language.

***** Labels are provided only for the so-called condition monitors in the American version, and since this language concept has been generalized in the Karlsruhe version, labels could be omitted.

****** The concept of processes in HELP comes close to the task concept (cf. Sect. 7.2).

Constants

There are two ways to define *constants*: implicitly, by writing down a number, as in an expression in AL

```
pound <- pennies / 100;
```

or explicitly, as in PASRO:

```
CONST divisor = 100;
```

The above assignment would be in PASRO:

```
pound := pennies/divisor;
```

(Please note the only formal difference in the assignment signs in AL or PASRO). The possibilities arising from explicit constant definition will be considered in detail in the next section.

Dimensions

All values must have dimensions, particularly as robot languages aim to describe the environment through various data types. This may be done by either fixing all lengths to be in millimeters (or inches) and all angles in degrees, as is implemented in VAL and ROBEX, or by allowing the user to combine dimensions and form new ones, as in AL. It supposes the following predefined dimensions of AL:

- DISTANCE in cm
- TIME in sec
- FORCE in gm
- ANGLE in deg

From these, new dimensions can be formed, such as acceleration:

```
DIMENSION acceleration = distance / (time * time);
```

New units of dimensions are obtained by macro definitions (Sect. 2.2.4):

```
DEFINE meter = ≠ 100 * cm ≠
```

This way, it is possible to introduce the units inches, oz, lbs and radians, which are additionally provided in the American implementation.

Furthermore, the following dimensions are predefined:

- TORQUE in gm/cm
- VELOCITY in cm/sec
- ANGULARVELOCITY in 1/sec
- DIMENSIONLESS

The AL compiler checks expressions where dimensioned variables are combined to see whether the combination is possible at all, whether the left- and right-hand sides have the same dimensions and whether the dimension is allowed in the particular instruction. This can lead to problems with frames (cf. Sects. 2.3.1 and 3.1.4), in which the vector part can be dimensioned. There is, for instance, a special frame – the "@", which returns the current arm position. What is its dimension? This problem – and others which will not be considered here – led to the combination of dimensionless and dimensioned frames (DISTANCE) in the Karlsruhe implementation.

Dimensioned variables are obtained by placing the dimension name in front of the data-type declaration. It is sufficient during implicit declarations to assign a dimensioned value to a variable. For example:

```
DISTANCE VECTOR gap,spacing;
acuteangle <- 30 * deg;
```

The variable acuteangle has been defined implicitly as a scalar with the dimension ANGLE.

In all VAL systems and in HELP, coordinates have to be entered in millimeters and all angles in degrees, whereas ROBEX allows the choice between metric (MM and DEGREES) and imperial (INCHES and DEGREES) measures.

```
TYPE vector = RECORD
                x, y, z : REAL
              END;
```

Program section 3.1: Predefined declaration of vector in PASRO

```
TYPE distancetype = RECORD
                      startpoint,
                      direction : vector;
                      length    : INTEGER
                    END;
```

Program section 3.2: Declaration of distancetype in PASRO

```
VAR     xmax,
        ymax,
        zmax    : INTEGER;
        xvect,
        yvect,
        zvect   : vector;
        cubex,
        cubey,
        cubez   : distancetype;
```

Program section 3.3: Declarations in PASRO

SIGLA does not recognize any dimensioned values but regards quantities which change the arm position as instructions for stepper motors (cf. Sect. 4.2.3).

Data types

So far, three data types have been introduced: INTEGER, CHARACTER and VECTOR, where VECTOR can be defined as a combination of three integer values. Certain data types are predefined elements in most languages - unfortunately, this is also true for most robot languages -, and apart from these, there are no possibilities for the programmer to define new, problem-oriented data types. This is only possible in SRL, PASRO and AML. In the example of program section 3.1, the data type vector, consisting of three components, x, y and z, is introduced. This data type can be used in further definitions, as in program section 3.2. The capital letters for symbols used here and in the remainder of the text serve only to give better legibility and are by no means mandatory.

Numerous variables can be declared with the predefined and programmer defined types as shown in program section 3.3 . Section 3.1.5 considers the possibilities for data type declarations in further detail.

Variables
Declarations of variables have already been dealt with in detail and are only mentioned here for completeness.

Macros
In general, program text in any language can be processed with a general macro generator, which itself is not part of that language. A macro generator is already contained in the compiler of some languages (AL and ROBEX). As already shown in Sect. 2.2.4.3, macro calls are a substitution of the possibly parameterized call by some text which has been defined in the course of the macro definition.

Labels
An instruction can be marked with a tag, also called a label. In nearly all languages, this is done by preceding the particular instruction with a tag terminated by a colon. In HELP an assignment with the label `aim2` would appear as follows:

```
aim2: xmax := 2 * zmax;
```

This assignment can be reached by a *jump instruction* (cf. Sect. 4.7). Explicit definitions are compulsory in some languages such as PASCAL (or PASRO) for methodological reasons; but most allow implicit definitions as in the previous example. Additionally, AML offers label variables to which the value of an implicitly declared label can be assigned. A following jump instruction can use this variable subsequently.

Subroutines
Subroutines (already treated in Sect. 2.2.4.1) are not declared explicitly in the simpler languages. They are marked by a preceding label and terminated by a RETURN instruction. The label can be reached by a *subroutine call* as well as by a conditional or unconditional jump instruction. The compiler is only interested in the start of the subroutine which is identified by the label during the definition and is otherwise treated just like any labelled instruction (cf. Sect. 6.1).

Procedures and Functions
As already mentioned in Sect. 2.2.4.2, procedures and functions are subroutines with local data scope accessible only within the procedure (or function). A further difference is the possibility of passing parameters during the procedure (or function) call. Both have always to be defined explicitly (cf. Chap. 6).

Interrupt Routines
Special program sections can be declared to be interrupt routines. These are only entered when a designated sensor has reached a particular threshold, either immediately or after the currently executed instruction. Further information can be obtained from Sects. 4.2.6 and 4.5.3.

Tasks
Tasks, as introduced in Sect. 2.2.6, can only be defined in SRL. HELP, in its process concept, allows a program construction which is very similar to the task idea. The AL version of the University of Karlsruhe has some very important commands for

task concepts available, in which only concluded programs can be defined as tasks. More about that in Sect. 7.2.

3.1.2 Definitions of Constants

Apart from the implicit definition of a constant in all languages, there is an explicit definition of a name as a constant, as in SRL, PASRO and AML/E but not in AML.

PASRO:
```
CONST divisor = 100;
or
CONST objectnumber = 10;
```

If a task has to be executed ten times - for example, the emptying of a pallet with ten objects - then a counting loop (Sect. 4.7.2.1) will be programmed. Should these loops appear several times with the object number as the repetition factor, then it is convenient to write the constant `objectnumber` instead of the digits 10 and define it to have a value of 10. So, if new pallets have to be manipulated with - for example - twelve objects, then it is not necessary to search through the whole program to find all the locations where the number of objects was specified, but it is sufficient to change the constant `objectnumber` in the definition part. Naturally, one could define a variable and initialize it with a value of ten, however, the definition of a constant makes it obvious that `objectnumber` is a program constant and does not change during program execution. Thus the clear differentiation between constant and variable program data increases **program security**. The feature allowing declarations of easy-to-change and legible data structures is an additional advantage. This is further treated in Sect. 3.1.5.1. Only SRL and PASRO offer these possibilities.

As already mentioned, the ROBEX version on hand does not recognize any variables in the sense described here. Various geometric objects, from points to bodies consisting of several objects, can be defined (cf. Sect. 3.1.4). They are defined with constant values, if necessary with already present constants. From these, the compiler generates numeric code with fixed values and without computational instructions for the interpreter driving the robot (cf. also Sect. 2.3.6, and Chap. 8). In this sense, ROBEX works only with constants - but very complicated constants, such as matrices. This procedure is derived from the numerical computations in the widespread languages APT and EXAPT and has the principal disadvantage of being unable to change values during program execution (and hence cannot allow modification of robot motion in response to sensory data).

3.1.3 Standard Data Types

Standard data types are simple data types which are present in many languages. Among them are:

INTEGER: whole, positive or negative numbers
REAL: fractional numbers in *floating point* notation
BOOLEAN: logical values of TRUE and FALSE

CHARACTER: characters are highly machine-dependent (the ASCII character set, for example)

STRING: predefined array of characters, such as a linear sequence of characters

EVENT: special counters *(semaphore variables)* for program synchronization (cf. Sect. 4.7.3)

One can regard digital input and output as events, even if this "data type" has not been declared and is not accessible to any operators (cf. Sect. 4.7.3).

Floating point notation separates the mantissa and exponent (standard in good pocket calculators) giving a presentation in normalized form. Example:

```
   5.3     represented as  0.53000000 E +01
-100       represented as -0.10000000 E +03
   0.0513  represented as  0.51300000 E -01
```

When using real values, one has to keep in mind that ordinary laws of algebra are not valid in all cases, because of a limited number of digits and the inaccuracies associated with this. For example, the expression

```
1/3 * 3 = 1
```

is valid in algebra.

If this is evaluated by the computer, probably by the expression (1/3)*3, it may be that 1/3 * 3 is only 0.99999999 E 00. A logical comparison of this result with the number 1 would be FALSE. Such *rounding errors* are not only relevant in logical operations but can accumulate to large errors in substantial calculations. This area of numerical mathematics will not be considered here in detail, but the reader is referred to relevant literature [3.1, 3.2]. Generally, the available accuracies should be sufficient for robot applications; it is enough to remember during practical programming that a logical test for equality with real numbers is only meaningful within a small range of tolerance. So, it is possible to write for the above example

```
|(1/3)*3 - 1| < 0.0001
```

and the comparison will deliver the correct result.

If a computer with a wordlength of 16 bits is assumed (2 bytes with 8 bits each), the memory demand and range of values can be taken from Table 3.4. It is notice-

Table 3.4: Range of values for the standard data types

Data type	Number of bytes	Range
INTEGER	2	-32768...+32767
REAL	4	0.15224277E-39...0.17014111E+38 **
BOOLEAN	at least 1 *	FALSE, TRUE
CHARACTER	at least 1 *	‹e.g. all ASCII characters›
EVENT	2	-32768...+32767

* If the processor can address bytes, one byte is sufficient to represent these data types.

** The range does not only depend on the number of bits, but also on the partitioning between mantissa and exponent and hence on the desired accuracy.

Table 3.5: Standard data types in the different languages

Data type	PASRO	SRL	AL	AML	VAL	VAL-II	HELP	SIGLA	ROBEX
INTEGER	INTEGER	INTEGER	SCALAR	INT	Integer	Scalar	Scalar	Counter	./.
REAL	REAL	REAL	SCALAR	REAL	./.	Scalar	Scalar	./.	./.
BOOLEAN	BOOLEAN	BOOLEAN	SCALAR	./.	./.	./.	./.	./.	./.
CHARACTER	CHAR	CHAR	./.	./.	./.	./.	./.	./.	./.
STRING	*	STRING	./.	STRING	./.	./.	./.	./.	./.
EVENT	./.	SEMAPHORE INTERRUPT	EVENT	./.	./.	Semaphore	Flags	./.	./.

* Many PASCAL implementations include strings, even though they are not standard.

Type identifiers of the relevant languages are given here in capital letters.

Table 3.6: Boolean values of AML data types

Type	Truth value
INT	True - if value is not zero False - if value is zero
REAL	True - if value is not 0.0 False - if value is 0.0
STRING	True
Aggregate (cf. Sect. 3.1.5.4)	True - if it has at least one true element False - if it has no true element or if it is empty

able at this point that data types are not only important for the interpretation of their content, but they also determine the necessary memory allocation for each variable.

Table 3.5 gives a summary of the standard data types in the various languages.

HELP and VAL-II combine the types INTEGER and REAL to form the type SCALAR, as does AL with INTEGER, REAL and BOOLEAN. Here integer values are real numbers without the fraction, and all numbers larger than zero are taken to be logical TRUE.

The representation of Boolean values in AML is an interesting feature. The value of the Boolean variables is given in Table 3.6.

The counters used as integer variables in SIGLA have predefined names, such as M1 to M1023 for global counters, P1 to P16 for passing of parameters in subroutines, and I1 to I16 for indirect addressing. The following explains the indirect addressing in SIGLA: The specification if I2 delivers the corresponding content of P2, which represents the number of a counter. Its content is then taken and substituted for I2. If, for example, the counter M100 in a program a was assigned the value 705 and the program b is called as a subroutine from a

```
EX/b,100
```

then this signifies firstly that the parameter P1 has the value 100 and secondly that there are no further parameters. Using the instruction I1 in program b, the content of P1, which is 100, is taken as the counter number and the contents of the counter evaluated with this number, which is the content of M100, 705.

3.1.4 Geometric Data Types

Most robot languages offer special data types for programming robot moves. The programmer can easily describe geometric relationships and frames with the help of these new data types. Unfortunately, only a few languages allow the combination of standard data types with geometric ones, or else there are no operators to combine the operands of the new data type (Sect. 3.2.1.2). Geometric data types supporting explicit motion programming (Sect. 4.2) are

- VECTOR (three-dimensional vectors)
- ROTATION (rotation defined by an angle about a vector)

```
TYPE
  vector      = RECORD
                  x, y, z: REAL
                END;

  rotmatrix   = RECORD
                  t, o, a: vector
                END;

  rotation    = RECORD
                  axis   : vector;
                  angle  : REAL;
                  matrix : rotmatrix
                END;

  frame       = RECORD
                  rot    : rotation;
                  transl : vector
                END;
```

Program section 3.4: Declaration of geometric data types in PASRO

```
TYPE
  vector        = ARRAY [1..3] OF REAL;
  rotmatrix     = ARRAY [1..3, 1..3] OF REAL;
```

Program section 3.5: Defining vectors and rotation matrices with arrays (PASCAL)

- ORIENTATION (rotation defined by three angles)
- FRAME (frame, defined by rotations and vectors)
- TRANS (transformation, defined like a frame)

Vectors describe *positions* and rotations describe *orientations*, and a frame represents a point of motion (see Sect. 2.3.1).

Using structured data types (Sect. 3.1.5), the geometric data types are predefined in PASRO (see program section 3.4). The declaration of the rotation matrix rotmatrix from three three-dimensional vectors is only necessary for internal representations of a 3x3 rotation, but it should be transparent to the user. Of course, vectors and rotation matrices can also be defined in arrays (see program section 3.5).

SRL:

```
conveyor: VECTOR;
partrotation: ROTATION;
approach_point, pickup, putdown: FRAME;
conveyor := VECTORC (0,35,0);
```

PASRO:

```
conveyor: VECTOR;
partrotation: ROTATION;
```

```
approach, pickup, putdown: FRAME;
MAKEVECTOR (conveyor,0,35,0);
```

AL:

```
DISTANCE VECTOR conveyor;
ROT partrotation;
FRAME approach,pickup,putdown;
conveyor <- VECTOR (0,35,0) * CM;
```

See also Fig. 3.2.

One distance vector, one rotation and three frames have been defined. The values x = 0 cm, y = 35 cm and z = 0 cm are assigned to the distance vector.

There is an implicit usage of geometric data types in AML; for example, a frame which is used as a parameter for some system procedures calls consists of an aggregate (cf. Sect. 3.1.5.4) of seven elements:

1. element: value of the x-axis
2. element: value of the y-axis
3. element: value of the z-axis
4. element: value of the roll axis
5. element: value of the pitch axis
6. element: value of the yaw axis
7. element: value of the gripper opening

A vector can be defined by an aggregate of three elements as well as a rotation.

AML:

```
CONV:  NEW 3 OF 0.0;
PARTRT:NEW <90, 0.0, 45>;
APPRO: NEW 7 OF 0.0;
PICKUP:NEW 7 OF 0.0;
PUTDOWN: NEW 7 OF 0.0;
CONV = <0, 35, 0>;
```

There are neither rotations nor vectors in VAL; it is only possible to assign the values of one frame to another with the SET instruction, or values can be specified ex-

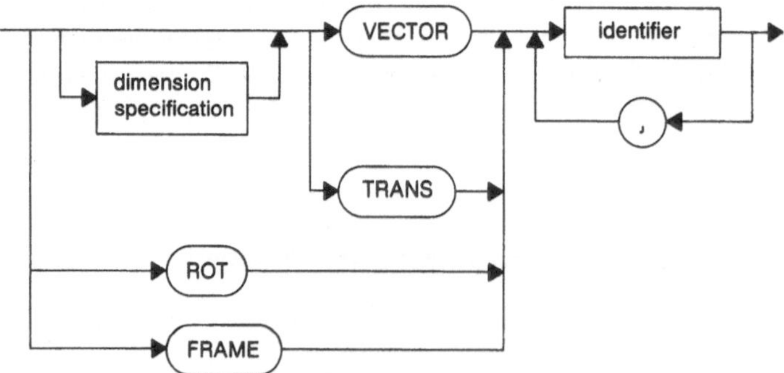

Fig. 3.2: Declaration of geometric data types in AL

plicitly with the POINT command (in immediate mode, but not during program execution). A *transformation* of a frame can be specified implicitly if a base frame is noted, followed by a transformation frame – separated by a colon. A small example, where only frame positions are considered for simplicity, is used to illustrate this point.

VAL:

```
POINT basis 300,550,250,0,0,0
```
Basis frame with x-, y-, and z-values defined, as well as Euler angles for orientation specification

```
POINT transform 0,0,150,0,0,0
```
Transformation frame defined

```
MOVE basis:transform
```
The robot will approach the point X = 300 mm, Y = 550 mm and Z = 400 mm.

Several transformations can be specified, each separated by a colon.

In VAL-II, frames can be defined explicitly with the TRANS instruction and assigned to a frame or transformation variable.

A vector can be defined as an array in HELP but cannot be used in a motion command (Sect. 4.2). Equally, a frame may be stored in a six-dimensional array (three coordinate values and three angle specifications), but no operators exist.

The language SIGLA does not have any geometric data types.

In ROBEX, a vector can be defined with the point declaration; rotations are missing as data types. Frames and transformations can be defined via MATRIX declarations, while the actual frame values are stored in point variables, where the frame or transformation has been engaged previously by the TRASYS instruction.

ROBEX:

```
frame = MATRIX/XYROT,-180,TRANSL,300,550,250
TRASYS/frame
aim = POINT/0,0,150
GOTO/aim
```
The same location as in the previous VAL example is approached.

Apart from these data types for *explicit motion programming*, ROBEX recognizes further geometric data types for implicit motion programming. These have mostly been taken from EXAPT or APT. They are

- POINT
- LINE
- CIRCLE
- PATTERN
- PLANE
- BEAM
- CYLNDR
- SPHERE

- CONE
- BODY (body defined from previous types)
- PART (parts defined from bodies)

The type vector should be mentioned in this context as well, although it cannot be used for explicit motion programming.

3.1.5 Structured Data Types

It has already been pointed out several times that complicated data types, such as frames and rotations, can be made up from simpler ones, where the components are accessible individually. This consideration leads to the idea of **structured data types**, like the ones contained in PASCAL. Four important concepts for structured data types - arrays, records, files and aggregates - are introduced in the following sections.

3.1.5.1 Arrays

An array is understood to be a collection of objects of the same data type into one data object, when the individual elements of an array can be identified by an index. So, a vector v can be written as an array with three elements:

```
SCALAR ARRAY v [1:3];
```

This way, the array v, consisting of three scalar values (Table 3.5) with indices 1, 2 and 3, is defined in AL. A Denavit-Hartenberg-matrix (see also Sect. 2.3.1) is easily constructed, too:

```
SCALAR ARRAY dhm [1:4,1:4];
```

In order to access single components, one writes

```
v [2]
```

or

```
dhm [2,4]
```

and obtains the y-component of the vector v or the y-component of the translation vector in the DH-matrix. The two-dimensional matrix dhm can be visualized as follows:

```
dhm[1,1] dhm[1,2] dhm[1,3] dhm [1,4]
dhm[2,1] dhm[2,2] dhm[2,3] dhm [2,4]
dhm[3,1] dhm[3,2] dhm[3,3] dhm [3,4]
dhm[4,1] dhm[4,2] dhm[4,3] dhm [4,4]
```

Arrays can be used in SRL, PASRO, AL and HELP, but are not available in VAL, SIGLA and ROBEX. In addition, VAL-II supports one-dimensional arrays with a maximum of 32767 elements. The size of an array does not have to be explicitly defined - an array automatically grows as its elements are referenced. This implementation of VAL-II is contrary to all principles of structured programming concepts and may cause unexpected run-time errors. SRL, PASRO, AL and HELP allow an arbitrary number of array indices (n-dimensional matrices). SRL, PASRO and AL permit an arbitrary range of indices, even negative values. The only condition is that the upper limit is to be larger than the lower limit. A definition in AL is shown by way of an example:

```
field1: ARRAY [-1:4, -10:-5] OF REAL;
```

whereas the line

```
field2: ARRAY [-1:-3, -2:7] OF REAL;
```

will lead to an error, since the lower limit of the first index is larger than the upper limit of -3.

In HELP, the indices have to begin with 1, which should, however, not impose any significant limitations in robot programming.

There is a difference in the way in which the array limits have to be specified. It has to be a numeric constant in HELP, whereas explicitly defined constants can be used in PASRO and SRL. AL even allows variables and arithmetic expressions. Hence very flexible and clear data structures are defined. For instance, if flanges with a varying number of screws have to be assembled, an array `flangeholes` (in AL) can be defined as:

```
FRAME ARRAY flangeholes [1:holenumber];
```

If the assembly operation is programmed as a loop with `holenumber` as the loop variable (see also Sect. 4.7.2.1), then only the variable (or in PASRO the constant) `holenumber` has to be varied in the program to adjust it for different flanges. `holenumber`, if it is a variable, can be input from a terminal, further increasing program flexibility.

Variables, constants and arithmetic expressions can be used to access each array

```
TYPE
  vectortype          = ARRAY [1..3] OF REAL;

  rotationtype        = RECORD
                          axis        : vectortype;
                          angle       : REAL
                        END;

  transformationtype  = RECORD
                          orient      : rotationtype;
                          pos         : vectortype
                        END;

  frametype           = transformationtype;
```

Program section 3.6: Declaring the data types of AL in PASRO

```
TYPE
  boxtype = RECORD
               coordsystem  : frametype;
               xlength,
               ylength,
               zlength      : REAL;
               gripposition : frametype;
               full         : BOOLEAN
            END;
```

Program section 3.7: Declaring a boxtype with types declared in Program section 3.6 in PASRO

element in all four languages. Hence the index can be calculated and the versatility of the data structure is further increased.

Arrays are only limited by the available main memory in each of the introduced languages.

3.1.5.2 Records

If different data types have to be collected into one, then the array is not suitable, and one has to refer to the more general type of the record. This language element is available in PASRO and SRL. Differing data types, which have been defined previously or are standard data types, can be combined into a record.

The data type `vectortype`, consisting of three integers, has been introduced in Sect. 3.1.1. Program section 3.6 defines the data types in the language PASRO with the help of records and arrays based on real numbers.

Hence a box-like object can be defined as shown in program section 3.7. The rela-

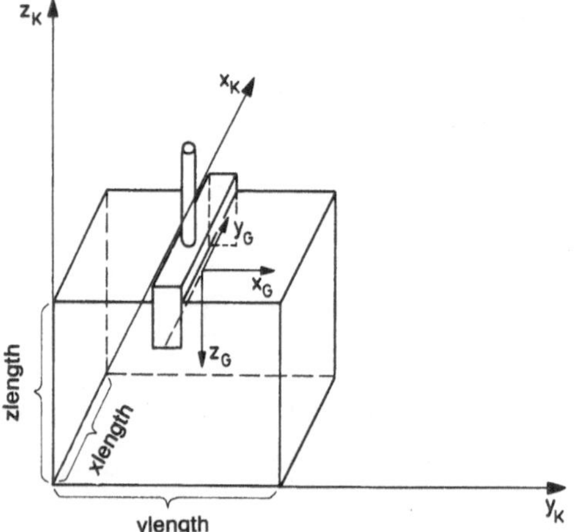

Fig. 3.3: Box with gripping position and gripper. The coordinate system of the boxframe is indexed with K, the one of the gripping position with G

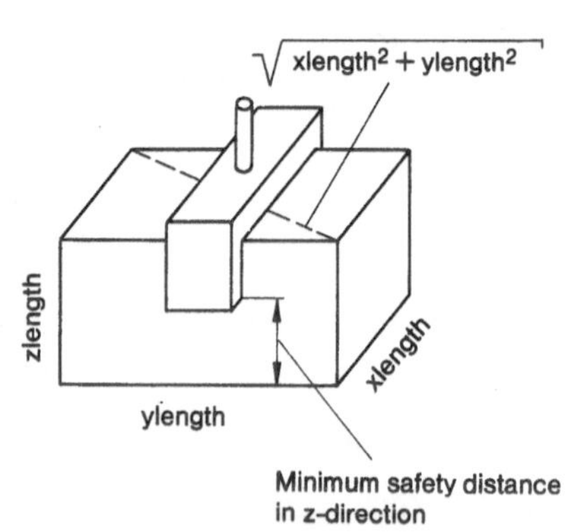

Fig. 3.4: Minimum safety distances when moving a parallelepiped block

tionships between the coordinate systems of the frames and the variables of the record are reflected in Fig. 3.3.

The now defined box could be open at the top, for example, so that it is possible to fill it. Consequently, the record element `full` is either TRUE or FALSE. The frame `gripposition` is specified for holding the box by a robot hand, and dimensions are given to ease the estimate of maneuvrability between obstacles. There has to be a certain amount of space on either side of the path, given by

```
SQRT (SQR (xlength) + SQR (ylength) ) / 2
```

if the box is to be turned arbitrarily, and clearance space below the box can be computed from `zlength` minus a part of the gripper length (compare Fig. 3.4; SQRT and SQR are standard functions for calculating square roots or squares).

Access to *record components* is obtained by specifying a variable name followed by a component name separated by a period. Let the variable box be declared as follows:

```
VAR box : boxtype;
```

then the status of filling is given by

```
box.full
```

and the origin of the coordinate system of the box

```
box.coordsystem.position
```

so that the front edge of the bottom left-hand side and the height of box above the base coordinate system are found by

```
box.coordsystem.position [3].
```

3.1.5.3 Files

The means of structuring introduced so far (arrays and records) are problem-orientated elements of a language, whereas a file has a technically motivated origin from data processing. Files serve to store large amounts of data, consisting of the same elements, arrays - for example. Generally, they are not in **main memory**, but reside on **peripheral memory devices**, like *hard disks, magnetic tapes* or *floppy disks.* Their elements are sequentially ordered, and their main difference from a one-dimensional array is the undeclared length of a file.

One distinguishes files with *random* or *sequential* access. The random access is equivalent to that in arrays, and file elements are usually numbered starting with 1, similar to indices in arrays. Random access makes the time to read or write to a file not (or only very little) dependent on the contents of the data index. This is only meaningful for hard or floppy disks. Tapes have to be run forward and backward, and this can lead to very long positioning times (in the order of seconds) in certain circumstances. Typical positioning times for disks are between 20 and 50 ms.

The more common access is sequential and starts after file opening with the first element, in order to take one after the other until the end of the file. A renewed access to already processed elements is only possible by resetting the file to the beginning and rereading all elements up to the one desired. AML contains sequential and random access files. It is implementation-dependent, whether or not SRL, PASRO (or PASCAL) also offer random access files.

The possibilities for file usage are greatest in PASCAL, PASRO and SRL. Elements can be of any type. The declarations

```
TYPE text  = FILE OF CHAR;
     boxes = FILE OF boxtype;
```

define a *text file* and a file for storing the records describing box-like objects defined in Sect. 3.1.5.2. The text file could contain problem-oriented program data.

The robot languages SRL, PASRO, AL, AML and HELP also support file handling. The file-management routines in PASRO and SRL for creating, opening and closing files as well as file input/output are similar to those in PASCAL [1]. SRL, PASRO and AML provide a general form of file handling, which is mostly hardware-independent.

Only text files are known in HELP, which have the purpose of storing HELP instructions created through the self-teaching program. Data accumulated in the teach-in part, the so called self-teach program, are written textually onto a file formatted as value assignments (in HELP), and compiled together with the actual program. HELP has the following functions available:

- CREATE to create and open a file
- OPEN to open an already existing file
- RECORD to write to a file
- CLOSE to close a file

Naturally, any text file can be be created in HELP, not only those with specific HELP texts. No further data types are catered to in HELP.

3.1.5.4 Aggregates

An aggregate is a collection of different data types into one. So far, they are like records, but the difference lies in the access to the components, which is done by an index mechanism known from arrays. Aggregates in AML are not explicitly defined as a new data type, which could be used later on. They are defined in the variable declaration statement using the AML reserved words NEW (which must not be confused with the procedure NEW from PASCAL, cf. Sect. 3.1.6) or STATIC (see Chap. 6).

Examples for valid declarations of the same frame are

```
f1: NEW <0.0,0.0,0.0,0.0,0.0,0.0,0.0>;
f2: NEW 7 OF 0.0;
```

Another example for a nested aggregate is

```
nest1: NEW <'nested',<'aggregate',1>,1.5>;
nest2: NEW <'aggregate',1>;
nest3: NEW <'nested',nest2,1.5>;
```

The third variable nest3 is equal to the first nest1. Access to the components is gained as follows:

Component access	Value
`nest1(1)`	`'nested'`
`nest1(2)`	`<'aggregate',1>`
`nest1(3)`	`1.5`
`nest1(2,1)`	`'aggregate'`
`nest1(2,2)`	`1`

The size of an aggregate is determined after the execution of its definition. As with arrays in AL, the size need not be known at compile time. Thus it is possible to write:

```
f2: NEW LENGTH OF REAL;
```

which results in an aggregate of LENGTH elements of type REAL.

3.1.6 Pointer Type and Word Model

Before the connections between pointer types and world model should be explained in greater detail they themselves are considered more closely.

Data types introduced up to now allow the declaration of variables, which are defined in one block, as shown in Sect. 2.2.3, claiming the memory allocation on block entry and releasing it again on block exit. On the other hand, it is feasible to create variables at any time independently from the block structure and also to release the memory at another time. Such variables are termed **dynamic variables**. As it is not known at the time of the program writing how many variables are needed, they cannot be declared in a (static) variable definition.

Dynamic variables are created in a procedure (NEW in PASRO and SRL) which returns a pointer to that variable. This pointer is nothing else than the address of the newly-created variable. It can be stored in a different variable of type pointer. The possibilities hence created are illustrated in the example program 3.8 (written in PASRO).

The pointer types are declared in PASRO, SRL and PASCAL by preceeding them with the symbol "^" (or "@" in SRL). The data object, which is addressed by the pointer, is then accessed by appending the pointer name with a "^" (or "@"). The pointer name is followed by the component name, which is indicated by a period, because we are dealing with records here. Hence

```
box^.next
```

names the component `next` of the record to which `box` points.

The program serves as a work basis for two examples. The first is to build up a linear list of a number of `box` records. The number is specified by the user from a terminal. Only already defined data types and `boxtype` have to be considered. For example, let the pointer `boxes` point to the start of the list, and let the number of the box records to be generated be entered during the program execution.

The program reads the number of the desired boxes and creates a record as illustrated in Fig. 3.5. For reasons of clarity, values are only assigned to the pointer variables relevant here: `boxes^.affixrel` and `boxes^.next` or `k.affixrel` and `k.next`. The remaining variables are not considered. After that, a loop is executed `boxnum-`

```
PROGRAM pointerexample (input);

TYPE
  boxpointer          = ^boxtype;
  framepointer        = ^frametype;
  affixpointer        = ^affixelement;
  transpointer        = ^transformationtype;

  vectortype          = ARRAY [1..3] OF REAL;

  rotationtype        = RECORD
                          axis         : vectortype;
                          angle        : REAL
                        END;

  transformationtype  = RECORD
                          orient       : rotationtype;
                          pos          : vectortype
                        END;

  frametype           = transformationtype;

  affixelement        = RECORD
                          affixedbox   : boxpointer;
                          transrelation: transpointer;
                          next         : affixpointer;
                        END;

  boxtype             = RECORD
                          coordsystem  : frametype;
                          xlength,
                          ylength,
                          zlength      : REAL;
                          gripposition : frametype;
                          full         : BOOLEAN;
                          next         : boxpointer;
                          affixrel     : affixpointer
                        END;

PROCEDURE affix (kx, ky: boxpointer);
(************************************************************************)
(*                                                                      *)
(*  Generates the affix relation between kx and ky, such that kx       *)
(*  depends on ky                                                       *)
(*  Comparable with the AL instruction  AFFIX kx TO ky NONRIGIDLY      *)
(*                                                                      *)
(************************************************************************)
```

```
  VAR
    affixelem  :  affixpointer;                (* new affix element      *)
    affixaux   :  affixpointer;                (* auxiliary pointer      *)

BEGIN (* affix *)
    NEW (affixelem);                   (* Generate a new affix element  *)
    affixelem^.next := NIL;            (* No element following          *)
    affixelem^.affixedbox := kx;       (* Point to kx                   *)
    IF ky^.affixrel = NIL              (* Tests whether ky already has  *)
                                       (* affix relations               *)
    THEN ky^.affixrel := affixelem     (* The generated affix element   *)
    ELSE                               (* is linked directly to ky      *)
    BEGIN (* linking the affixelement *)
      affixaux := ky^.affixrel;        (* Point auxiliary pointer to    *)
                                       (* first affix element           *)
      WHILE affix^.next <> NIL DO      (* Look for last affix element   *)
        affixaux := affixaux^.next;    (* Auxiliary points to it        *)
      affixaux^.next := affixelem;     (* The generated affix element   *)
                                       (* is appended to the list       *)
    END (* of linking affix element *);
END   (* of affix *);

  VAR
    boxes              : boxpointer;   (* Start of link pointer         *)
    boxnumber,                         (* Number of list elements       *)
    i                  : INTEGER;      (*      Auxiliary                *)
    k                  : boxpointer;   (*      variables                *)

  BEGIN (* example program *)
    (*******************************************************************)
    (******** PROGRAM - PART 1:                         ****************)
    (******** Generation of list according to Fig. 3.6  ****************)
    (*******************************************************************)

    READ (boxnumber);                  (* Read number of boxes to be    *)
                                       (* generated from user's terminal*)
    NEW (boxes);                       (* Make a new box                *)
    boxes^.next := NIL;                (* This box has no successor     *)
                                       (* For result see Fig. 3.5       *)
    FOR i := 2 TO boxnumber DO         (* Loop from 2 to boxnumber to   *)
                                       (* make and link boxnumber-1     *)
                                       (* box records                   *)
    BEGIN (* of loop *)                (* Start of loop                 *)
      NEW (k);                         (* Make a new box                *)
      k^.next := boxes;                (* New box links to the first    *)
                                       (* box in list                   *)
      k^.affixrel := NIL;              (* There is no affix relation    *)
```

```
    boxes := k;                          (* Start pointer boxes points to *)
                                         (* first (new) element in list   *)
  END (* of loop *);                     (* loop end                      *)
  (*********************************************************************)
  (******** End of linked list according to Fig. 3.6     ***************)
  (*********************************************************************)

  (*********************************************************************)
  (******** PROGRAM - PART 2:                             ***************)
  (******** Generating affix relations according to      ***************)
  (******** Fig. 3.7                                      ***************)
  (*********************************************************************)
  k := boxes^.next^.next;               (* k points to box record of k2 *)
  affix (k^.next, k);                   (* Affix k1 depending on k2     *)
  affix (boxes^.next, k);               (* Affix k3 depending on k2     *)
  affix (k, k^.next);                   (* Affix k2 depending on k1     *)
  affix (boxes^.next, k^.next);         (* Affix k3 depending on k1     *)
END (* of example program *).
```

Program 3.8: Example program for the usage of pointer types in PASRO

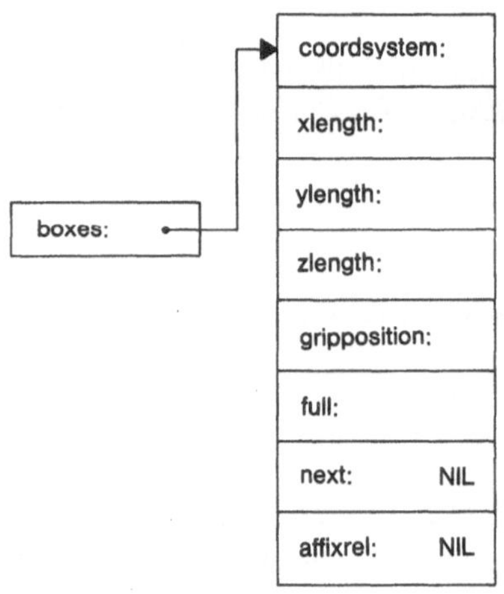

Fig. 3.5: The list boxes with an element of type boxtype. No assignments have been made to the record components `coordsystem`, `xlength`, `ylength`, `zlength`, `full` and `gripposition`

`ber-1` times, and each time a record is created and inserted at the beginning, so that a linear list is generated (cf. Fig. 3.6). The order of creation is marked K1 to K4 in the figure.

Without pointers, this problem can only be solved by allowing fields of dynamic lengths, as in AL, and then only if the number of records does not change during program execution. Even a field of dynamic length cannot be changed during the program run after it has been declared, and in contrast to the linked list, an addition of further field elements is not possible. Only SRL, PASRO and AML contain pointer types for solving the problem of *dynamic data structures.*

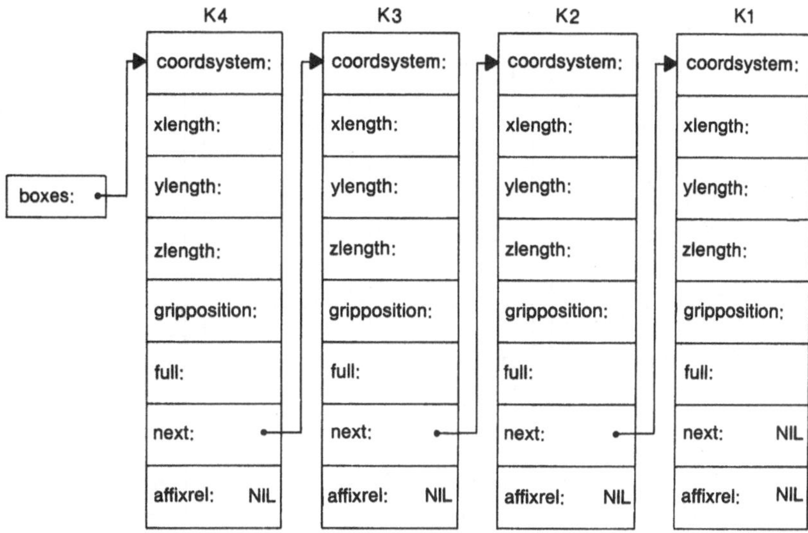

Fig. 3.6: A simple, linked and linear list consisting of four elements of type boxtype. No assignments have been made to the record components `coordsystem`, `xlength`, `ylength`, `zlength`, `full` and `gripposition`

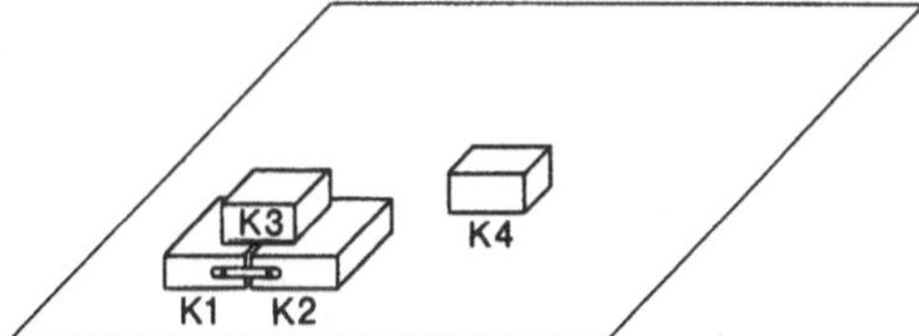

Fig. 3.7: Relative positioning of the boxes on the work table

In AML a dynamic variable is generated by the STORE system function, for example:

```
NEXTP: NEW PTR;
LIST1: NEW STORE (<NEXTP,<1,2,3>>);
```

First, the empty pointer `NEXTP` is declared. After that `LIST1` points to the newly generated aggregate `<NEXTP,<1,2,3>>`. The following new statements generate a new aggregate, which is linked to the first one.

```
NEXTP = NEW STORE (<'another aggregate', PTR>);
!LIST1(1) = NEXTP;
```

The second statement uses the dereferencing operator "!", similar to "^" in PASCAL or "@" in SRL. Another method which differs from PASCAL to generate a pointer is the "&" operator. It produces a pointer to the variable to which it is applied, for example:

```
A: NEW <1,2,3>
NEXTP = &A;
```

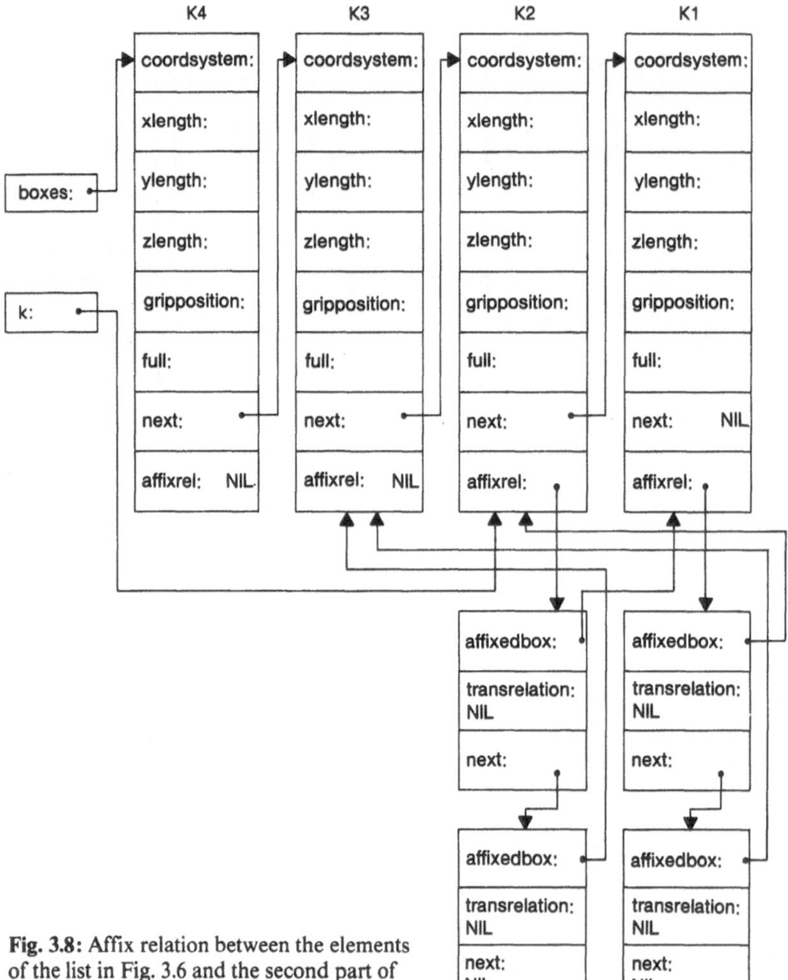

Fig. 3.8: Affix relation between the elements of the list in Fig. 3.6 and the second part of program section 3.7 and Fig. 3.7

Let us proceed to the PASRO example of program 3.8. As it is shown in Fig. 3.7, two boxes are fixed to each other with screws, and a third is stacked on top because of previous manipulations, whereas a fourth is still awaiting some manipulation. If K1 or K2 are moved, K3 and K2 or K1 are moved, too. However, the movement of K3 leaves K1 and K2 unaltered. How can relations which frequently alter during program execution according to the manipulations already performed be represented in data structures? This world model not only has to contain the relationships between the objects (here boxes), but also relative positions, which are important for the movements of boxes fixed to one another.

To demonstrate these connections - called **affix-relation** in AL - a record `affix-element` is given with the following components:

`affixedbox` (points to the box to which it is coupled, and hence has to be moved as well);

`transrelation` (points to the transformation giving the relative position of both boxes);

`next` (points to the next affix element, if there are several relations).

The data model in Fig. 3.8 can be constructed to demonstrate the affix relation with the help of program 3.8. This is the purpose of the second section in connection with the procedure `affix` (compare Sects. 2.2.4.2 and 4.1). Let `k1` and `k2` be pointer variables of type `boxpointer`. Then the procedure call

```
affix (k1,k2)
```

makes `k1` and `k2` dependent on one another, so that a movement of `k1` changes the position of `k2`, but not vice versa. This is comparable with the AL instruction

```
AFFIX k1 to k2 NONRIGIDLY;
```

(where `k1` and `k2` have to be frames). SRL and AL only allow frames to form an affix relation, while the PASRO example enables dynamically handled objects to do so. It is easily expanded to let any dynamically generated records form an affix relation.

If `NONRIGIDLY` is replaced by `RIGIDLY` in AL, a symmetric relationship is created. This would have to be done as follows in program 3.8:

```
affix (k1,k2);
affix (k2,k1);
```

The order of the procedure calls is obviously not important. The complicated calculations of the transformations between the different reference frames of `k1` and `k2`, i.e., `k1^.coordsystem` and `k2^.coordsystem`, has been left out for the sake of clarity, since they do not contribute to the presentation of pointer structures. It is just pointed out that the calculations with the internal DH-matrices are done as follows:

$$DH_{\text{Transformation}} = \begin{pmatrix} R_1 & l_1 \\ 0 & 1 \end{pmatrix} \begin{pmatrix} R_2 & l_2 \\ 0 & 1 \end{pmatrix}^{-1}$$

This corresponds to the AL expression

```
DH_Transformation <- k2^.coordsystem -> k1^.coordsystem;
```

(cf. Sect. 2.3.1). The second part starts with the assignment of `k` as an auxiliary pointer to the box record "K2". Then the affix relations between "K1", "K2" and "K3" (see Fig. 3.8) are defined by calling `affix` four times.

Although AL has the `AFFIX` instruction available to describe such relations (Sect. 4.1), it would certainly be sensible and useful if AL were to have such a versatile and easily implemented concept as that of pointer types.

3.2 Manipulation of Data

A large variety of **operators** and **functions** exist for manipulation of data objects, and they can be used to build up **formulae**. The power of a language depends equally on the data structures and the possibilities of manipulating them.

3.2.1 Operators

Operators are taken to be the usual monadic and dyadic ones, in contrast to functions like `SIN(x)` or the AL function `POS(f)`, which delivers the position vector of a frame. The compiler has to check the validity of a particular operation. For example, 5+7 is a valid operation in contrast to `5+VECTOR(1,0,3)`. This is obvious so far, but what about `5+7.1`, the addition of integers and reals? Here a type conversion is done automatically and the integer "5" is made into a real. This problem is dealt with in more detail when the individual operators are described.

3.2.1.1 Arithmetic Operators

Arithmetic operators are concerned with the data types INTEGER and REAL, where AL combines them to the one type of SCALAR. In Table 3.7, i stands for integers, r for real and s for scalars. ROBEX is not contained in this table, since the version on hand does not provide variables and hence no operators for calculations during program execution. PASRO and SRL execute a type conversion of integer to real, if the four basic arithmetic operations combine different data types. Integer division truncates the result. Neither AL nor VAL specifies whether the fraction of the result is rounded or truncated, and hence this depends on the language implementation.

As shown in Table 3.7, SRL, PASRO, AL and AML allow a large variety of operations and hence enable the operator to formulate arithmetic calculations relatively easily. The versatility decreases in HELP and VAL-II, VAL allows only the most important operations for the type INTEGER, and SIGLA offers the least opportunities.

3.2.1.2 Geometric Operators

Special geometric operators are needed in order to combine the geometric data types introduced in Sect. 3.1.4 with simple types as well as with elements of structured data types. SRL, PASRO and AL offer a broad spectrum of geometric operators, and hence these are described in detail and their application demonstrated with examples. AML includes special functions for geometric calculations on a lower level because there are no predefined geometric data types like frame or vector. However, as shown in Sect. 3.1.4, the user can simulate a vector type, for example, with the help of aggregates, and can also write a function which adds two variables of type vector.

Many of the geometric operators are represented by the same symbol as the arithmetic ones, because of pragmatic reasons and also because most keyboards contain only a limited number of special characters. Table 3.8 lists the operators available in

Table 3.7: Arithmetic operators in various languages

Operation	Type of result in	Representation in						
	SRL and PASRO	SRL and PASRO	AL	AML	VAL	VAL-II	HELP	SIGLA
Addition	INTEGER REAL REAL REAL	i+i r+r i+r r+i	s+s	i+i r+r i+r r+i	i+i	s+s	s+s	IC/i,i
Subtraction	INTEGER REAL REAL REAL	i-i r-r i-r r-i	s-s	i-i r-r i-r r-i	i-i	s-s	s-s	IC/i,-i
Multi-plication	INTEGER REAL REAL REAL	i*i r*r i*r r*i	s*s	i*i r*r i*r r*i	i*i	s*s	s*s	./.
Division	INTEGER REAL REAL REAL	i/i r/r i/r r/i	s/s	i/i r/r i/r r/i	i/i	s/s	s/s	./.
Integer division	INTEGER	i DIV i	s DIV s	r IDIV r i IDIV i	i/i	./.	./.	./.
Modulo	INTEGER	i MOD i	s MOD s	**	i % i	s MOD s	./.	./.
Exponent	REAL	./.	s ^ s	**	./.	./.	./.	./.
Modulus	INTEGER REAL	ABS(i) ABS(r)	|s|	**	./.	yes	./.	./.
Negation	INTEGER REAL	-i -r	-s	-i -r	-i	yes	-s	NE/-i

** Assumed (no information available)

The type of the result is always SCALAR in AL and HELP, where the result is truncated in the case of integer division. The result in VAL and SIGLA is always an integer.

SRL, PASRO and AL, ordered into monadic and dyadic operators and according to the result delivered by the operation. By combining the standard functions for the geometric data types of Sect. 3.2.2 and the standard arithmetic functions, even complicated paths and search patterns can be programmed explicitly. A comparatively simple example in SRL will illustrate this. SRL has been implemented to program several robots which do not have path control but only point-to-point motion (cf. Sect. 2.3.2). Let the robot execute a straight-line motion between start and target frame because of safety reasons, and, in addition, change the orientation linearly. Since this is not a frequent demand, an implementation in the robot control is too expensive and the motion is allowed to be slow. The problem is solved by an SRL procedure realizing a *linear Cartesian interpolation* (program 3.9). The distance between the calculated intermittent frames is to be 0.5 cm (Fig. 3.9).

Table 3.8: Geometric operators in SRL, PASRO and AL

Operation representation			Operation description	Operation result (using AL standard routines)
SRL	PASRO	AL		
s := VLENGTH(v)	VABS(s,v)	s <- |v|	Vector length	s = SQRT(SQR(x) + SQR(y) + SQR(z))
s := r.angle	s := r.angle	s <- |r|	Angle of rotation	s = alhpa
s := v1 DOT v2	VDOT(s,v1,v2)	s <- v1.v2	Scalar product	s = x1*x2 + y1*y2 + z1*z2
v := s*v	VMUL(v,v,s)	v <- s*v	Vector expansion	v = (s*x, s*y, s*z)
v := v/s	VDIV(v,v,s)	v <- v/s	Vector contraction	v = (x/s, y/s, z/s)
v := v1+v2	VADD(v,v1,v2)	v <- v1+v2	Vector addition	v = (x1+x2, y1+y2, z1+z2)
v := v1-v2	VSUB(v,v1,v2)	v <- v1-v2	Vector subtraction	v = (x1-x2, y1-y2, z1-z2)
v := v1*v2	VCROSS(v,v1,v2)	v <- v1*v2	Vector cross product	v = (y1*z2 - z1*y2, z1*x2 - x1*z2, x1*y2 - y1*x2)
v := r*v	VROT(v,r,v)	v <- r*v	Vector rotation	v is rotated around the rotation vector vr by an angle alpha
./.	./.	v <- t*v	Vector transformation	v = rt* (vt+v)
v := f*v	./.	v <- f*v	Vector transformation	v = ft* (vf+v)
./.	./.	v <- v WRT f	Vector transformation with respect to (WRT) frame orientation	v = rf*v
v := ROTAXIS(r)	ROTAXIS(v,r)	v <- AXIS(r)	Calculation of the rotation axix	v = rotation axis
r := r1*r2	ROTROT(r,r2,r1)	r <- r1*r2	Linking of two rotations	Multiplication of the rotation matrices
f := f+v	FRAMETRANSL (f,f,v)	f <- f+v	Frame translation	f = FRAME(rf, vf+v)
f := f-v	./.	f <- f-v	Frame translation	f = FRAME(rf, vf-v)
f := r*f	FRAMEROT(f,f,r)	./.	Frame rotation	f = FRAME(r*rf,vf)
./.	./.	f <- t*f	Frame transformation	f = FRAME(rt*rf, vt+vf)
f := f*f	TRANSFRAME (f,f1,f2)	f <- f*f	Frame transformation	f = FRAME(rf*rf, vf+vf)
f := f1 FREL f2	FRAMEREL (f,f1,f2)	t <- f1 -> f2	Construct transformation	solution of the equation f2 = t*f1
fi := FINV(f)	FRAMEINV(fi,f)	ti <- INV(t)	Frame (transformation) Inversion	nilframe = fi*f
./.	./.	t <- t*t	Link 2 transformations	t = TRANS(rt*rt, vt+vt)

s = SCALAR s or REAL in SRL and PASRO
v = VECTOR (x,y,z)
r = ROT (vr, alpha)
f = FRAME (rf, vf)
t = TRANS (rt, vt)

A call to reach a point of deposit would be:

```
smove (deposit);
```

where the frame `deposit` has been defined explicitly or via the teach-in procedure. It is remarkable in this solution of SRL that apart from the problem-orientated formulation, the orientation is interpolated linearly as well. Should the programmer specify the change in orientation by several rotations about different rotation axes, only one rotation matrix is calculated internally, from which *one* rotation axis or *one* rotation vector can be deduced at any time with *one* unambiguous rotation angle. The start frame is turned around this axis successively until the orientation of the target frame is reached. This procedure also enables an interpolated motion to tar-

```
PROGRAM srlsmove (INFILE,OUTFILE);
VAR
  goal : FRAME;

PROCEDURE smove (target_frame: FRAME);
VAR
  posnum, i                    : INTEGER;
  deltas, delta_rotangle       : REAL;
  translation, delta_rotaxis   : VECTOR;
  start_frame, delta_frame     : FRAME;

BEGIN_PROCEDURE
  start_frame := ARM(2);                  (* Store current robot position  *)
                                          (* position in start frame       *)
  deltas := 0.5;                          (* Interpolation distance. Nor-  *)
                                          (* mally global program constant.*)
  translation  := target_frame.TRANSL -   (* Translation vector            *)
                  start_frame.TRANSL;
  delta_frame  := start_frame ->          (* Transformation frame          *)
                  target_frame;
  delta_rotangle:=delta_frame.ROT.ANGLE;  (* Angle of orientation change   *)
  delta_rotaxis :=delta_frame.ROT.AXIS;   (* Rotation vector of orientation*)
                                          (* change                        *)
  posnum := ROUND(LENGTH(translation / deltas));   (* Number of inter-   *)
                                          (* mediate frames                *)
  IF posnum > 1 THEN
    FOR i := 1 TO posnum - 1 DO
      PTPMOVE ARM (2) TO                  (* Motion to intermediate frame  *)
        FRAMEC (start_frame.TRANSL + i*translation/posnum,
                start_frame.ROT * ROTC (delta_rotaxis,
                                        i*delta_rotangle/posnum))
    END_FOR
  END_IF;
  PTPMOVE ARM (2) TO target_frame;        (* Goto final position           *)
END_PROCEDURE;

BEGIN_ROGRAM
  goal.ROT := ROTC (VECTORC(0, 10, 90), 75);
  WRITELN (' Give position X Y Z :');
  READLN (goal.TRANSL.X, goal.TRANSL.Y, goal.TRANSL.Z);
  smove (goal);
  WRITELN (' smove interpolated explicitely in SRL done !');
END_PROGRAM.
```

Program 3.9: Generation of a linear Cartesian interpolation in SRL

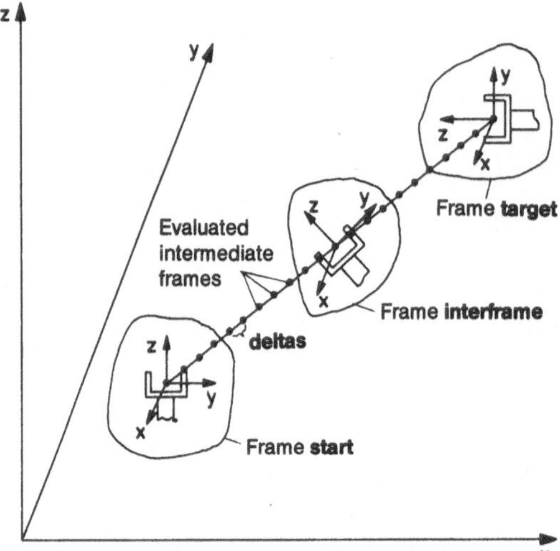

Fig. 3.9: Representation of linear Cartesian interpolation of the AL procedure `smove`

Table 3.9: Part of AML system functions

BOXTRANS	Computes the transformation with respect to the used robot joints associated with the set of joint positions. The input parameters are the joints used, the position vector and the roll, pitch and yaw angles and an offset of the TCP. The result is the corresponding transformation matrix in the form translation vector and 3 by 3 rotation matrix (corresponding to the DH-matrix, see Sect. 2.3.1). This is useful for computing the coordinate transformation of the tool offset.
TRANSBOX	The inverse operation of BOXTRANS
EULERTRANS	Calculates the rotation matrix from the given Euler angles.
TRANSEULER	The inverse operation of EULERTRANS
TRANSPOSE	Transposition of any matrix
INVERSE	Inverting a 3 by 3 matrix
CROSS	Crossproduct of two vectors
DOT	Dotproduct of two vectors
MAG	Magnitude of a vector

get frames, which have been defined after the teach-in procedure or by sensor data.

Besides the standard arithmetic operations, AML includes matrix operations and functions for coordinate transformation. Table 3.9 shows a number of these special functions in AML.

Although VAL contains the frame concept, no geometric operators are available to the programmer. Not even individual coordinate values of frames are accessible, for example, to read the z-coordinate after the robot might have been stopped by a contact sensor. Only the SHIFT instruction can be used to perform a frame transla-

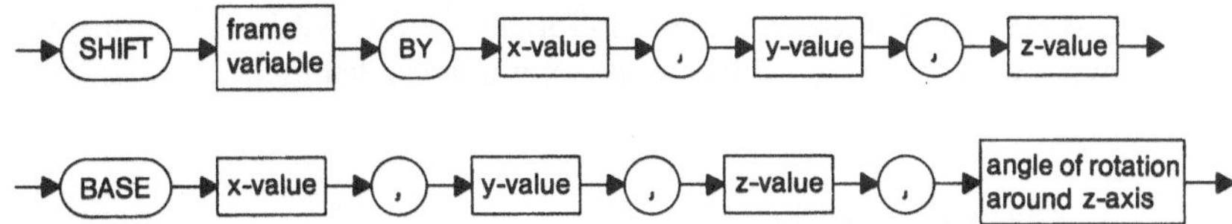

Fig. 3.10: Frame translation in VAL

```
        SETI counter = 0
100     SHIFT targetframe BY 1,0,0
        SETI counter = counter + 1
        IF counter LE xvalue THEN 100
        SETI counter = 0
200     SHIFT targetframe BY 0,1,0
        SETI counter = counter + 1
        IF counter LE yvalue THEN 200
```

Program section 3.10: Frame translation in VAL

tion. Should all frames be shifted or rotated around the z-axis of the base coordinate system, the BASE instruction can be utilized (Fig. 3.10).

Since the SHIFT and BASE instructions do not allow variables for the x-, y- and z-values, but only constants, a trick has to be employed if the frame translation is, for example, determined by sensory input. Program section 3.10 shows the cumbersome solution in VAL, which is realized simply with the SRL command:

```
targetframe := targetframe + VECTORC (xvalue,yvalue,0);
```

This restriction in VAL is really quite unnecessary, as the implementation of variables in the SHIFT instruction demands only very little effort (program section 3.10).

Table 3.10: Operators for comparisons

Operator	SRL and PASRO	AL	AML	VAL and VAL-II	HELP	SIGLA	ROBEX **
Less than	x < x	s < s	x LT x	i LT i	s < s	*	LT
More than	x > x	s > s	x GT x	i GT i	s > s	*	GT
Equal	x = x	s = s	x EQ x	i EQ i	s > s	*	EQ
Less than or equal	x <= x	s<=s	x LE x	i LE i	./.	./.	LE
Larger than or equal	x >= x	s>=s	x GE x	i GE i	./.	./.	GE
Not equal	x <> x	s <> s	x NE x	i NE i	./.	./.	NE

x represents a real or integer value, s a scalar value and i is an integer.

* SIGLA does not recognize logical comparisons apart from special instructions for conditional jumps.

** ROBEX only caters to comparisons in conditional branching (Sect. 4.7.1).

The restrictions mentioned above are no longer valid in VAL-II, because it is possible to have access to frame components, joint angles, distances between frames and so on.

3.2.1.3 Comparing Operators

The operators listed in Table 3.10 are contained in the languages for comparisons between integer, real and scalar values.

The comparing operators allow the construction of simple logical expressions like: `s1 < s2`. They are only allowed for the control of program flow in VAL and HELP. The result can be assigned to variables in SRL, PASRO, AL and AML. Operators can be used on constants, variables and arithmetic expressions in SRL, PASRO, AL, AML, VAL-II and HELP, but only on constants and variables in VAL. As AML does not include Boolean variables, the result of comparisons returns -1 for TRUE and 0 for FALSE.

The test for equality is only of limited use with reals (and hence scalars), since the calculations of the values to be compared can deliver different values because of rounding errors, even in the case of identical algebraic expressions (see Sect. 3.1.3).

3.2.1.4 Logical Operators

These are included for the logical combination of Boolean variables, constants and simple logical expression as they are formed by comparing operators. As an example, the expression

```
a > 5 AND a < 7
```

is only true if the value of a lies between 5 and 7. Such expressions can be formed in PASRO, SRL, AL, VAL-II and HELP. Table 3.11 gives the information about the availability of different operators.

In HELP, available logical expressions control only the program flow (cf. Sect. 4.7).

Table 3.11: Logical operators

Operators	SRL and PASRO	AL	AML	VAL-II	HELP
Logical AND	b AND b	s AND s	i AND i	yes	expr AND expr
Logical OR	b OR b	s OR s	i OR i	yes	expr AND expr
Logical NOT	NOT b	NOT s	NOT i	yes	NOT expr
Exclusive OR	b <> b	s XOR s	i XOR i	./.	./.
Equivalence	b = b	s EQV s	./.	yes	./.

i is an integer constant, variable or expression;
b is a Boolean constant, variable or simple logic expression;
s is a scalar constant, variable or logical constant;
expr is a simple logical expression.

AL does not distinguish any Boolean variables. Instead, all scalar values greater than zero are taken to be TRUE.

The logical operators of AML operate on the binary value of integer (or aggregate of integer) operands. The result is also an integer.

3.2.2 Standard Functions

Section 3.2.1 distinguishes operators from certain functions which are predefined in the various languages and hence termed standard functions. A nucleus of functions has evolved which is common to most languages, consisting of the most important

Table 3.12: Standard arithmetic functions

Function	PASRO	SRL	AL	AML	VAL-II	HELP	ROBEX
sine	SIN	SIN	SIN	SIN	yes	SIN	SIN
cosine	COS	COS	COS	COS	yes	COS	COS
tangent	./.	TAN	TAN	TAN	./.	./.	./.
arctangent	ARCTAN	ATAN2	ATAN2	ATAN	yes	ARCTG	ATAN
arcsine	./.	ASIN	ASIN	ASIN	./.	./.	./.
arccosine	./.	ACOS	ACOS	ACOS	./.	./.	./.
conversion REAL ->INTEGER with							
- truncation	TRUNC	TRUNC	INT	****	yes	./.	./.
- rounding	ROUND	ROUND	./.	****		./.	./.
exponentiation	EXP	EXP	EXP	./.	./.	./.	EXP
natural logarithm	LN	LN	LOG	./.	./.	./.	NLOG
logarithm	./.	./.	./.	./.	./.	./.	LOG
square root	SQRT	SQRT	SQRT	SQRT	yes	SQRT	SQRT
square	SQR	*	*	./.	yes	./.	./.
modulus	ABS	**	**	ABS	yes	ABS	ABS
Input of a							
- scalar value	***	INPUT	INSCALAR	***	yes	ASK	./.
- Boolean value	***	INPUT	QUERY	***	./.	ASKN	./.
- vector value	vread	INPUT	./.	./.	./.	./.	./.
- rotation value	rread	INPUT	./.	./.	./.	./.	./.
- frame value	fread	INPUT	./.	./.	./.	./.	./.

* The square can be generated by the more general procedure of exponentiation.

** The modulus is formed in SRL and AL with the operator |s|.

*** In PASRO and AML, values are read in with standard procedures for input and output.

**** When real values are used in situations, where integer type is required, an implicit rounding is performed.

trigonometric functions, at least one conversion from REAL to INTEGER, the e-function, the natural logarithm and the square root.

Arithmetic standard functions are contained in HELP in two versions: Firstly, the base version with trigonometric functions, where the angles have to be specified in radians, and without the square root, and secondly the expanded version containing - apart from the SQRT function - trigonometric functions allowing angle specification in degrees.

AL also contains a number of functions for geometric data types (see Table 3.13) apart from the standard functions listed in Table 3.12.

Since there are no functions in VAL, the definitions of frames and their inverses are realized through special commands:

```
FRAME <name> = <frame 1.>,<frame 2.>,<frame 3.>
INVERSE <name> = <frame>
```

It is possible in VAL-II to define frames explicitly by setting position (x, y, z) and rotation angles (o, a, t).

Table 3.13: Standard functions for geometric data types in SRL, PASRO, AL and VAL II

Function call				Type of result	Description
SRL	PASRO	AL	VAL-II		
VECTORC(r,r,r)	MAKEVECTOR(v,r,r,r)	VECTOR(s,s,s)	./.	VECTOR	Forms a vector
./.	./.	UNIT(v)	./.	VECTOR	Forms unit vector of v
*	*	POS(f) POS(t)	yes	VECTOR	Position vector of a frame or transformation
ROTAXIS(r)	ROTAXIS(v,r)	AXIS(r)	./.	VECTOR	Rotation axis of r
ROTC(v,r)	MAKEROTATION(r,v,s)	ROT(v,s)	./.	ROT	Forms a rotations around v by an angle s
*	*	ORIENT(f) ORIENT(t)	yes	ROT	Orientation of a frame or translation
FRAMEC(v,r)	MAKEFRAME(f,r,v)	FRAME(r,v)	yes	FRAME	Forms a frame with orientation r and position v
FCONSTRUCT (f1,f2,f3)	./.	CONSTRUCT (v1,v2,v3) CONSTRUCT (f1,f2,f3)	similar	FRAME	Generates a frame with v1 (or POS(f1)) as the position vector. v1 or POS(f2)) points into the x-direction and v3 (or POS (f3)) is a point in the x-y plane
./.	./.	TRANS (r,v)	./.	TRANS	Forms a transformation with rotation r and translation v
FINV(f)	FRAMEINV(f,f)	INV(t)	yes	FRAME (TRANS in AL)	Forms the inverse transformation of f

s is a scalar f is a frame
v is a vector t is a transformation
r is a rotation
* no function needed, because the value can be selected as a record component

SRL, PASRO and AL offer a large number of possibilities to manipulate geometric data types by using the standard functions from Table 3.13 and the operators described in Sect. 3.2.1.2.

The lack of variables in ROBEX leaves the standard functions together with the operators +, -, * and / to be evaluated during the compilation. It is possible to write

```
POINT = POINT/100 * COS(30), SIN(30), 0
```

and hence a location POINT has been defined which lies in the x-y plane with a distance of 100 mm from the origin and forms an angle of 30 degree with the x-axis. It is equally feasible to calculate the values with the help of a pocket calculator and enter them as program constants. On the other hand, one might find it useful to integrate the capabilities of pocket calculators into the compiler, particularly if it is kept in mind that the NC language EXAPT, on which ROBEX is built, was developed at a time when pocket calculators were not to be imagined.

The languages SRL, PASRO, AL and AML enable the user to define further functions for the evaluation of application specific formulae. This has already been mentioned in Sect. 2.2.4 and will be dealt with in further detail in Sect. 6.3.

3.2.3 Complex Expressions

The operators introduced in previous sections permit the construction of formulae or complex expressions, immediately raising the problem of **operator priorities**. This only holds for SRL, PASRO, AL, AML, VAL-II and HELP which allow formulae. VAL and SIGLA recognize only integer expressions with two elements such as

```
SETI A = B <op> C
```

or

```
IC/a,b
```

where B and C may be integer variables or numbers. <op> stands for one of the operators listed in Table 3.7. IC/a,b adds the content of variable b to variable a. More complex expressions have to be broken down into the individual dyadic commands. The PASRO expression

```
x := (a + b) DIV c; (* Integer division *)
```

(where a, b and c are integer values) becomes in VAL:

```
SETI X = A + B
SETI X = X/C
```

The division itself has to be programmed explicitly in SIGLA, as shown in program section 3.11.

The current version of ROBEX does not give any information about the order of evaluation, and it is to be assumed that functions are executed first, followed by * and /, and finally the + and -. Table 3.14 gives an overview of the priorities of the different operators in SRL, PASRO, AL, AML and HELP, but note that the +, - are monadic as well as dyadic. Operators with the same priority are evaluated from left

```
                  Meaning:
SE/M1,M2;         X := A
IC/M1,M3;         X := A + B
SE/M5, 0;         Clear loop counter
NU/10;            Loop label
IC/M5,1;          Increment loop counter
IC/M1, -M4;       Subtract c from a+b
BG/M1, 0, 10;     Test loop end (M1 <= 0)
BE/M1, 0, 11;     Test for zero remainder
IC/M5, -1;        Correct result of division if remainder non-zero
NU/11;            Jump label, if remainder zero
SE/M1, M5;        Store result in X
```

Where:
M1 contains X
M2 contains A
M3 contains B
M4 contains C
M5 contains result of division

Program section 3.11: Programming of "(a+b)/c" in SIGLA (integer division)

Table 3.14: Operator priorities in SRL, PASRO, AL and HELP

Priority	SRL and PASRO	AL	AML	HELP
1	function call (), NOT	function call (), ||, NOT	=	standard function call (), NOT
2	*, ., /, DIV, MOD, AND	WRT, ->, ^	NOT	*, /, AND
3	+, -, OR	*, /, ., DIV, MOD, MIN, MAX	*, /, IDIV	+, -, OR
4	=,<,>,<=,>=,<>	+, -	+, -	=, <, >
5		=,<,>,<=,>=,<>	ROTL	
6		AND	EQ, NE, LT, LE, GT, GE	
7		OR, XOR	AND	
8		EQV	OR, XOR	
9			OF	
10			IS	
11			=	

The highest priority is 1.
The dot operator does not exist in PASRO.

to right in SRL, PASRO, AL and AML, whereas this is not defined in HELP. This is particularly important in AL, too, as is shown in the following example:

```
v1.v2 * v3
(v1.v2)*v3   or
v1.(v2 * v3)
```

It is possible to perform an assignment in AML at any place in an expression. The interpretation of an assignment is position-dependent. Therefore, the operator "=" has two priorities: 1 and 11. The following examples illustrate this:

```
5 + A = 3;
```

Here "=" has priority 1 (compared to any other operator on the left hand side of "="). The effect is an assignment of 3 to A, and the expression finally has a value of 8 because of the addition of A and 5.

```
A = 3 + B;
```

Now the "=" sign has priority 11 (compared to any other operator on the right hand side of "="). First the addition is executed and the result assigned to A.

The ROTL operator performs a bit rotation on a 16-bit word, and the IS operator is used for data conversion.

The capability to write complicated arithmetic formulae in familiar mathematical conventions increases the legibility quite significantly, as demonstrated in the next example from PASRO:

```
IF ABS (arm.transl.z - workpiece.transl.z) < safedistance
THEN {instruct collision avoidance}.
```

This instruction monitors the approach of the arm in the z direction. A relative movement in the z-direction can be expressed in SRL as:

```
SMOVE arm TO arm + VECTORC (0,0,distance);
```

where `distance` is a real.

3.3 Assignments

The values obtained from expressions or functions can be assigned to variables, and consequently the previous content of that variable is lost.

3.3.1 Value Assignments

The simplest form of assignment is the value assignment, where a value is stored in a variable of the same type. This assignment has very different forms in the various languages (Table 3.15).

SRL, PASRO, AL, AML, VAL-II and HELP may contain expressions of arbi-

Table 3.15: Syntax of assignments

SRL:	a := 5;
PASRO:	a := 5;
AL:	a <- 5;
AML:	A = 5;
VAL:	SETI A = 5
VAL-II:	A = 5
HELP:	A := 5;
SIGLA:	SE/M1,5
ROBEX:	A = 5

trary complexity on the right-hand side. The attribute "arbitrary" has to be viewed in context, since the actual complexity is obviously dependent on the implementation of the compiler. Note that an assignment in AML could be part of an expression (see examples in Sect. 3.2.3).

ROBEX assignments are such that the compiler takes the value 5 instead of the symbol A in all following instructions.

Data types have to be matched in all assignments and in every language. The only exception in many high-level languages is the assignment of integers to reals, where type conversions are performed implicitly. SRL, PASRO and AML overcome this problem in such a manner. AL, VAL-II and HELP have no distinction for INTEGER and REAL values, and the remaining languages do not allow any variables of type REAL.

AL also demands dimensions apart from the matching of the data types, and this can lead to rather complicated expressions. Additionally, the equality of dimensions is considered during addition and subtraction. These checks for consistency are meant to increase **program security**, which is almost certainly true for SCALAR and VECTOR data types. It is, however, doubtful whether legibility and security of a program is aided by dimensioned expressions for rotation, frames and transformations. Further statements toward this can only be made after a certain phase of industrial use of AL.

Examples for the use of value assignments are contained in program section 3.10. Other assignments, apart from the ones introduced here, change values without appearing to do so, like the input of a value (Sect. 3.3.2) or the motion instruction of AL (Sect. 3.3.2.2). An example of this is the special form of the AFFIX instruction in AL (AFFIX...AT <trans>), which will be discussed in Sect. 4.1.

3.3.2 Inputs

Data entered during program execution can be classified according to its origin: There are entries resulting from a dialogue with the user, possibly from a VDU, or readings from instruments, like the arm position of a robot or limit switches in the robot environment. The *comfort* and *flexibility* of a robot language depends principally on the variety of those inputs and the associated possible outputs, and any program can adjust itself to the *working environment* only to the same degree as the language supports the possibilities for communication between the technical and human environment.

Table 3.16: Input routines

Type of result	AL	HELP
SCALAR *	QUERY (‹printlist›)	ASK ('‹text›')
SCALAR	INSCALAR	ASKN ('‹text›')

‹text› any character string without the apostrophe.
‹printlist› consists of parameters separated by commas. They may be algebraic expressions, variables or "‹text›".
* The scalar value is interpreted as a Boolean value.

The next sections concern themselves with assignments through user input and entry of world coordinates and arm positions, while the entry of data from interrupts or sensors is dealt with in the technical context of the robot itself in Chap. 4.

3.3.2.1 User Inputs

SRL, PASRO, AL, AML, VAL-II and HELP contain standard instructions to input Boolean and numeric values. Table 3.16 summarizes these functions for AL and HELP.

The <text> - or the content of the <printlist> - is displayed on the terminal, together with a prompt for the desired data type. The run-time system asks the user to enter Boolean values as Y for YES (= TRUE) or N for NO (= FALSE). The user can then type his response, and the program is halted until the data is read in. Unfortunately, the implementation of INSCALAR does not contain a <printlist>. Instead, INSCALAR announces "Scalar, please" on the terminal, and it is left to the user to specify by a previous PRINT (<printlist>) which actual value is to be entered.

Since input routines are functions, the result has to be evaluated immediately, as shown in program section 3.12. Let the variable height be a DISTANCE SCALAR (see also Sect. 3.1.1).

It is up to the routines to check for the validity of the data entered. For example, a number may only consist of certain characters, such as digits and decimal points, and it has to fall within the limits of the representation capability of the computer. Actions following any mistakes are implementation-dependent and may, for exam-

```
num <- 0;
WHILE QUERY (num, " Box processed, continue? ") DO
BEGIN
  PRINT ("Enter box height: ");
  height <- INSCALAR * cm;
  num    <- num + 1;
          .
          .
          .
END;
```

Program section 3.12: Example for an input routine in AL

ple, repeat the data request after an error message has been conveyed. The programmer, too, should check the plausibility of entered data, since the entry of a height of 10000 cm, for example, may be represented by the computer but is certainly not sensible.

PASRO offers a general form of the input procedure with the instruction READ. It contains an unlimited number of parameters which have to be of type CHAR, INTEGER or REAL, to which the value is assigned. Data can be read from files as well as from terminals (as a special file) with this instruction. The WRITE instruction permits output to terminals (or, generally, files).

In SRL and PASRO, the user can read and write the geometric data types VECTOR, ROTATION and FRAME directly. For this purpose, PASRO offers the system procedures vread, vwrite, rread, rwrite, fread and fwrite.

The statements INPUT and OUTPUT of SRL can be used to read data from or send data to any device, like a terminal or a sensor. The compiler checks the input/output, to which the variable is connected, with respect to the system specification part of the user program. For example, if the variable switch was connected to the digital port 2, then the statement

```
INPUT (switch);
```

sets switch to the value TRUE if a voltage is asserted at port 2 or to FALSE otherwise. It is possible in SRL to handle input and output of values of numbers, characters and strings on any device.

AML supports a general form of file handling such as SRL, PASRO or PASCAL. The user can read aggregates, representing, for example, geometric data types, from a terminal or a file with the help of the system procedure READ.

VAL-II supports the input and output from/to terminals or sensors for real values.

3.3.2.2 Reading of Current Robot Position and Orientation

Most robots, particularly those used for assembly, are equipped with a position measuring system, which allows interrogation of the current robot position. Hence most languages provide an instruction to read the joint angles or displacements. Usually, these are transformed into Cartesian coordinates and assigned to a frame variable (Fig. 3.11).

In SRL, the current robot position and orientation can be obtained by applying the system function robotframe (‹robot-identifier›), while in AL these data are stored in system variables ARM, ARM1, ..., ARM7. The user can refer to these frames in all assignments and expressions where "normal" frames are permitted. If the specified robot is not in motion, the actual position and orientation are already in a robot frame, otherwise the coordinates are read, transformed and entered into the robot frame.

```
SRL:    current_position := robotframe(puma).transl;

AL:     current_position <- POS(ARM);
```

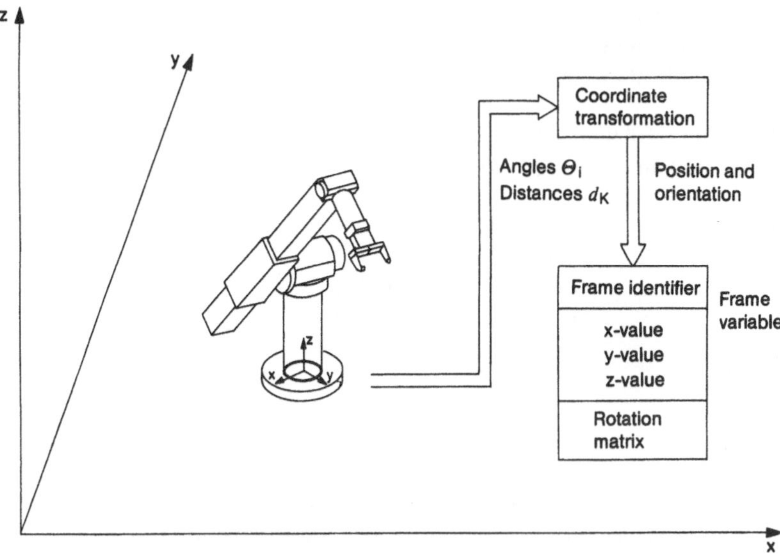

Fig. 3.11: Reading of the current robot position and orientation

SRL:

```
IF robotframe(puma).transl.z > 50 THEN
   WRITELN ("Arm too high!");
```

Robot coordinates in PASRO are stored in the system variables `robotjoints` and `robotframe`.

The system subroutine QGOAL in AML returns the current position of the robot in Cartesian coordinates, such as X, Y, and Z position values and O, A and T Euler angles. The seventh value of the aggregate represents the gripper opening (see Sect. 3.1.4).

VAL offers a special instruction to read the robot position and orientation. The HERE instruction enters the coordinate values into a frame after they have been read.

VAL:

`HERE targetframe`	The current robot position and orientation are entered into the `targetframe`
`HERE #precisepos`	Storage of the frame as a precision frame, storing the actual values of the joint variables and not the Cartesian coordinates.

HELP performs the reading of coordinates via the COORD instruction. Access to the entered coordinates is gained through the arithmetic functions AX, AY, AZ, AR, AP and AYW, where

AX x-coordinate
AY y-coordinate
AZ z-coordinate

AR angle of orientation (roll)
AP angle of orientation (pitch)
AYW angle of orientation (yaw)

HELP does not contain any frame variables so that the coordinate values have to be assigned to an array explicitly.

HELP:
```
COORD(1);
framef (1) := AX(1);
framef (2) := AY(1);
framef (3) := AZ(1);
framef (4) := AR(1);
framef (5) := AP(1);
framef (6) := AYW(1);
```
Position and orientation of robot 1 are assigned to the array framef after they have been read.

As SIGLA is based on stepping motors without position transducers and ROBEX does not contain variables, neither robot language provides instructions to read in the current position and orientation.

3.3.3 Assignments to Pointers

Program section 3.8 uses pointer variables and assignments to pointers. The value transferred in this transaction is the *address* of a data object. The address contained previously on the left-hand side is overwritten. This may result in the loss of pointers to objects creating inaccessible data space in memory, and its elimination and reclaiming is only possible in SRL, PASRO (or PASCAL) and AML.

The compilers in SRL and PASRO or PASCAL check the consistency of types in the same way as the assignments dealt with so far. Hence only those assignments with pointers referring to the same data type are permissible. The only exception to this is the empty pointer which is called NIL in SRL and PASRO. The program section 3.13 may clarify this, assuming the type declarations from program 3.8.

```
VAR boxes, currentbox   : boxpointer;
    frameaddress        : framepointer;

BEGIN
  NEW (boxes);                       (* Make a data object of type    *)
                                     (* box                           *)
  currentbox := NIL;                 (* Valid assignment              *)
  frameaddress := boxes;             (* Invalid assignment            *)
  boxes := NIL;                      (* Data object is not accessible *)
                                     (* any more
END
```

Program section 3.13: Pointer assignments (PASCAL)

The first line generates an object of type boxtype, and the variable boxes contains the pointer to that object. After the correct assignment in the second line, an incorrect one is carried out in the third line because they are of different types and consequently point to different objects. The last line is an example for the loss of a data object which is now inaccessible but still exists and occupies memory space.

Since the AML compiler does not check the consistency of pointers and the data types of objects to which they refer, there are no limitations to the programmer and it is possible to write very ambiguous and unsafe programs.

3.4 Output of Text and Numbers

The instructions for output serve to communicate with the environment just like the input. There are:

- Output in a form readable by the human on a VDU or a printer (or to a text file)
- Output of data to an external memory to store data amounts in files not readable by humans (see Sect. 3.1.5.3)
- Output to the robot or other components in the technical environment (see Sects. 4.2, 4.3 and 4.4).

The first form of output is discussed in this section. The second form is mostly usable in a problem-oriented manner and is generally accessible to the programmer in SRL, PASRO or PASCAL and AML. An example for special file output may be found in the frame stores of PASRO. Chapter 5 takes a closer look at this.

All languages except ROBEX allow the output of text and numbers during program execution. In addition, AL and VAL offer the possibility combining the output of text with a program halt and letting it wait for a special input. In particular, the following instructions are available:

SRL:
```
OUTPUT (<channel>, <parameter list>);
WRITELN (<parameter list>);
```

PASRO:
```
WRITE (<parameter list>);
```

AL:
```
PRINT (<parameter list>)
PROMPT (<parameter list>)
```

AML:
```
WRITE (<i/o channel>, <parameter list>);
DISPLAY (<parameter list>);
PRINT (<parameter list>);
```

The `<parameter list>` may contain scalar variables and expressions or strings. The latter is enclosed in quotation marks. While the program run continues after the execution of the PRINT instruction, the PROMPT instruction displays "Type P to proceed" after the parameter list has been output and waits until the user types a "p". The WRITE command of AML is the more general form of an output command and applicable to files, too, while DISPLAY is for output to VDU only. PRINT uses the currently open i/o channel.

VAL:
```
TYPE [<text>]
TYPEI <integer>
PAUSE [<text>]
```

The TYPE instruction prints the text, just as PAUSE does, but PAUSE interrupts the program execution, which can be continued by typing "PROCEED". TYPEI contains a peculiarity: it types out the values together with the name of the variable.

The TYPE instruction of VAL-II accepts a series of arguments which may be any combination of string constants, arithmetic expressions or format control.

HELP: PRINT (‹parameter list›)

This instruction corresponds to the one in AL, except that no scalar expressions are allowed, only variables. Strings are enclosed by a '-sign, instead of quotation marks, and special control characters are available such as line feed, bell, etc.

SIGLA: NT/‹text›,‹counter›

The <text> is limited to 8 characters and <counter> specifies the register whose content has to be printed.

The general WRITE instruction of SRL and PASRO or PASCAL has already been mentioned in Sect. 3.3.2.1.

4 Instructions

Instructions in robot languages describe activities which are executed either with program data or with industrial robot(s) and effector(s). Usually, instructions are obeyed in the order in which they have been written. Exceptions to this are jump instructions and commands invoking parallel program execution.

Those instructions contained in "normal" programming languages have been dealt with in Sects. 3.3 and 3.4 (assignments, input/output), because they are closely linked to data structures. In addition, general instructions for program flow control are introduced in this chapter in Sect. 4.7. However, the detailed treatment of the instructions necessary for *special robot programming* is completely new, particularly those for the world model, motion, effector and sensor commands and dealing with exceptional situations.

As in the previous chapter, various solutions and forms of representations in the programming languages are illustrated by numerous programming examples. The syntax is described by syntax diagrams.

In addition to the generally treated languages SRL, PASRO, AL, AML, VAL, HELP, SIGLA and ROBEX, the instructions of RAIL for vision systems are introduced in Sect. 4.5.4 as they represent a comprehensive starting point for the sensor instructions necessary for communication with vision systems.

4.1 Instructions for World Models

Apart from **world models** which are generated and administered offline, some solutions exist for online models which are built up during execution of a user program and adjusted according to real changes in the workspace of the robot. Such operations on models may be carried out automatically via sensors or via special instructions in the user program. However, the first case of updating the world model with sensory information has hardly been realized. The second possibility is considered here, changing the internal model explicitly through instructions. Only SRL and AL include an attempt at an offline model (see also PAUL [4.1]). In PASRO, it is possible to program structures explicitly and access a world model via records and file input/output. Geometric models in ROBEX are dealt with offline (cf. Sect. 3.1.4). AML allows the explicit construction and management of data for a world model with the help of its pointers (see Sect. 3.1.6), aggregates (Sect. 3.1.5.4) and its general file handling (see Sect. 3.1.5.3).

The SRL-system is designed to be an interface to higher level modules for implicit programming. One of the features needed for world modelling is the robot data base RODABAS, which holds the relevant data describing the world model. Part of this world model should be provided at run-time, for which the language SRL provides an interface to a world model. The underlying logical structure of the world model is based on two features:

```
object:    consists of (name, attribute)
attribute: consists of (name, value)
```

Some of the SRL statements dealing with the access to the robot data base are listed below.

SRL:

`EXISTENCE (part1);`
The programmer can ask about the existence of an object such as `part1`, and the function returns a Boolean value.

`weight_part1 := ATTRIBUTE weight OF part1;`
The value of the attribute `weight` of object `part1` is assigned to the variable `weight_part1`.

`ATTRIBUTE length OF stick := 5.3;`
The attribute `length` of the object `stick` in the data base is given the value 5.3.

```
AFFIX object1 TO object2;
UNFIX object1 FROM object2;
```
The effect of these affix statements, which are described later, is the same as in AL.

Within the context of pointer types, Sect. 3.1.6 contains an example showing the capability of records in PASCAL to build up a data structure for world models, a feature also applicable in PASRO. Therefore, the extended frame concept of AL is given in PASCAL notation at this point without extensive introductions, and the corresponding AL instructions are also shown. Program section 4.1 demonstrates the definition of geometric data types on the basis of the extended frame concept as well as operations for world models. In comparison with program section 3.4, the frame in Sect. 3.1.4 contains a pointer `next` to a frame list, pointers `approach` and `depart` for the approach or departure frames (Sect. 4.2.4) as well as an element for an affixment. An affixment in AL is the concatenation of frames observing mutual spatial relationships. The spatial relation consists of a translation and rotation, with which values of one frame can be referenced to another (Sect. 2.3.1). For example, if frame A is fixed to frame B, all values of A are changed automatically, should B change, and in such a manner that the relation calculated during an affixment is always maintained. For simplicity in this example, the affixment is carried out only with the pointer `affixframe` of the frame to be connected, and the relationship is entered in `affixrel`. This is flagged for later reference in the element `affixedon`. A frame is created by the procedure `newframe` and filled with empty pointers. If one frame is to be affixed to another, the procedure `affix` is called with the following parameters

```
PROGRAM model;

TYPE
  vector      = RECORD
                  x, y, z: REAL
                END;

  rotmatrix   = RECORD
                  t, o, a: vector
                END;

  rotation    = RECORD
                  axis  : vector;
                  angle : REAL;
                  matrix: rotmatrix
                END;

  trans       = RECORD
                  translation: vector;
                  rot        : rotation
                END;

  framepointer=^frame;

  frame       = RECORD
                  next        : framepointer;
                  orient      : rotation;
                  pos         : vector;
                  approach    ,
                  departure   : framepointer;
                  affixedon   : BOOLEAN;
                  affixframe  : framepointer;
                  affixrel    : trans
                END;

VAR
  nilvector   : vector;
  nilrot      : rotation;
  affixvariable, dummytrans : trans;
  box, hole, gripp, appropos : framepointer;

PROCEDURE newframe ( VAR framevar : framepointer;
                         xf, yf, zf : REAL);
BEGIN (* newframe*)
  NEW (framevar);
  WITH framevar^ DO
  BEGIN
```

```
    next        := NIL;
    orient      := nilrot;                  (* for simplification *)
    pos.x       := xf;
    pos.y       := yf;
    pos.z       := zf;
    approach    := NIL;
    departure   := NIL;
    affixedon   := FALSE;
  END;
END (* newframe *);

PROCEDURE affix (VAR atframe     : framepointer;
                     targetframe : framepointer;
                 VAR affixtrans  : trans );
BEGIN
  WITH atframe^ DO
  BEGIN
    affixedon   := TRUE;
    affixframe  := targetframe;
    affixrel.translation.x      := pos.x - targetframe^.pos.x;
    affixtrans.translation.x    := affixrel.translation.x;
    affixrel.translation.y      := pos.y - targetframe^.pos.y;
    affixtrans.translation.y    := affixrel.translation.y;
    affixrel.translation.z      := pos.z - targetframe^.pos.z;
    affixtrans.translation.z    := affixrel.translation.z;
    affixrel.rot                := nilrot;      (* for            *)
    affixtrans.rot              := nilrot;      (* simplification *)
  END;
END (* affix *);

PROCEDURE unfix (VAR atframe    : framepointer;
                     tragetframe: framepointer);
BEGIN
  atframe^.affixedon  := FALSE;
  atframe^.affixframe := NIL;
END (* unfix *);

BEGIN (* main section *)
  WITH nilvector^ DO

BEGIN
  x := 0;
  y := 0;
  z := 0;
END;
nilrot.axis   := nilvector;
nilrot.angle  := 0;
```

```
        .
        .
        .
  newframe ( box,        55, 30, 22 );
  newframe ( hole,       45, 30, 28 );
  newframe ( gripp,      45, 50, 34 );
  newframe ( appropos,   45, 50, 40 );
        .
        .
        .
  affix  ( hole,        box,    dummytrans );
  affix  ( gripp,       box,    affixvariable);
  affix  ( appropos,    gripp,  dummytrans );
        .
        .
        .
  unfix  ( gripp, box );
        .
        .
        .
END (* main section *).
```

Program 4.1: Definitions and operations for environmental models, formulated in PASCAL

1. a frame to be connected,
2. a frame to be connected to,
3. and a transformation to be entered into the frame relation.

The procedure call

```
affix (hole, box, dummytrans);
```

causes `affixedon` of the frame to be set, the pointer `affixframe` to point to the frame box and the values of the frame relation or transformation to be calculated. For the sake of clarity, only the evaluation of the translation is given here. The translation and a "null rotation" is entered into `affixrel` of the frame `hole` with the following values:

```
affix.translation.x = 45 - 55 = -10
affix.translation.y = 30 - 30 =   0
affix.translation.z = 28 - 22 =   6
```

Suppose the frame box is given the new values x = 25, y = 35 and z = 22 because the box has been moved to that position, then the run-time system automatically calculates the new values of the frame `hole` according to `affixrel`:

```
hole.pos.x = box.pos.x + hole.affixrel.translation.x
           = 25 - 10 = 15
```

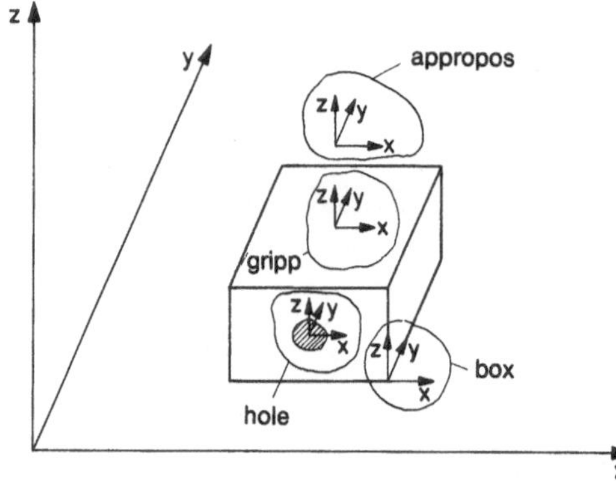

Fig. 4.1: Geometric representation of frame relations in the environment model

```
hole.pos.y = box.pos.y + hole.affixrel.translation.y
           = 35 +  0 = 35
hole.pos.z = box.pos.z + hole.affixrel.translation.z
           = 22 +  6 = 28
```

Hence, the frame `hole` now has the position x = 15, y = 35 and z = 28. The value of the frame relation is then assigned to the transformation variable `affixvariable` in the second procedure call

```
affix (gripp, box, affixvariable);
```

which is referenced more easily. The breaking of a frame relation is carried out by the procedure `unfix`, in which `affixedon` is reset and the pointer `affixedframe` cleared. Fig. 4.1 illustrates the geometric representation of frame relations in a world model. As already mentioned, treatment of the rotation has been left out since it increases the size of the program significantly and hence diminishes clarity.

The corresponding program section 4.2 in AL is obviously significantly shorter because definitions of the geometric data types VECTOR, ROT, FRAME and TRANS are missing, as well as the procedures `newframe`, `affix` and `unfix`. The language AL offers these data types as standard, and the procedures are realized by assignments to frame variables as well as the instructions AFFIX and UNFIX. Figure 4.2 illustrates the syntax of AL instructions for world models.

Just as in the PASCAL example, the instruction

```
AFFIX hole TO box;
```

causes a pointer from the frame `hole` to point to the frame `box` and the values of the frame relation to be calculated. Contrary to PASCAL, the programmer has no explicit access to pointers or values of frame relations. The latter is only possible if a transformation variable is specified after the keyword BY:

```
AFFIX gripp TO box BY affixtrans;
```

Thereafter, access to the values x = -10, y = 20, z = 12 of the relation can be gained via the vector part of `affixtrans`. In the PASCAL example, both frame values had to be specified for one affixment. This is not necessary in AL. The keyword

```
FRAME box, hole, gripp, appropos;
TRANS affixtrans;
box   <- FRAME (NILROT, VECTOR (55, 30, 22) * CM);
hole  <- FRAME (NILROT, VECTOR (45, 30, 28) * CM);
gripp <- FRAME (NILROT, VECTOR (45, 50, 34) * CM);
     .
     .
     .
AFFIX hole      TO box;
AFFIX gripp     TO box BY affixtrans;
AFFIX appropos TO gripp AT TRANS (NILROT, VECTOR (0, 0, 6) * CM);
     .
     .
     .
UNFIX gripp FROM box;
```

Program section 4.2: Instructions for the environment model in AL

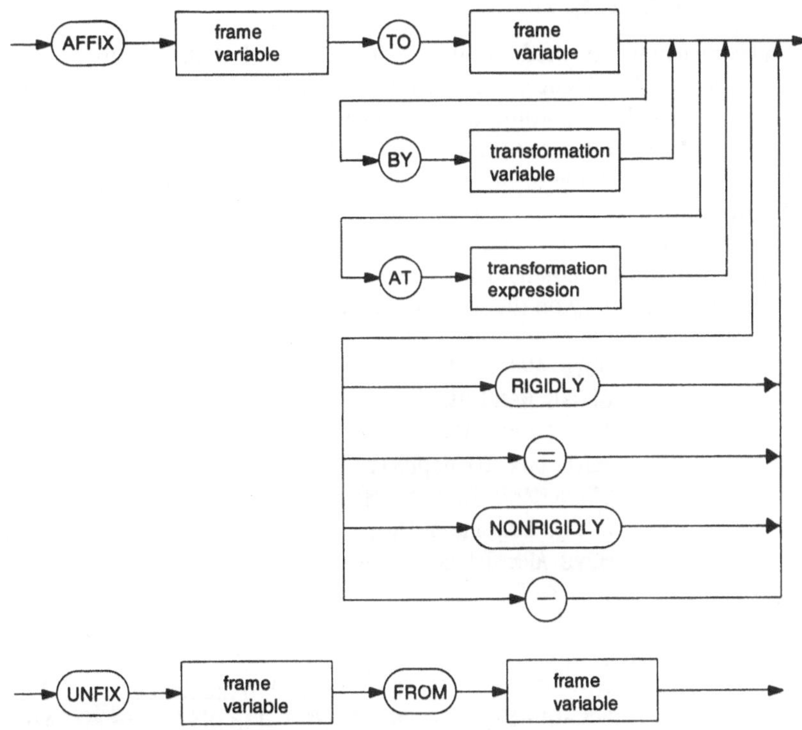

Fig. 4.2: Instructions for the environment model in AL

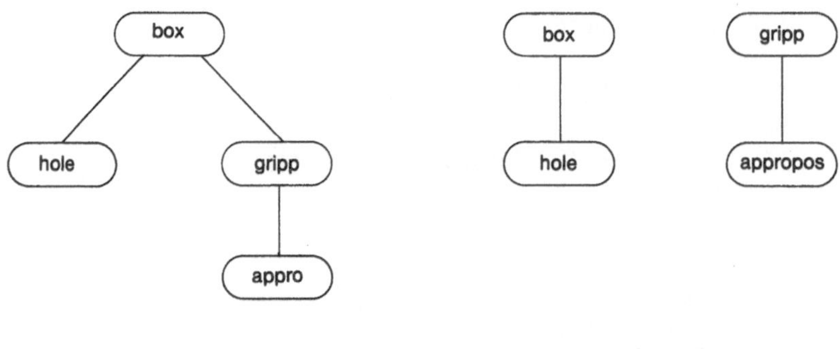

Fig. 4.3: Structure of affixments in the AL example

AT causes the frame, which is to be concatenated, to be evaluated relative to the referred frame by means of the given transformation expression:

```
AFFIX appropos TO gripp AT TRANS(NILROT,VECTOR(0,0,6)*CM);
```

After this instruction, **appropos** contains the values x = -10, y = 20, z = 40 according to the position of **gripp** and the transformation vector x = 0, y = 0, z = 6 cm. UNFIX cancels that affixment, and Fig. 4.3 illustrates the affixment structures of the example before and after the instruction

```
UNFIX gripp FROM box;
```

A frame can be attached not only to other frames but also to those of the robot or ARM frames in AL. Hence, apart from the arm frame, all values of concatenated frames are calculated automatically according to their frame relations. This is obviously useful if the robot has gripped an object and moved it to a different place. The relevant gripping frame of the part is corrected accordingly, so that the robot does not grip thin air when it tries to pick the part up again. However, the programmer must not forget to break the link after the part has been put down because otherwise the frame values are changed with every move instruction, although in reality the part is lying stationary on the work surface.

AL:

```
MOVE ARM TO box;
CLOSE HAND TO 5 * CM;
AFFIX box TO ARM RIGIDLY;
MOVE box TO deposit;
OPEN HAND TO 8 * CM;
UNFIX box FROM ARM;
MOVE ARM TO continue;
.
.
.
```

After the robot has been moved to the frame box and the gripper is holding the part, box is affixed to the arm frame. After the box has been moved to the place of **deposit**, the link is disconnected by the UNFIX instruction. The frame box now contains the new gripping position of the box after the motion.

The keyword RIGIDLY - or the character "=" - causes a mutual affixment, i.e., changes in one frame are automatically transferred to changes in the other and vice versa. If NONRIGIDLY is specified, only the dependent frame is changed with changes in the reference frame, but not the other way round. In SRL every affixment is rigid.

The example above only contains the moveable frame box instead of the robot arm. It is moveable because it is fixed firmly to the arm frame.

4.2 Motion Instructions

However an instruction for robot moves is defined, it represents the **character of a robot language**, it is contained in all of the robot languages known to the authors (compare with BLUME [2.5]), and it is given by the symbol MOVE almost without exception. The programmer controls the movements of robots through these motion instructions. They can be divided roughly into *explicit* and *implicit* move commands. With explicit commands, the programmer specifies position and orientation coordinates of the gripper or tool (usually in frames). Frames are not necessarily defined textually; their definitions are also possible by the teach-in procedure (see Chapt. 5). An implicit move instruction assumes that the programming system itself is able to evaluate the necessary coordinates for any movement with the help of a data base.

Execution of motion instructions during the run-time of the program is realized by an appropriate **robot controller**. Whatever kind of control it is capable of, the path of the effector is either straight (linear Cartesian interpolation), or circular (circular interpolation) or it depends on the kinematic of the robot, and hence is hardly predictable by the programmer (point-to-point motion). Consequently, the effect of even a simple motion instruction like

```
<move robot arm to point x>
```

depends entirely on the **run-time system**. Currently, there are only a few languages which allow a specification of the motion by the programmer. The reason for this can be found in the special kind of motion control around which most languages have been developed. Transfer to other systems has not been foreseen at all and thus, control strategies and hence paths are contained *implicitly* in each language and cannot be specified explicitly by the programmer. On the other hand, ROBEX does not contain any instructions for the kind of motion, although it has been designed as a system-independent language.

4.2.1 Implicit Motion Instructions

In the languages dealt with here, only ROBEX contains an implicit motion command, which exists additionally to the explicit one (cf. Sect. 4.2.3). It is implicit because the coordinate values of frames are not given. The programmer only specifies

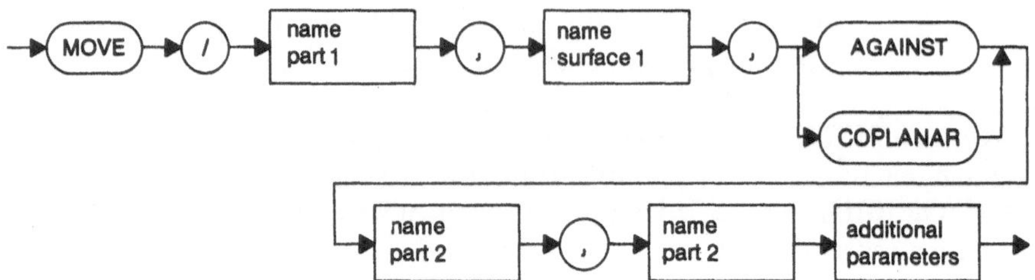

Fig. 4.4: Implicit motion instructions in ROBEX

the description of the part geometry and the relation to be generated during each movement of one part; for instance, part 1 is parallel to part 2. The implicit motion instruction in ROBEX has the structure illustrated in Fig. 4.4 (see also AMBLER [4.2]).

Parts and surfaces have to be defined previously by explicit coordinates (cf. Sect. 3.1.4). The compiler then evaluates the gripping points and paths from the geometric description of the manipulated parts, the robot and the working environment. The generated code contains explicit motion instructions which may contain evasive actions to avoid collisions. Additional parameters are concerned with the exact approach of predefined target frames and the inclusion of external signals.

Here is an example in ROBEX:

```
MOVE/box,underside,AGAINST,construct,top,EVENT,2,ELSE,message
```

The compiler generates any necessary motion and gripping commands, in order to put the `box` with its `underside` onto the `construct`. If there is no signal from channel 2 during the motion - as may be triggered by a force sensor when depositing the box - program execution jumps to the label `message`.

4.2.2 Explicit Motion Instructions

Coordinate values for frames of motion can be specified explicitly by text or via a teach-in procedure (Chapt. 5). *Textual definition* allows distinction between different coordinate systems and allows for possible sensor integration. In any case, the programmer has to know the position and corresponding orientation of the manipulated objects as well as the gripper or tool, and has to be clear about the geometric relations of the objects so that the programmed path does not cause collisions and can be calculated by the controller.

Explicit coordinate definitions refer usually to Cartesian coordinates, but cylindrical, polar and joint coordinates are also common. The use of Cartesian coordinates is nearly always linked to the frame concept (cf. Sect. 2.3.1), where orientations are specified as rotations around the coordinate axes or any other vector as rotation axis (SRL, PASRO and AL) or as Euler angles (VAL). Another common use is that of "roll", "pitch" and "yaw" (AML, HELP), having aeronautic and maritime origins. Formally defined, the three Euler angles in the VAL system refer to a rotation around the world z-coordinate axis, a further rotation around the new y-axis and finally a rotation around the new z-axis of the Cartesian coordinate system. This

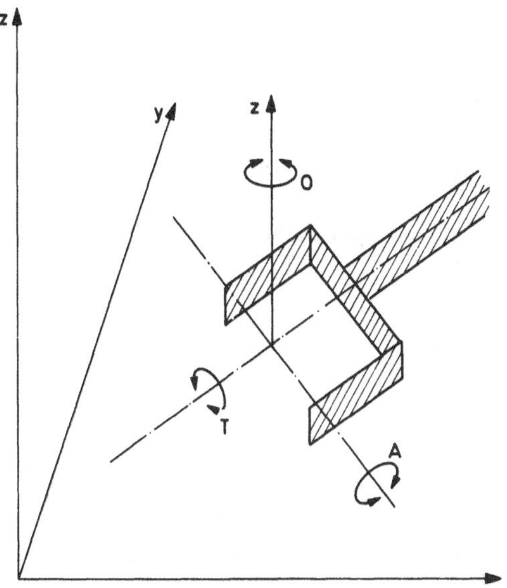

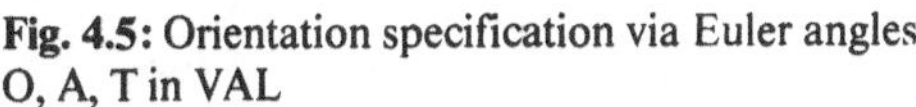
Fig. 4.5: Orientation specification via Euler angles O, A, T in VAL

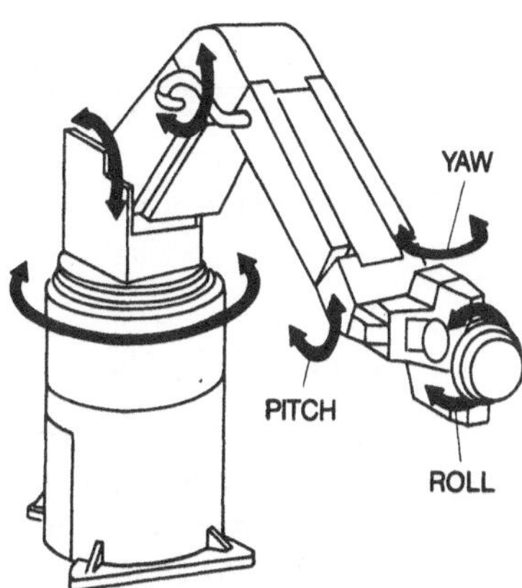

Fig. 4.6: Roll, pitch and yaw as orientation specification

corresponds to a rotation of the gripper axis around the z-axis, a rotation around the y-axis of the gripper coordinate system and lastly a rotation around the gripper axis itself, since this coincides with the z-axis of the gripper coordinate system (Fig. 4.5). *Roll* corresponds to the turning of a screw by a human hand, *pitch* to the bending of the wrist and *yaw* to the swivelling motion from left to right. Figure 4.6 shows these motions on a Cincinnati Milacron robot.

4.2.3 Simple Motion Instructions

A simple motion instruction is understood to be a command with the specification of just a target frame *without parameters* like intermediate frames, speed or duration. Naturally, a simple motion can be executed with sensory information, event and time-out checks, but these will be treated in Sects. 4.2.5 to 4.2.7.

A direct *"parameterization"* is present if the programmer describes the positions of targets by the number of *motor steps* which each robot axis has to execute in order to reach the point. For example, the SIGLA instruction

```
MO/1,1317,3,M2,P1,158
```

commands motor 1 to go to the absolute position of 1317 steps (relative to the index position), and motor 3 to go to the corresponding value of counter M2, and the motor specified in P1 to step to position 158.

Relative step demands are also possible with respect to the current robot position. This is indicated in SIGLA by the preceding command II, so for example

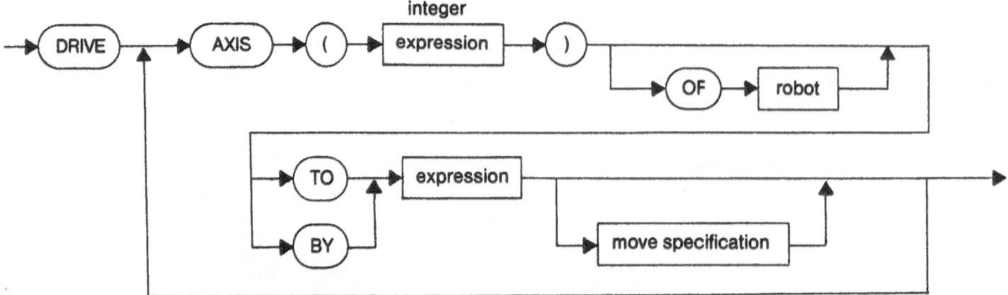

Fig. 4.7: DRIVE statement in SRL

```
II
MO/1,357,2,-182
```

moves motor 1 into the positive direction by 357 steps and motor 2 by 182 steps into negative direction. Although each motor in SIGLA is assigned unambiguously to a Cartesian coordinate axis, a programmer may have difficulty in visualizing the exact position after 357 steps.

The languages SRL (see Fig. 4.7), PASRO, AL, ROBEX and VAL contain the "robot-related" command DRIVE. It allows individual robot axes to be moved by specified angles (in case of revolute joints) or distances (in case of prismatic joints). The following instructions in SRL, PASRO, AL and ROBEX rotate the second joint of the robot to a position of 110 degree:

SRL: `DRIVE AXIS(2) OF arm1 TO 110;`

PASRO: `drive (2, 110);`

AL: `DRIVE JOINT (2) OF ARM1 TO 110;`

ROBEX: `DRIVE/2,110;`

VAL provides only relative rotations, e.g., 20 degrees, together with a speed demand in percent (Fig. 4.8).

SRL: `DRIVE AXIS(2) OF arm1 BY 20;`

PASRO: `drive (2, robotjoints[2] + 20);`

AL: `DRIVE JOINT (2) OF ARM1 BY 20;`

VAL: `DRIVE 2, 20, 100`

ROBEX: `DRIDLT/2,20`

In AML, joint moves correspond to the x-, y- and z-directions of the base coordinate system. AML therefore does not include a DRIVE statement; however, it is pos-

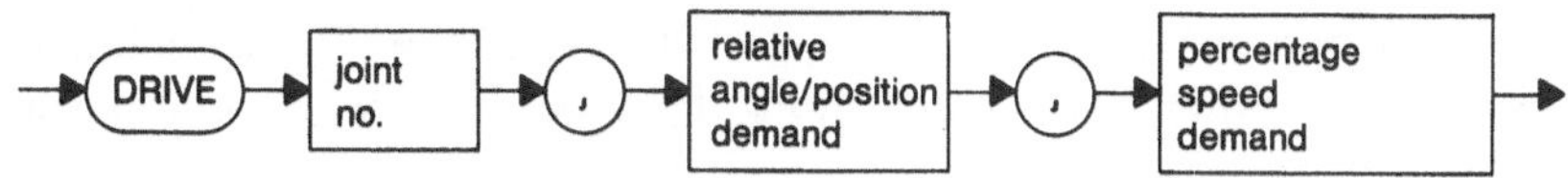

Fig. 4.8: DRIVE command in VAL (relative angle/distance demands)

sible to move joint 2 (y-direction), for example, to a value of 10 inches with the MOVE instruction.

AML:

```
MOVE (JY, 10);
```

JY is the reserved name in AML for the joint moving in y-direction. A relative translational move is performed by DMOVE:

```
DMOVE (JY, 0.2);
```

Some languages permit the specification of target points in robot coordinates. The DRIVE command in SRL can be used to specify all values of the axes for a goal position and orientation (see Fig. 4.7).

The PASRO system includes a predefined array type thetai, which consists of six real values representing the robot axis (see Sect. 2.3.2). An array of this type can be used as a parameter for the JDRIVE system procedure. The programmer has access to the system procedures of coordinate transformations, allowing the calculation of Cartesian coordinates for learning or control purposes.

Due to the special kinematics in AML, every point specified in Cartesian coordinates is also given in robot coordinates.

VAL stores robot coordinates in *precise points.* A precise point is marked by a preceeding " # " sign. The definition of such a point is carried out explicitly (Sect. 4.2.2) or via the teach-in procedure. The programmer drives the robot to the desired position and orientation. The precise point is then defined by a system command, and the angle or distance values are stored in a frame list under a symbolic name. The MOVE instruction allows the robot to be positioned at that frame during program execution.

PASRO:

```
jdrive (axisvalues); (* axisvalues is of type *)
                     (* thetai                *)
```

VAL:

```
1st step: POINT #P1
          0,110,30,0,60,0  (system command)
2nd step: MOVE #P1         (program instruction)
```

A specification in **Cartesian coordinates** is, however, much more suitable for human comprehension. This is directly possible in SRL, PASRO, AL, AML, HELP and ROBEX. Orientation is specified in SRL and PASRO by specifying a vector as the rotation axis and an angle as the rotation angle. In AL, rotations around the coordinate axes are used, in AML and HELP the "roll", "pitch" and "yaw" refer to the "wrist", and in ROBEX rotations are specified around three mutually orthogonal planes in space. An example is given in the languages to illustrate the movement of the robot to the point x = 95 cm, y = 20 cm and z = 40 cm. Let the orientation be as follows: the gripper points along the x-axis relative to a null position in the z-direction.

```
SRL:        SMOVE puma600 TO FRAMEC(VECTORC(95, 20, 40),
            ROTC(yaxis, 90));

PASRO:      setframe (goal, 95,20,40, yaxis, 90);
            smove (goal);

AL:         MOVE ARM TO FRAME (ROT (YHAT,90*DEG),
            VECTOR(95,20,40)*CM);

AML:        MOVE (<JX,JY,JZ,JP>, <95,20,40,90>);

HELP:       MOVE (1, #1,950, #2,200, #3,400, #5,90);

ROBEX:      GOTO/95,20,40,ZXROT,90
```

SRL and PASRO contain differing identifiers for the move command because they invoke different path interpolations (see Sect. 4.2.4). It is noticeable that ROBEX uses the keyword GOTO for the (explicit) motion command, whereas this is normally reserved for an unconditional jump in programs (in ROBEX JUMP). In later versions of ROBEX the keyword has been changed to MOVE, and also some concepts of normal computer science - such as variables - have been introduced in this version called ROBEX-M (M for microcomputers).

Figure 4.9 illustrates the approach to a position and orientation by explicit specification of coordinate values.

Such explicit demands cannot be programmed directly in VAL. However, the programmer can define a point or frame through explicit coordinate values with a system command. This frame is then entered automatically into a global frame list (see also Chapt. 5). The symbolic name can be referred to in the MOVE instruction and the movement to the corresponding position and orientation can be carried out. The example from above is implemented in VAL as follows:

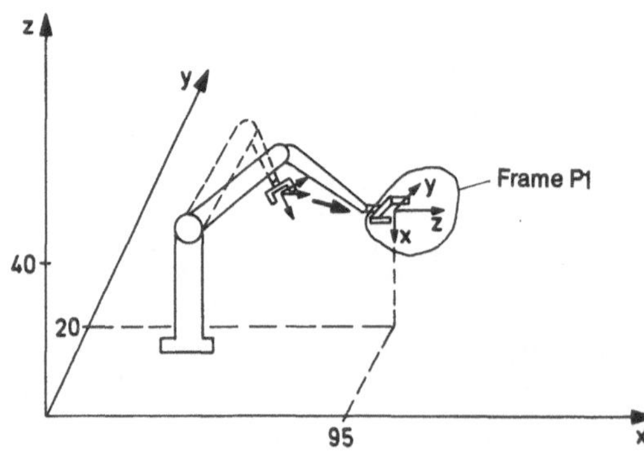

Fig. 4.9: Movements specifying explicit coordinates

```
SCALAR externsignal;
FRAME  targetframe;
     .
     .
     .
IF externsignal THEN
   targetframe <- FRAME (NILROT, VECTOR (30, 50, 8) *CM)
ELSE
   targetframe <- FRAME (NILROT, VECTOR (30, 50, 20) *CM);
     .
     .
     .
MOVE ARM TO targetframe;
     .
     .
     .
```

Program section 4.3: Motion instructions with frame variables in AL

VAL:

```
1st step: POINT P1
          95,20,40,0,90,0  (system command)
2nd step: MOVE P1          (instruction)
```

Orientation in VAL is specified as Euler angles.

The languages SRL, PASRO, AL, AML, VAL and ROBEX permit the definition of points or frames via teach-in procedures (Chapt. 5).

Out of the eight languages introduced here, only SRL, PASRO and AL contain frame variables as data types in their own right. An extension to frame variables is planned in ROBEX. Hence it is only possible in these languages to carry out a movement to a frame specified by a variable. The definition of an aggregate with the values of a frame is possible in AML, as well as its usage in a move statement, just as a frame variable. This, for example, enables the definition of a target frame depending on external signals, and further movements can be modified.

If `externsignal` in program section 4.3 has been set, the z-coordinate has a value of 8 cm, otherwise 20 cm.

In HELP, a simulation of frame variables is possible in which the X, Y, Z, roll, pitch and yaw values are stored in scalar variables. SRL, PASRO and AL also permit the definition of rotations which are evaluated during program execution. In ROBEX, the programmer could specify the point P1 and recall it in the motion instruction.

ROBEX:

```
P1 = POINT/17,25,15
.
.
.
GOTO/P1
```

With the inclusion of EX, an exact approach to the demanded position and orientation is carried out (see also Sect. 4.2.4).

In addition to movements to absolute frames, relative motions in Cartesian coordinates can be realized in SRL, PASRO, AL, AML, VAL, HELP and ROBEX. A movement of 20 cm in the x-direction and 10 cm in the z-direction relative to the current robot position is specified as follows:

SRL:

```
SMOVE pragma TO ROBOTFRAME(pragma) + VECTORC
(20,0,10);
```

PASRO:

```
makevector (relvect, 20, 0, 10);
frametransl (goal, robotframe, relvect);
smove (goal);
```

AL:

```
MOVE ARM1 TO @ + VECTOR(20,0,10) * CM;
```

AML:

```
DMOVE (<jx,jy,jz>, <20,0,10>);
```

VAL:

```
DRAW 20,,10
```

HELP:

```
MOVE (1,#1,X+200, 2,Y,#3,Z+100);
```

ROBEX:

```
GODLTA/20,0,10,0,0
```

Frame, vector or scalar expressions may be combined with the robot frame in SRL, PASRO and AL.

SRL, PASRO and AL do not provide a special command for relative motions; for this purpose, the target frame is evaluated relative to the current robot position. HELP does not have a special command, either. Here current positions have to be interrogated and stored in the scalar variables X, Y and Z.

4.2.4 Motion Instructions with Parameters

Apart from target frames themselves, a programmer may wish to influence the control of motions by further *parameters*. They specify

- Speed of movement
- Acceleration and deceleration of movement
- Duration of motion
- Robot configuration at target frame
- Precision of control at target frame
- Type of control
- Execution of parallel operations
- Superimposed weaving motion
- Intermediate frames or positions
- Approach and departure frames

Thus **paths** (trajectories) can be demanded which are geometrically complicated and may be altered dynamically. With all these different parameters, a controller will only be able to realize movements more or less approximately.

Different parameters as well as different combinations of parameters are allowed in various languages (see also Table 4.1). In AML, parameters are specified by calling system procedures. They automatically refer to the total movement with simple motion instructions without intermediate positions or frames, or approach/departure frames. This case is considered in the following section.

Speed can be set globally or only for the next move. Usually, it is not specified in physical quantities but in percentages relative to the maximum or medium speed calculated by the system.

SRL:

```
SMOVE rm55 TO targetframe
WITH VELOCITY = 20;
```

The move is executed with 20 cm (or mm) per second.

```
SPEEDFACTOR := 50;
```

All following moves are carried out with half the maximum speed.

PASRO:

```
SPEEDFACTOR := 5;
smove (goal);
```

All following moves are carried out with half the maximum speed.

AL:

```
MOVE ARM TO targetframe
WITH SPEED_FACTOR = 3;
```

Motion is executed with a third of the maximum speed.

```
SPEED_FACTOR <- 2;
```

All following movements are carried out with half the maximum speed.

AML:

```
SPEED (0.8);
```

All following moves are carried out with 80% of the maximum speed.

```
targetframe: NEW <x,y,z,roll,pitch,yaw>;
MOVE (ARM, targetframe);
```

VAL:

```
SPEED 200
MOVE targetframe
```

Movement with twice the normal speed.

```
SPEED 80 ALWAYS
```

All following movements are carried out with 80% of the normal operating speed.

Table 4.1: Comparison between motion instructions in AL, AML, PASRO, ROBEX, SIGLA, SRL and VAL

Language	AL	AML	PASRO	ROBEX	SIGLA	SRL	VAL
Simple motion instructions:							
- Specification of motor steps	-	-	-	-	MO/	-	-
- Specification of axis joints/ distances	DRIVE	-	DRIVE DRIVE	DRIVE	-	DRIVE	DRIVE precise points
- Explicit Cartesian coordinates	FRAME	REAL	FRAME	FRAME	-	FRAME	(FRAME)
- Frame variables	yes	-	yes	only at compile time	-	yes	yes
- Relative motion	FRAME + FRAME FRAME + VECTOR	DMOVE	via computation	GODLTA	-	FRAME expression	DRAW
Motion instructions with parameters:							
- Velocity	SPEED-FACTOR	SPEED	SPEED-FACTOR	RAPID FEDRAT	-	VELOCITY	SPEED
- Duration	DURATION	-	*	-	-	DURA-TION	-
- Robot configuration	-	-	-	-	POS-TURE		RIGHTY, LEFTY, ABOVE, BELOW
- Control accuracy	NULLING NO-NULLING	SETTLE	-	EX	-	EXACT	COARSE FINE NONULL, NULL INTOFF, INTON
- Parallel task execution	yes, generally	AMOVE WAIT-MOVE	*	**	yes via tasks	yes via SECTION	MOVET, MOVEST (gripping only)
- Superimposed weaving motion	WOBBLE	-	-	-	-	WOBBLE	WEAVE
- Intermediate frames	VIA	-	*	-	-	VIA-FRAMES	CP
a. With change in speed	VELOCITY	-	-	-	-	VELOCITY	SPEED
b. With duration	DURATION	-	-	-	-	DURA-TION	-
- Approach frames	APPROACH	-	calculated	simulated	-	APPRO	APPRO
- Departure frames	DEPAR-TURE -		calculated	simulated	-	DEP	DEPART
Motion instructions with sensory monitoring :							
- Force at gripper	FORCE	in MOVE	*	-	RP	ALWAYS WHEN	-
- Torque at gripper	TORQUE	in MOVE	*	-	-	ALWAYS	-

Table 4.1 (continued)

Language	AL	AML	PASRO	ROBEX	SIGLA	SRL	VAL
- Any other sensors	-	-	-	-	-	WHEN ALWAYS WHEN	-
Motion instructions with event monitoring: - Interrupts	EVENT-variable	-	*	EVENT-specification	-	ALWAYS WHEN interrupt	REACT (IGNORE)
Motion instruction with time monitoring	DURATION	-	*	-	-	ALWAYS WHEN	-
End effector control - Gripping	OPEN, CLOSE	in MOVE	GRIPOPEN, GRIPCLOSE, GRIPWIDTH	OPEN, CLOSE	-	OPEN, CLOSE, GRIPWIDTH	OPEN, CLOSE
- Execution of operations	-	-	-	-	-	OPERATE	-

* Implementation-dependent, e.g., in CONCURRENT PASCAL, MicroPower/Pascal or MODULA-2
** Planned

HELP: `MOVE(1, #1, 30, #3, 10, #11, 80)`
Motion with 80% of the maximum speed in x-direction.

`SPEED(1,80)`
All following movements of arm 1 are carried out with with 80% of the maximum speed.

ROBEX: `RAPID`
Fast speed for the following movement.

`FEDRAT/80`
80% of normal speed set.

Accelerations and decelerations can be specified only in AML.

AML: `ACCEL (0.5);`
The accelerations of following motions are carried out at half the normal acceleration.

Duration of motion can only be specified in AL.

AL:
```
MOVE ARM TO targetframe
WITH DURATION = 5 * SEC;
```
The motion lasts about 5 seconds.

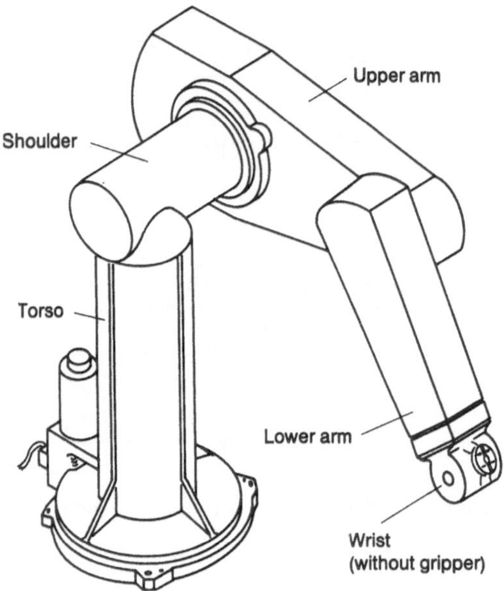

Fig. 4.10: PUMA robot as an imitation of the human torso and arm

Some industrial robots allow the approach to target frames in different *robot configurations*, because of their kinematics. This ambiguity can be resolved by the system or is determined by the predefinition of the programmer. Of those languages we are concerned with here, only SRL and VAL contain choices for the programmer. In SRL, the programmer can specify the configuration of the robot by a logical number which depends on the special kinematics of the robot. The VAL system is specially tailored to the PUMA robot of Unimation, which closely resembles the human arm with waist, shoulder and elbow (Fig. 4.10).

SRL:

```
SMOVE puma600 TO goalframe
  WITH POSTURE(1);
```

The posture 1 configuration may correspond to the human left arm.

VAL:

`RIGHTY` Configuration corresponds to the right arm of the human.

`LEFTY` Configuration corresponds to the left arm of the human.

`ABOVE` The "elbow" of the PUMA points upwards.

`BELOW` The "elbow" of the PUMA points downwards.

Because an exact approach to a position (and also to intermediate frames) is more time demanding than one which has a tolerance associated with it, the programmer may specify whether an *exact control* is necessary or not.

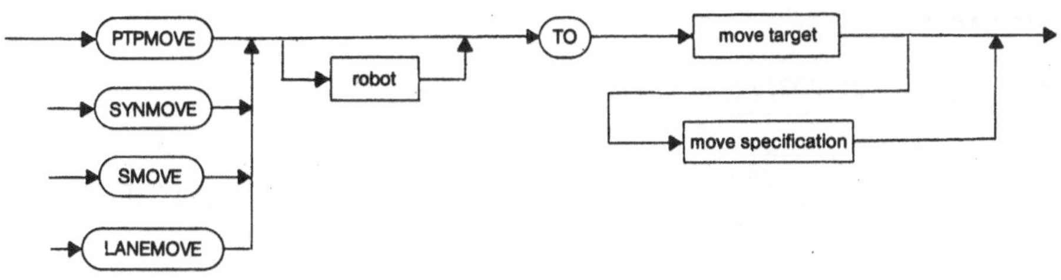

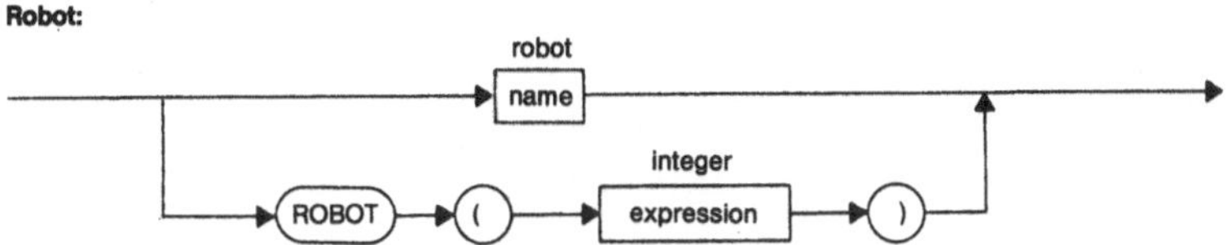

A robot name must be defined in the system specification before.

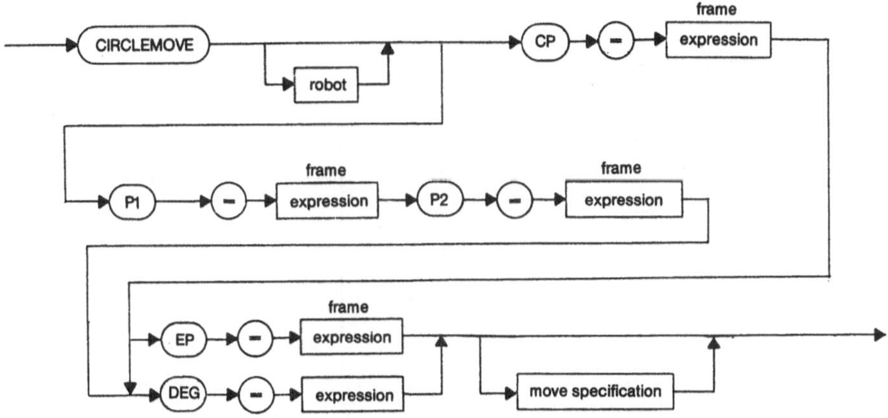

If the centre point resp. frame is given (CP) the plane of the circle is the x-y plane of the frame. The move target is given by an end point (EP) or by specifying the degrees of a circle segment. The degrees are defined with respect to the z-axis of the center frame. It is also possible to define the circle move by three points on the circle segment (P1, P2 and EP).

Fig. 4.11: Some move statements in SRL

SRL:

```
SMOVE r55 TO goalframe
WITH NULLING;
```

Small positioning errors at the end of the motion are corrected.

AL:

```
MOVE ARM TO targetframe
WITH NULLING;
```

Small positioning errors at the end of the motion are corrected.

AML:

```
SETTLE (OFF);
MOVE (ARM, targetframe);
```

The system returns control to the user immediately after the last interpolated move command has been passed to the servo-control.

Specification:

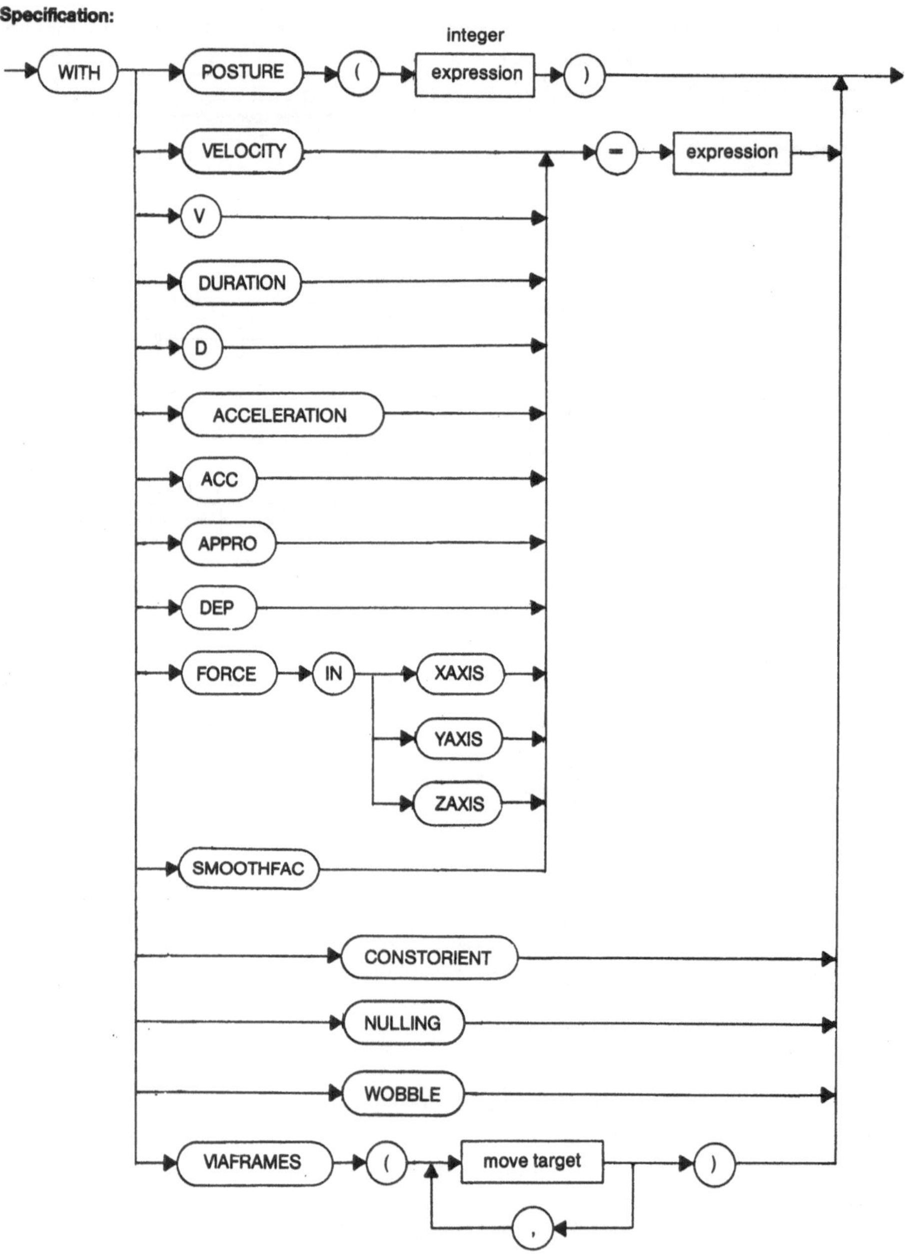

Fig. 4.12: Move specification in SRL

VAL:	COARSE	Set larger tolerances on the servo-control
	FINE	Small tolerances demanded.
	NONULL	Motion is terminated without confirmation of all axes.
	NULL	Confirmation about the reaching of the position must be given by all axes.

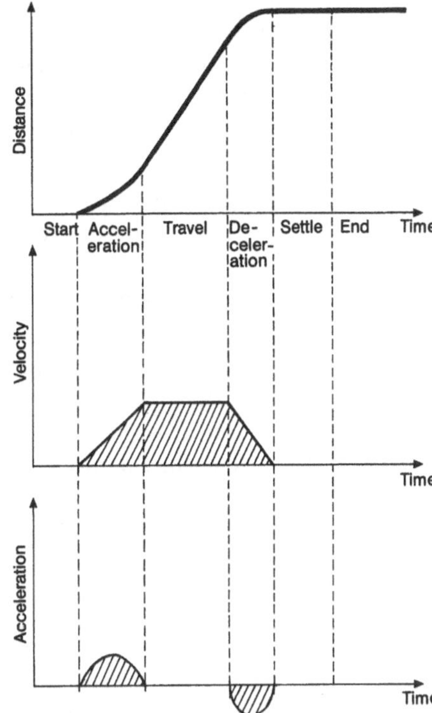

Fig. 4.13: Motion phase for one robot joint

	INTOFF	Error integration during path control is switched off.
	INTON	Error integration is enabled.
ROBEX:	GOTO/P1,EX	
	Exact servoing to frame P1.	

SRL and PASRO have been developed as much as possible independent of the hardware. Therefore they include not only one move command, but several; consequently, there is a move statement for every different kind of interpolation. Thus it is possible for the programmer to use motion commands for special kinds of interpolation and trajectory calculations which are not included in the simple robot control. Examples are shown in Figs. 4.11 and 4.12.

In the normal version of AML, there is no trajectory interpolation. All moves are executed as joint interpolation, that is, all joints start and stop their movements at the same time. Since the IBM robot has translational axes to move in x-, y- and z-direction, the resulting movements are straight lines. Each motion is divided into four phases: an acceleration phase, a travel phase, a deceleration phase and a settling phase. The user can influence the duration of each phase by setting the various parameters of the system subroutines for motion (for example, speed and acceleration). The motion starts with an acceleration (see Fig. 4.13), and after reaching the maximum velocity or the calculated velocity for synchronized moves, the travel phase is executed without acceleration. The motion then stops after the deceleration phase and it ends with a settling phase in which the system ensures that all joints in motion have stopped, before returning control to the calling routine.

Since the VAL system contains joint interpolation as well as linear Cartesian interpolation motion, the programmer may choose the *type of control* with the type of MOVE command.

SRL:

```
PTPMOVE asea TO goalframe;
```

The robot is moved to the goal frame by point-to-point control mode, that is, all the joints are moved to their final positions with maximum speed and without synchronization (see Sect. 2.3.3).

```
SYNMOVE t3 TO goalframe
  WITH DURATION = 5;
```

Move with linear joint interpolation, hence all joints start and end their motions at the same time.

```
SMOVE r106 TO goalframe
  WITH V = 25
  WITH ACC = 20
  WITH CONSTORIENT;
```

Move with linear Cartesian interpolation, that is the gripping point or tool center point is moved along a straight line from the current robot position to the given goal frame. The parameters specify the speed and acceleration, and the motion is executed with constant orientation.

```
CIRCLEMOVE r3 CP = centerframe DEG = -45;
```

The Mantec R3 robot performs a circular move, where the centerframe describes the center point of the circle by the position vector and the move direction by its direction of the frame z-axis. The move will start at the current robot position and result in a motion along the segment of an 1/8 of a circle.

```
LANEMOVE r56 TO goalframe
  WITH VIAFRAMES (savepos1, savepos2);
```

The move of the Jungheinrich robot r56 is performed by using polynomials for the trajectory calculations. The viaframes savepos1 and savepos2 describe a trajectory which is neither a straight line nor an asynchronous move (see below).

PASRO:

```
pmove (goalframe);
```

The effect is the same as for the PTPMOVE in SRL (see above).

```
jmove (goalframe);
```

Same as SYNMOVE in SRL.

```
smove (goalframe);
```

Same as SMOVE in SRL.

VAL:

```
MOVE targetframe
```

Joint interpolation motion

```
MOVES targetframe
```

Motion is executed with linear Cartesian interpolation.

Some languages which support an appropriate control permit specifications of *effector operations* during the robot movement. AML and VAL contain this directly in the motion command, whereas SRL and AL permit it indirectly.

SRL:

```
OPEN hand TO 1 DURING MOVE;
SMOVE arm TO targetframe;
```

AL:

```
COBEGIN
  MOVE ARM TO targetframe;
  OPEN HAND TO 1 * CM
COEND
```

Both instructions are executed in parallel.

AML:

```
MOVE (ARM, <x,y,z,r,p,yaw,1>);
```

VAL:

```
MOVET targetframe, 10
```

Joint interpolation motion with closing of gripper to 10 mm.

```
MOVEST targetframe, 10
```

Motion in linear Cartesian interpolation with closing of gripper to 10 mm

If the motion is to be neither in a straight line nor on an undefined path, **intermediate frames** are necessary. Most programming systems do not pass through these frames exactly (it is indeed not possible to do so on straight lines between frames without stopping), but the path follows a transition, i.e., it is rounded.

SRL:

```
SMOVE rs1 TO targetframe
  WITH SMOOTHFAC = 0.5
  WITH VIAFRAMES (interframe1, interframe2);
```

The resulting trajectory consists of straight lines: from the current robot position to `interframe1`, `interframe2` and the `targetframe`. The robot control calculates a curve to come from one straight line to another, because otherwise the move will stop for a short moment at the "corners" of the trajectory. The smoothing depends on the smooth factor SMOOTHFAC. The resulting trajectory is shown in Fig. 4.15.

```
LANEMOVE arm TO targetframe
  WITH VIAFRAMES (interframe1, interframe2);
```

The LANEMOVE statement with its intermediate frames is calculated so that the controller passes the endeffector through the posi-

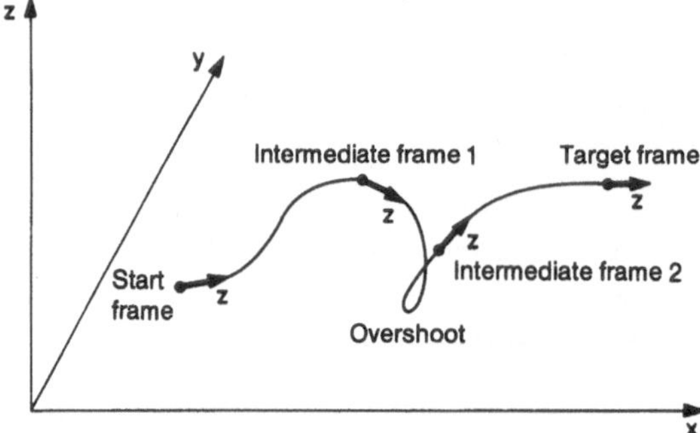

Fig. 4.14: Motion path with intermediate frames in AL

tion of the intermediate frame with the specified orientation. These paths are evaluated with polynomials which specify the interpolated positions between each intermediate frame. The start parameters (like velocity or orientation) of a segment between two intermediate frames and the output parameters to the next segment (constituting the input parameters of the next segment) are used to determine the factors of the polynomials for the segment. Segments between intermediate frames consist of positions and orientations which are interpolated and depend on time (for example, every 28 ms). Such resulting points are then transformed into robot coordinates.

```
VIAMOVE WITH SMOVE puma TO interframe;
```

This specific move command works as follows: The robot is moved to `interframe` while the controller awaits a new move command. The next move statement can be either another VIAMOVE or a finishing move. It will be calculated and executed without stopping the motion at the intermediate frame. If no other move is specified, the movement will stop with an error message. The execution of the VIAMOVE requires fast trajectory calculations.

```
MOVEDEF complex_handling;
BEGIN_MOVEDEF
  PTPMOVE TO above_pos;
  SMOVE TO bow_beg
    WITH CONSTORIENT;
  CIRCLEMOVE CP = bow_center DEG = 90
    WITH V = 150;
END_MOVEDEF;
MOVEDO complex_handling;
```

In SRL, a complex move can be defined, consisting of several single move statements which have been described before. During run-time, the trajectory is calculated first using the different interpolation modules (here PTP-control, straight line and circle interpolation). After the calculation of the whole trajectory, the

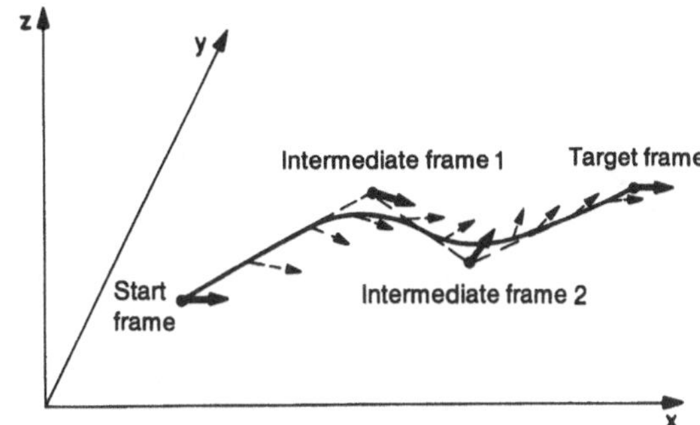

Fig. 4.15: Continuous path motion in SRL and VAL

MOVEDO statement causes the execution of the move `complex_handling`. Of course, this is only possible if the robot controller contains the different interpolation modules. Splitting the move specification and move execution can also be used for run-time optimization. For example, a move specification can be noted earlier in the program, where the control computer may be waiting for an input signal. Thus it has enough time available to calculate the trajectory and store the robot output values. When the MOVEDO statement is executed later on, no actual calculation time is needed before the move is started.

AL:

```
MOVE ARM TO targetframe
  VIA interframe1, interframe2;
```

In AL, motions with intermediate frames are calculated in the same way as for LANEMOVE in SRL. Due to this mode of calculation, overshoot may occur, i.e., the effector may execute a loop before the intermediate frame (Fig. 4.14). While the system can indicate this, the path itself is predictable by the programmer only within limits.

VAL:

```
ENABLE CP
MOVES interframe1
MOVES interframe2
MOVES targetframe
DISABLE CP
```

Here the continuous path approach to intermediate frames is switched on and off by the system switch CP. The resultant motion is shown in Fig. 4.15, in which the *transition* is particularly accentuated. The orientation of the effector is rotated evenly according to the values of the intermediate frames.

HELP:

```
SMOVE(1,#1,xz1,#2,yz1,#3,zz1,#4,
      rz1,#5,pz1,#6,yawz1,#7,400);
SMOVE(1,#1,xz2,#2,yz2,#3,zz2,#4,
      rz2,#5,pz2,#6,yawz2,#7,300);
```

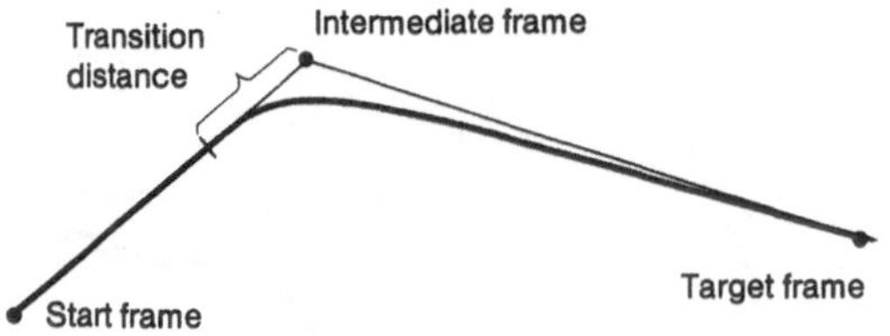

Fig. 4.16: Continuous path in HELP

```
MOVE (1,#1,x,#2,y,#3,z,#4,
      r,#5,p,#6,yaw);
```

Since HELP has no frame variables, the x, y, z, roll, pitch and yaw values are stored in scalar variables. Transitions are similar to those in VAL, only an additional parameter #7 in SMOVE determines the transition distance (Fig. 4.16).

In order to modify paths dynamically, several of the parameters described here - such as time or speed - may also be given for *intermediate frames* in some languages.

SRL:

```
SMOVE ir600 TO targetframe
  WITH V = 50
  WITH VIAFRAMES (interframe)
  WITH V = 15;
```

The move is first executed with a velocity of 50 cm/s, and after passing the intermediate frame the velocity is set to 15 cm/s.

AL:

```
MOVE ARM TO targetframe
  VIA interframe WHERE VELOCITY = 15 * CM/SEC,
    DURATION = 4 * SEC;
```

On passing the interframe, the controller is to set a velocity of approximately 15 cm/s for the effector. The time taken to reach the intermediate frame is to be four seconds.

VAL:

```
SPEED 100                   normal speed
ENABLE CP
MOVE interframe1
SPEED 20                    20% of normal speed
MOVES interframe2
MOVEST targetframe,5        Close the gripper
DISABLE CP
```

Changing of speed is possible after an intermediate frame or the simultaneous opening/closing of the gripper.

HELP:

```
SMOVE(1,#1,x,#2,y,#3,z,#4,r,#5,p,#6,yaw,#11,
      xspeed,#14,rollspeed, 16,yawspeed, 7,300);
```

A change in speed of individual axes is possible after an intermediate frame in HELP, because of the specification of speed in the SMOVE instruction.

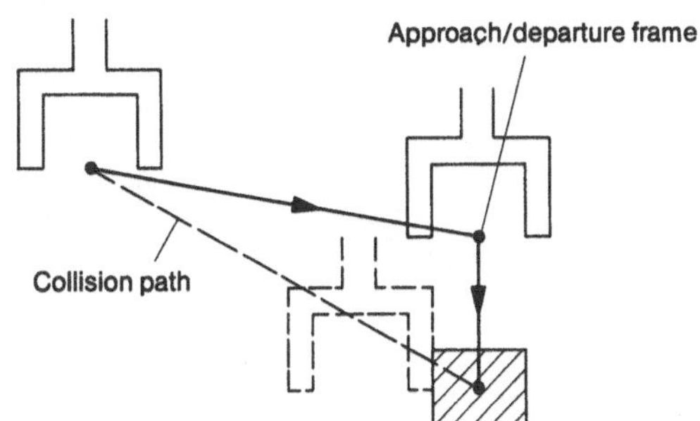

Fig. 4.17: Avoidance of object collision with approach/departure frames

Approach and *departure* frames represent a special kind of intermediate frames. They are defined relative to the target or start frame, so that the effector moves toward the target or away from the start frame in a predefined direction. This is useful, for example, to avoid collisions of object and gripper during the approach or departure phases (Fig. 4.17).

The possibility of defining approach and departure frames relatively is given in SRL, AL and VAL. An approach/departure frame in SRL is defined by a translation in the negative z-direction of the target or start frame.

SRL:

```
SMOVE hds36 TO targetframe
  WITH APPRO = 20
  WITH DEP = 20
  WITH SMOOTHFAC = 3;
```

The robot first moves up 20 cm (if the gripper points down), then the move goes to a point 20 cm above the target frame, and finally the robot moves down to it. Smoothing is present between the moves.

In AL, an approach/departure frame can be defined by a *frame relation* (i.e., translation and rotation), a *vector* (i.e., a translation only), or a *distance* in the direction of the z-axis of the target/start frame.

AL:

```
MOVE ARM TO targetframe
  WITH APPROACH = FRAME (rotation, vector)
  WITH DEPARTURE = FRAME (rotation, vector);
```

Figure 4.18 illustrates the geometric representation of frame relations, target/start frames and approach/departure frames. The approach/departure frame `aframe` could also be evaluated explicitly:

```
  aframe <- targetframe * FRAME(rotation,vector);
```

```
MOVE ARM TO targetframe
  WITH APPROACH = VECTOR (25, 9, 0)
  WITH DEPARTURE = VECTOR (0, 20, 12);
```

As shown in Fig. 4.19, the approach/depart frames have the same orientation as the target/start frames in this case. This can also be evaluated explicitly by

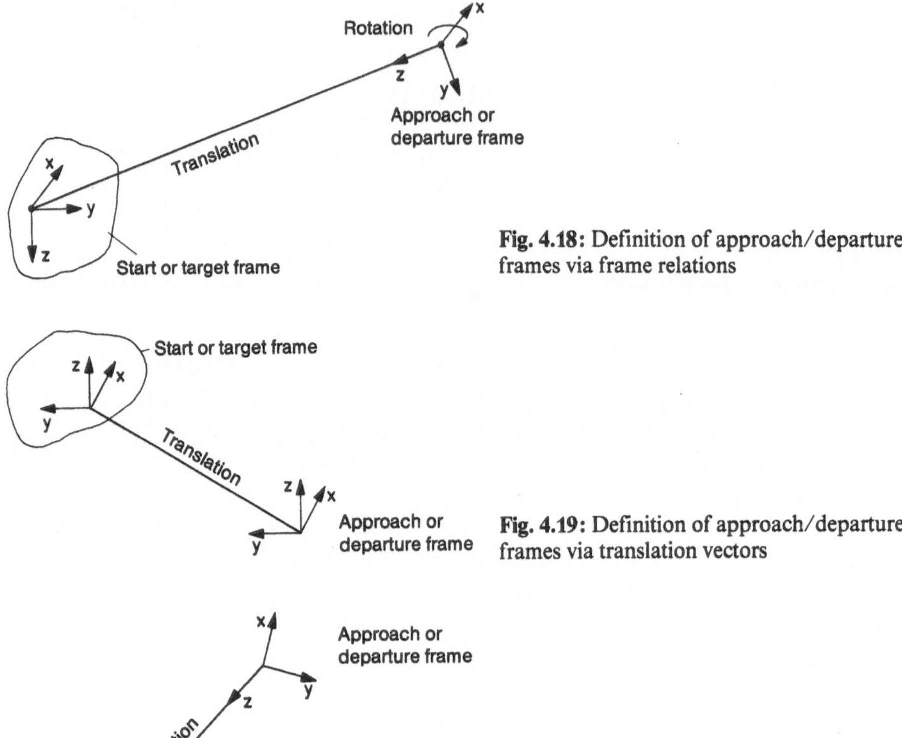

Fig. 4.18: Definition of approach/departure frames via frame relations

Fig. 4.19: Definition of approach/departure frames via translation vectors

Fig. 4.20: Definition of approach/departure frames via atranslation in the z-direction

```
aframe <- targetframe + VECTOR (25,9,0)
          WRT targetframe;
```

The coordinate values of the translation refer to the target or start coordinate system.

```
MOVE ARM TO targetframe
  WITH APPROACH = 20
  WITH DEPARTURE = 20;
```

As shown in Fig. 4.20, `aframe` can also be evaluated by a translation of the target/start frame in z-direction:

```
aframe <- targetframe + (20 * ZHAT)
          WRT targetframe;
```

The language VAL offers special instructions for reaching approach or departure frames. The movement to an approach frame is programmed by the command APPRO (joint interpolation) and APPROS (linear Cartesian interpolation).

Fig. 4.21: Weaving motion in AL and SRL

Fig. 4.22: Weaving motion in VAL

VAL:

```
APPRO targetframe, 50
APPROS targetframe, 50
MOVE targetframe
```

The distance of 50 mm is a translation with respect to the z-direction of the targetframe, which is reached itself only by the MOVE command.

```
DEPART 50
DEPARTS 50
```

Departure from a frame by 50 mm in the z-direction with the appropriate interpolation.

The languages SRL, AL and VAL allow the possibility of superimposing a weaving motion onto the robot movement. This is useful, for example, if objects are picked up by magnets mounted on the robot, moved to a dumping position and dropped. Some parts may stick because of the remaining magnetism, and this may be overcome by a shaking motion.

SRL:

```
SYNMOVE rs1robot TO targetframe
  WITH WOBBLE;
```

The robot is moved with an overlayed wobble motion.

AL:

```
MOVE ARM TO targetframe
  WITH WOBBLE = 2;
```

Figure 4.21 shows the addition of a sinusoidal motion onto the robot movement.

VAL:

```
WEAVE 25, 5, 2
MOVES targetframe
```

As shown in Fig. 4.22, the WEAVE instruction defines a sawtooth motion with an amplitude of 25 mm, a cycle time of 5 seconds and a dwell time of 2 seconds at the tip of the sawtooth pattern. It is executed during the following MOVE or MOVES instruction.

The language SRL includes a general MOVE statement with parameters which are identified by a logical number, and their semantic meaning is not predefined. Therefore this statement can be used for future extensions, like a possible module for ellipsoidal moves in the robot controller.

SRL:
```
MOVE a1robot TO corner
  WITH PARAM (1) = a,
  WITH PARAM (2) = b,
  WITH PARAM (3) = 30,
  WITH PARAM (4) = plane_frame;
```
The parameters 1 to 4 specify input values for a move done by ellipsoidal interpolation.

4.2.5 Motion Instructions with Sensor Integration

Monitoring and *control* of motion are carried out mainly by force, torque, proximity or touch sensors. Thus, some languages allow the presetting of **sensor thresholds**, which are monitored during the movement. Unfortunately, this is the only possibility in many robot languages for integrating sensory information into the software; there are no special instructions for sensors. Again, a reason for this disadvantage can be found in the fact that only available hardware or sensors are taken into consideration even in development systems, and language constructions are dedicated to these special system configurations only. So, for example, the (American) AL can communicate with force and torque sensors within a motion instruction, but not with vision systems, which in turn is possible in the VAL system (VAL 11V). However, even in VAL only one system is supported.

It is possible to distinguish whether sensor thresholds serve only for monitoring, that is, for checking the limits of sensor values, or whether the sensor value represents an argument. In the first case, the program branches accordingly or activates a task which exeecutes in parallel to the main program, and in the second case movements may be modified. An example for the latter case is the motion of an effector with a preprogrammed force in one direction.

The influence of sensory information on robot motions can be programmed in SRL with the help of the more general task concept (see Chapt. 7). In particular, the ALWAYS WHEN statement enables a clear and flexible specification of sensor thresholds, interrupts and/or limits on values of variables.

4.2.5.1 Motion Instructions with Sensory Monitoring

Most languages allow monitoring of movements of industrial robots. As soon as a predetermined threshold is exceeded, a branch in the program follows or the motion can be stopped. Besides the general ALWAYS WHEN statement, SRL includes a special instruction to perform a move in a specified direction until a monitored condition becomes true (here no endpoint of the move is known), or the end of the working envelope is reached and an error occurs. Possible conditions are not only

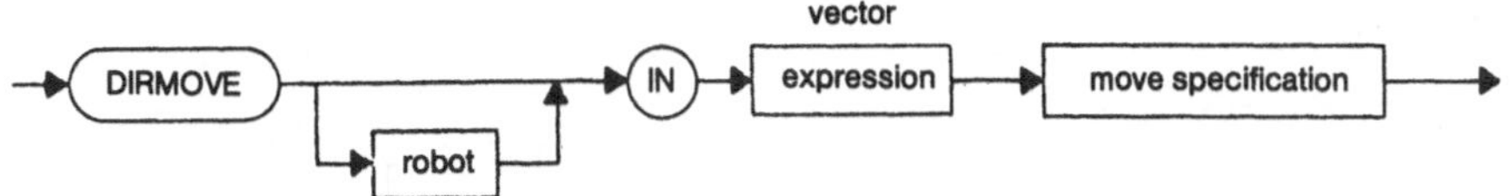

A robot move is started along the direction of the given vector until the specified move condition becomes true.

Move specification:

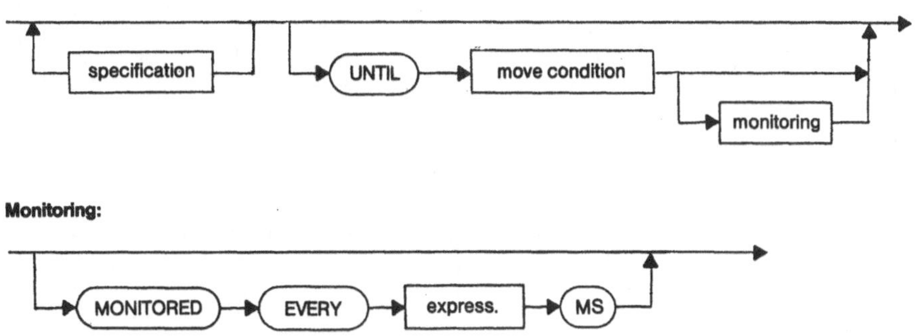

A condition is evaluated cyclic by the given cycle time.

Fig. 4.23: Special move in a direction in SRL

sensor thresholds but also interrupts or the distance to a target or from a startpoint (see Figs. 4.23 and 4.24).

SRL:

```
DIRMOVE rm501 IN ZAXIS
  WITH V = 20
  UNTIL DISTANCE TO START > 50
        OR SENSOR switch1 = TRUE
           MONITORED EVERY 100 MS;
```

The Mitsubishi robot is moved along the direction of the z-axis with a velocity of 20 cm/s until the distance to the startpoint is more than 50 cm or a switch is set. The condition is monitored every 100 ms.

Applying the `ALWAYS WHEN` statement to a robot move enables the robot controller to react to sensory information, such as a force sensor.

SRL:

```
WHEN SENSOR z_force >= 100
  MONITORED EVERY 75 MS
DURING
  SMOVE r56_robot TO box
    WITH VELOCITY = 0.5;
END_DURING
DO STOP r56_robot;
```

Move condition:

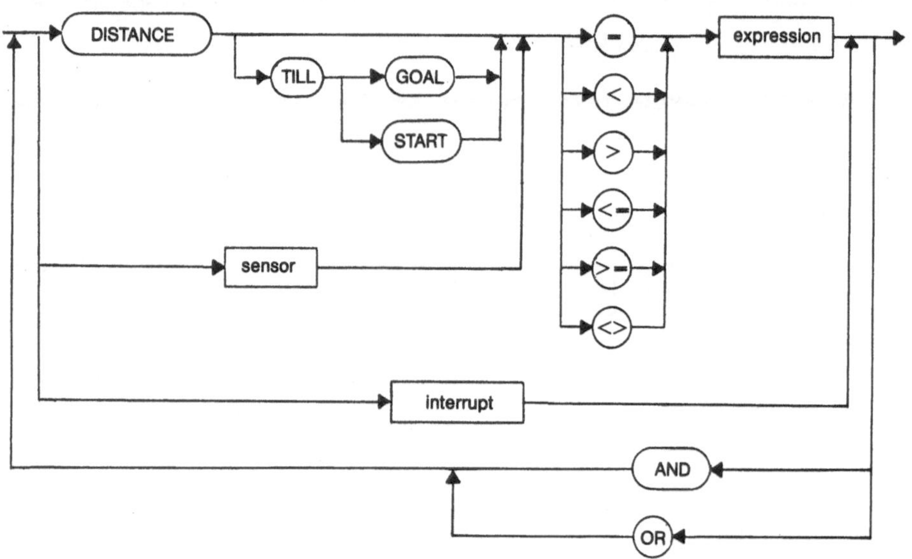

Fig. 4.24: Move conditions in SRL

If the sensor value of a force sensor in the z-direction is greater or equal to 100 g during the straight line move to the box, the robot will stop. The condition is monitored every 75 ms.

AL permits the specification of a threshold for a *force* or *torque touch sensor* mounted at the front of the gripper. The touch sensor measures forces in x-, y- and z-direction of the ARM frame (i.e., gripper) which - in turn - can be converted into the coordinate system of any frame. The frame to which the forces are referred is predefined.

Force frames can be defined in relation to the base coordinate system (`IN WORLD`) or gripper frame (`IN HAND`). In the latter case, the measurement direction changes during movements according to the current orientation of the gripper (see also Fig. 4.25). The slightly more complicated syntax will be explained with the help of some examples.

AL:

```
MOVE ARM1 TO targetframe
  ON FORCE (ZHAT) >= 100 * GM DO
    STOP &;
```

The explanation of the conspicuous "&" sign is left until the fourth example. This instruction causes the robot arm 1 to move to the point `targetframe`. Should a force of 100 g or more appear along the z-axis of the coordinate system during this action, the motion is stopped immediately. The coordinate system or frame to which "z-axis" refers is not specified in this instruction. It is assumed from a previous force frame command.

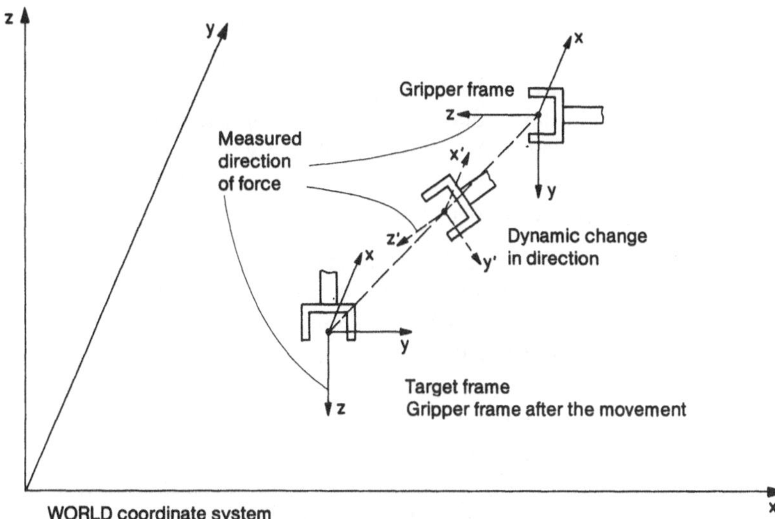

Fig. 4.25: Force monitoring in z-direction of the gripper frame in AL

```
MOVE ARM2 TO targetframe
  WITH FORCE_FRAME = gripperframe IN HAND
  ON FORCE >= 200 * GM ALONG ZHAT DO
    BEGIN
      STOP;
      PRINT ("Table reached");
      table <- ARM2;
      flag <- TRUE
    END;
```

First, the frame `gripperframe` is defined by `WITH FORCE_FRAME` as the coordinate system to which the following vector specifications of the force directions are referred (also in further `MOVE` instructions). By including `IN HAND`, `gripperframe` has been related to the current orientation of the effector, such that the measured direction of force changes together with the robot movement (Fig. 4.25). If a force equal to or larger than 200 g occurs in the z-direction of `gripperframe`, the motion is stopped, the text "Table reached" is printed, the newly reached position and orientation are assigned to the frame `table`, and the Boolean variable `flag` is set.

```
MOVE ARM3 TO targetframe
  ON FORCE < 30 * GM ALONG XHAT OF edge IN WORLD DO
    ON FORCE >= 500 * GM ALONG YHAT DO
      PRINT ("Attention") &;
```

In this example, a force in the x-direction is monitored to ensure that it does not fall below a previously defined value. The direction in which the force is measured is fixed, since the frame has been related to the base coordinate system. If the force is below the threshold, it is checked to see whether a force larger than or equal to 500 g is present along the y-axis of the previously defined WITH FORCE_FRAME. Only if this is true, is the message "Attention" printed. This is an example of nested monitoring of forces in two directions.

```
MOVE ARM4 TO targetframe
  ON FORCE >= 150 * GM ALONG ZHAT DO
    MOVE ARM2 TO deposit &
  WITH DURATION = 5 * SEC;
```

This last example shows why a MOVE instruction has to be terminated with a "&" after the symbols ON FORCE DO. If it were missing, it would not be clear to which arm, ARM2 or ARM4, WITH DURATION 5 * SEC refers. In this case, the motion to the targetframe lasts 5 seconds, but if the movement to deposit should take 5 seconds, the "&" sign would have to be placed before the semicolon.

AML contains a system procedure called MONITOR, which allows the monitoring of sensors for exceeding predefined limits (see Sect. 4.5.3). A number is assigned to such a monitor, and this may serve as a parameter for the MOVE instruction. Any motion will stop as soon as the condition becomes true.

AML:

```
MONITOR (LED, 2,0,0,1.5, 'passed');
MOVE (ARM, fgoal, LED);
```

The monitor checks the cross-fire sensor mounted in the gripper jaws every 1.5 seconds. If the light beam is interrupted by an object and the monitor detects the condition during the movement towards fgoal, the robot is stopped immediately.

In VAL, there is no way of specifying sensor monitoring explicitly in the motion instruction. However, it is possible to utilize a hardware feature in which the passing of a sensor threshold triggers a binary input. This signal then causes a call to the appropriate subroutine, which has been assigned previously (Sect. 4.2.6).

HELP allows the programmer to set a threshold with the FORCE instruction, and, in case it is exceeded, the current robot movement is stopped. This instruction has the syntax illustrated in Fig. 4.26. The FORCE instruction is noted after the SMOVE command and executed when the transition begins or after the specified distance from the target is reached (see also Sect. 4.2.4). If no distance is given, monitoring of the force starts immediately.

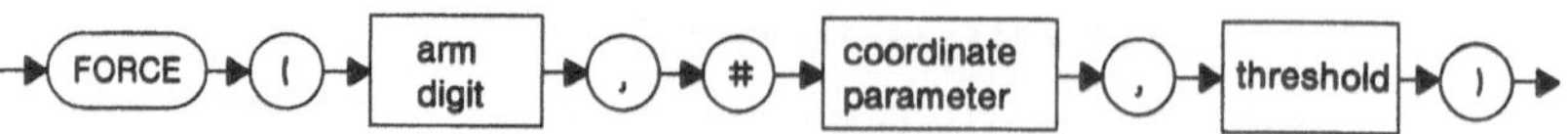

Fig. 4.26: Setting a force threshold to monitor motions in HELP

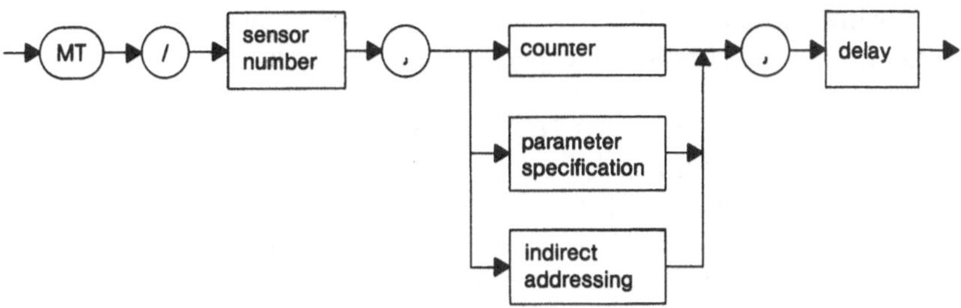

Fig. 4.27: Input of sensor values in SIGLA

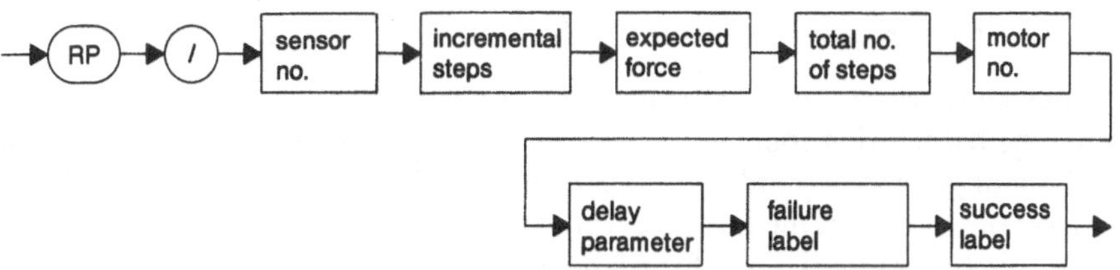

Fig. 4.28: Search motions monitored by sensors in SIGLA

HELP: `SMOVE(1,#3,150);`
`FORCE(1,#3,60);`
The industrial robot is moved to a position of 150 mm in the z-direction. Should the force measured during the movement equal or exceed the value of 60 (range from 0 to 255), the movement is stopped. A subsequent enquiry whether z = 150 mm may determine the reaching of the threshold and the program may branch accordingly.

Although SIGLA has neither frames nor variables, the current value of the force sensor can be read and a movement of one motor can be executed with force monitoring. Fig. 4.27 shows the instruction to interrogate sensor values. The wait state between setting a sensor and reading the current value is only necessary if several sensors are mounted on the robot. This value (from 0 to 255) is stored in a counter, parameter or indirectly addressed memory location (see Sect. 3.1.3). It may then be interrogated and branches executed accordingly.

SIGLA: `MT/3, M5, 1;`
The value of sensor 3 is stored in counter `M5`.

The `RP` instruction shown in Fig. 4.28 causes the specified motor to be moved by the total number of steps, whereby it is checked after each increment, whether or not the force sensor has reached the expected value. If this has been reached, the program jumps to the success label, otherwise to the failure label. This command is mainly present to move the SIGMA robot vertically downwards in the z-direction, or to turn the effector until an expected force/torque is exerted.

SIGLA:

```
RP/2,7,120,70,3,-10,4,5;
```

Motor 3 is moved 70 steps. After every 7 steps, sensor 2 is checked for exceeding a force of 120 (range from 0 to 255) and the program waits. If this does not happen, the program continues at label 4, otherwise at label 5.

In ROBEX, there are no possibilities to specify sensory monitoring in the motion instruction. It is possible, however, to generate an interrupt - similar to VAL - which stops the movement (Sect. 4.2.6).

4.2.5.2 Motion Instructions with Sensor-Monitored Parameters

It may be necessary for an effector to exert a particular force onto a work piece or onto the base during some of the various manipulation tasks like grinding, grating, drilling or glueing. According to this *force demand*, the movement is accelerated or slowed down, so that the force measured by tactile sensors mounted on the effector is reasonably constant. Out of the languages considered here, SRL and AL include a force or torque demand within a motion instruction. It may also be desirable to demand a constant distance in one direction relative to the object, as it may be necessary in welding applications using proximity sensors. Hence any arbitrarily curved surface could be followed.

Another possible parameter for a move in SRL is a force in the x-, y- and z-direction of the effector frame. Of course, some other parameters (like velocity and acceleration) are not permitted in this case.

SRL:

```
SMOVE puma600 TO left_side
  WITH FORCE IN XAXIS = 500;
```

A device for grinding the surface of a workpiece is moved in the x-direction with a force of 500 g.

In AL, force or torque demands which are monitored by sensors have the same syntax as the motion instruction with sensory monitoring described in Sect. 4.2.5.1. A frame referring to the force direction(s) is again defined by a force frame declaration (Fig. 4.29). The HAND/WORLD keywords have the same meaning as in Sect. 4.2.5.1.

AL:

```
MOVE grinder TO left_side
  WITH FORCE (XHAT) = 500 * GM;
```

A device for grinding the surface of a work piece is moved in the x-direction of a previously defined force frame with a force of 500 g.

```
MOVE drill TO below
  WITH FORCE = 700 * GM ALONG ZHAT
    OF drilltip IN HAND
```

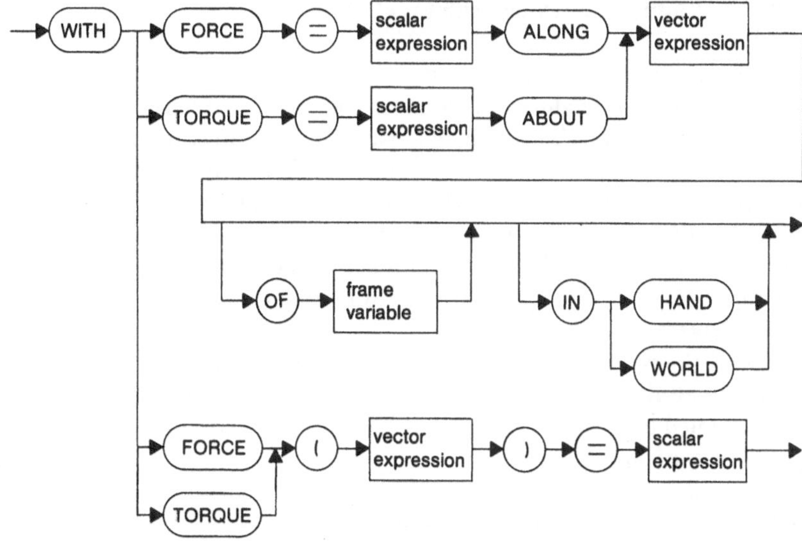

Fig. 4.29: Force and torque demands in AL

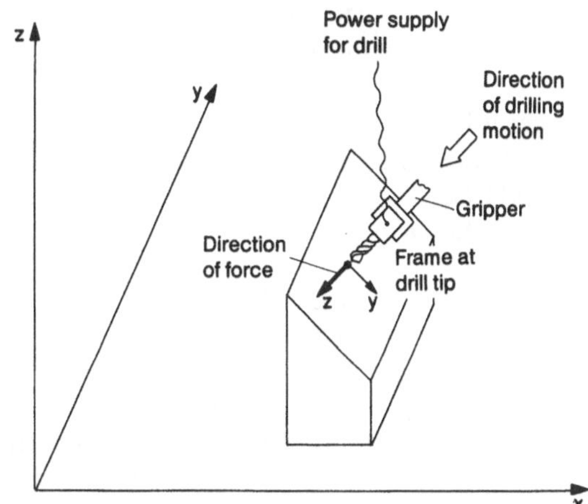

Fig. 4.30: Motion of a drill with force demands in AL

```
WITH FORCE = 0 * GM ALONG XHAT
WITH FORCE = 0 * GM ALONG YHAT;
```

As shown in Fig. 4.30, the special drill, which has been picked up previously, is guided by the industrial robot in such a manner that the drilling is carried out with a force of 700 g. The direction of force is relative to the drill, so that it is always measured along the drill.

4.2.6 Motion Instructions with Event Monitoring

Usually an industrial robot works in a space where there are other robots, machines, conveyors and/or feeding installations. Thus it may be necessary for certain **external events** not only to be polled, but also to have a direct influence on the program execution during the movement. For example, the motion of the robot may have to be stopped if the gripper arrives at a cross fire sensor in front of a machine and the robot has to wait for the output of the machine to be ready. While many languages allow an interruption of the motion because of an event, some, SRL, AL and HELP among them, allow parallel execution of program sections during motions after events have happened.

Again, the general ALWAYS WHEN statement can be used to perform motions with event monitoring. The events may originate from peripheral devices or they may be programmed using semaphores. External events are specified in the system specification part of an SRL program.

SRL:

```
WHEN machine_stop
  DURING
    SMOVE arm1 TO targetframe;
  END_DURING
  DO STOP arm1;
```

The interrupt machine_stop during the robot move terminates the motion immediately.

An event in AL is represented by a variable of type EVENT (see also Sects. 3.1 and 4.5.3). These event variables are normally used by the semaphore operations SIGNAL and WAIT (see Sect. 2.2.7). If an event is to represent an external interrupt, a special implementation and a predefined name are necessary.

AL:

```
MOVE ARM1 TO targetframe
  ON arm2halt DO
    STOP ARM1;
```

The robot is moved to the target frame. Should the event variable arm2halt be set during this move - in a parallel program, for example - ARM1 will stop.

VAL can also react to an external signal arriving during the robot motion. Figure 4.31 shows the notation of the REACT and REACTI instruction. The REACT command

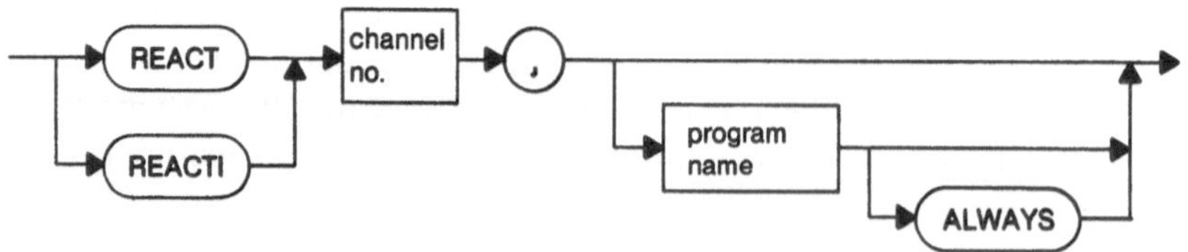

Fig. 4.31: Definition of a subroutine as a reaction to an external event in VAL

```
REACT 2, dummyprogram
MOVE targetframe
IGNORE 2 ALWAYS
     .
     .
     .
REMARK ***************************************
REMARK dummyprogram
RETURN 1
```

Program section 4.4: Motion instruction with event monitoring in VAL

causes the permanent interrogation of a specified signal during the next move and checks whether it is set. Should this be the case, the subroutine is called after the termination of the move, and when the subroutine is completed, the program continues with the instruction following the one during which the subroutine was called. If the program name is missing, the next instruction is skipped. The addition of ALWAYS causes the channel to be checked permanently and not only during the next movement. Should the signal be set, the appropriate subroutine is called and the process repeated until the IGNORE command.

VAL: `REACT 2, interrupt`
`MOVE deposit`
If input channel 2 is set during the movement to frame `deposit`, the subroutine `interrupt` is called.

The REACTI instruction differs from REACT by calling the subroutine immediately, when the signal is set during the execution of an instruction.

VAL: `REACTI 2, stopmessage ALWAYS`
As soon as channel 2 is set, the current instruction is terminated and the subroutine `stopmessage` is called. This happens on each setting of the channel, until the next IGNORE command.

`IGNORE 2 ALWAYS`
Channel 2 is no longer examined for a high potential.

Program section 4.4 shows how a motion may be stopped because of an external event. The IGNORE instruction after the motion command limits the checking of channel 2 to the MOVE instruction. The `dummyprogram` consists only of a comment line and the return instruction. The specification of RETURN 1 instead of RETURN or RETURN 0 has the effect of not continuing the movement, but executing the IGNORE instruction after the return jump.

Since it is possible to execute instructions in parallel with moves in HELP, event monitoring with program branching can be implemented with explicit polling. The events can be set in other program sections by the programmer (as **semaphores**, see also Chapt. 7), as well as by external devices connected to input lines. Program sec-

```
        SMOVE (1, 2, 150);
loop:   IF TESTB (5) THEN

            ! Instructions executed in parallel to the
            ! arm motion after input 5 has been set

          EOM (1)
        END;
        COORD (1);
        IF AY (1) < 149 THEN
          GOTO loop
        END;
        EOM (1);
```

Program section 4.5: Event monitoring during a robot motion in HELP

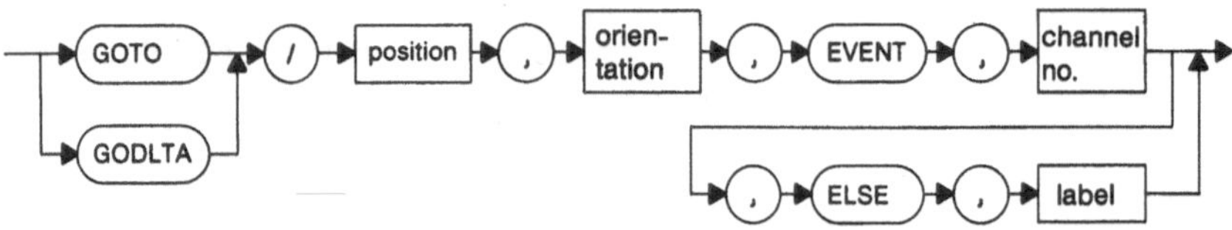

Fig. 4.32: Motion monitored by events in ROBEX

tion 4.5 shows the technical realization. After testing channel 5, the current robot coordinates are read and it is checked whether the final value of the move has nearly been reached. If this is not the case, an unconditional jump to the label `loop` follows and the whole process starts again. If the motion is nearly finished, the `EOM` instruction waits for confirmation of all robot axes and only then does the program continue. If channel 5 is set, the corresponding instructions within the `IF` statement may be executed several times. Again, this avoids an `EOM` instruction for the robot arm 1.

According to information of the WZL in Aachen, the semantics of an instruction of motion with event monitoring in ROBEX has not yet been finalised, however, the syntax is defined according to Fig. 4.32. During the motion of the robot, the specified channel is monitored for an event, if the `ELSE` part is left out. In case of an event, the robot motion is stopped. If the `ELSE` part is included and that particular event has not happened, the program jumps to the given label after the movement has stopped.

ROBEX: `GOTO/p3,XYROT,180,EVENT,2,ELSE,nobox`
The industrial robot arrives at the position p3 with corresponding orientation. During this move, channel 2 is monitored for an event. Should it happen, the robot stops and the next instruction is executed. If it does not, the program jumps to the label `nobox`.

4.2.7 Motion Instructions with Time-Out Monitoring

It is important with some movements that the **duration of motion** does not exceed particular values, because otherwise coordination with conveyers or similar systems may not be possible. Thus the motion instruction has a time parameter.

If a general time function is available, the duration of the move can be monitored by reading the move duration in parallel. The inclusion of such a time function in a SRL implementation on a robot controller can then limit the time for a move. Additionally, SRL includes a time-out check, so that the user can identify a task in the specification part which is to be executed in the event of a time-out.

SRL:

```
WHEN start_time <= SYSTIME-move_duration
  MONITORED EVERY 20 MS
DURING
  SMOVE e440 TO targetframe;
END_DURING
  DO START next_action;
```

The time function SYSTIME is called every 20 ms and the expression is evaluated to check whether the time limit for the move duration has been exceeded. In that case, the task `next_action` is started.

In AL, it may be possible to utilize the time monitor to execute a program section in *parallel* to the motion after a certain delay. The start of the robot also resets a time counter, thus initiating some action if it expires before the motion has stopped.

AL:

```
MOVE ARM1 TO targetframe
  ON DURATION >= 3 * SEC DO
    BEGIN
      MOVE ARM2 TO center
        WITH DURATION = 2 * SEC;
      PRINT ("Start process");
    END;
```

Three seconds after the start of `ARM1`, the second arm `ARM2` is moved to the frame `center` with a duration of 2 seconds and some text is printed.

In HELP, a motion with time monitor can be programmed explicitly, as shown in the structure of program section 4.6. `TIME()` returns the value of the system clock (in seconds), so that the time difference `dt` between the current time and the time of start can be compared with the given `timeduration` in a loop parallel to the movement. As soon as the time difference is equal to or larger than the preset time duration, the instructions of the `IF-THEN` part are executed and/or the movement terminated.

Since a real-time clock can be interrogated in AML at intervals of 20 ms (via the function CLOCK), an explicit monitoring is possible, similar to HELP.

4.2.8 Parallel Processing

The execution of a move command is very time-consuming compared to arithmetic operations, program branching or even input/output to disks and, within reason, printers, too. Thus AML and HELP enable the programmer to execute a move command and a sequence of other instructions in parallel. (This is apart from the more general task concept in HELP, SRL and other languages).

The AMOVE (asynchronous move) and WAITMOVE routines are for this pupose in AML.

```
AML:        AMOVE (ARM, targetframe);
              .
              .              -- Instructions specified here are
              .              -- executed immediately after the
              .              -- initialization of the move com-
              .              -- mand.
              .
            WAITMOVE;        -- Wait until movement is fin-
                             -- ished.
```

The SMOVE instruction in HELP can be used in a similar fashion, and it serves additionally for the overshoot (see Sect. 4.2.4). The waiting for the motion to be finished is accomplished by an EOM (End Of Move) instruction, in which an identification of the robot is also necessary (cf. program section 4.6).

```
       duration := 3;
       SMOVE (1, 2, 120);
       starttime := TIME ();
loop:  dt := TIME () - starttime;
       IF dt = duration OR dt > duration THEN

            !  Instructions executed after 3 seconds,
            !  e.g. HALT (1) to stop arm 1

         EOM (1)          ! waits for movement end
       ELSE
         COORD (1)
         IF AY (1) < 119.9 THEN
           GOTO loop
         END
       END;
```

Program section 4.6: Explicit monitoring of motion duration in HELP

4.2.9 Moving to a Home Position

Many robot systems define a special position (and to some extent orientation), which can be reached from any position and orientation of the robot by a special instruction. This *home position* or first position may be a hardware feature, where each axis may contain an endstop, or a software feature with look-up tables for the coordinate values of the robot. The arm nearly always points upwards, so that human interference in the workspace can be carried out more easily and other machines can operate without danger of collision. Out of the languages considered here, SRL, PASRO, AL, VAL and ROBEX know such a resting position. The arm always points up in SRL, PASRO, AL and VAL.

SRL: `FIRSTPOS rm501;`

PASRO: `nullpos;`

AL: `PARK ARM1;`

VAL: `READY`

The programmer may define a home position in ROBEX (Fig. 4.33), where either x-y-z coordinate values are given or only a single z-value. The latter case is equivalent to a safety surface above the workspace, onto which the robot retires. The current x- and y-coordinates are preserved. The move to the previously defined resting position follows the SAFP instruction.

ROBEX:

```
SAFPOS/50,20,35
.
.
.
GOTO/SAFP
```

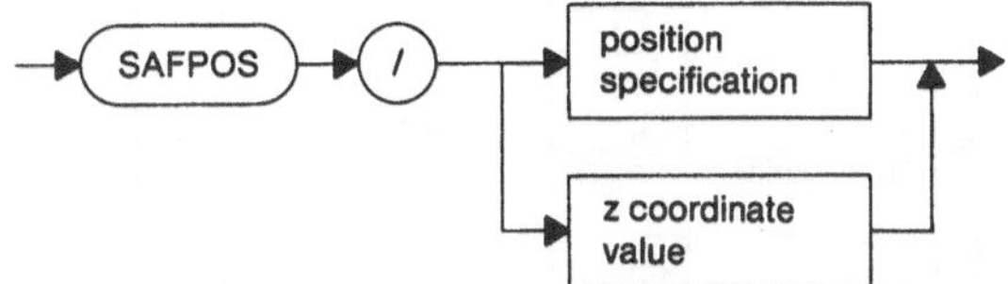

Fig. 4.33: Definition of a home position in ROBEX

4.3 Effector Instructions

At the end of a robot, a gripper, tool or some other means of production is mounted; these are collectively called **endeffectors** or **effectors**. Although it is part of an industrial robot, the effector itself combines elements of mechanics, drives, kinematics and control (see Fig. 4.34 and LUNDSTROEM [4.3]). Assembly usually requires **grippers**, for which some of the robot languages, such as SRL, PASRO, AL, VAL, and ROBEX, contain special instructions. In AML, the gripper is programmed by the MOVE instruction. Other languages (like HELP and SIGLA) classify grippers in the same manner as other tools and production means, namely as *peripheral devices*

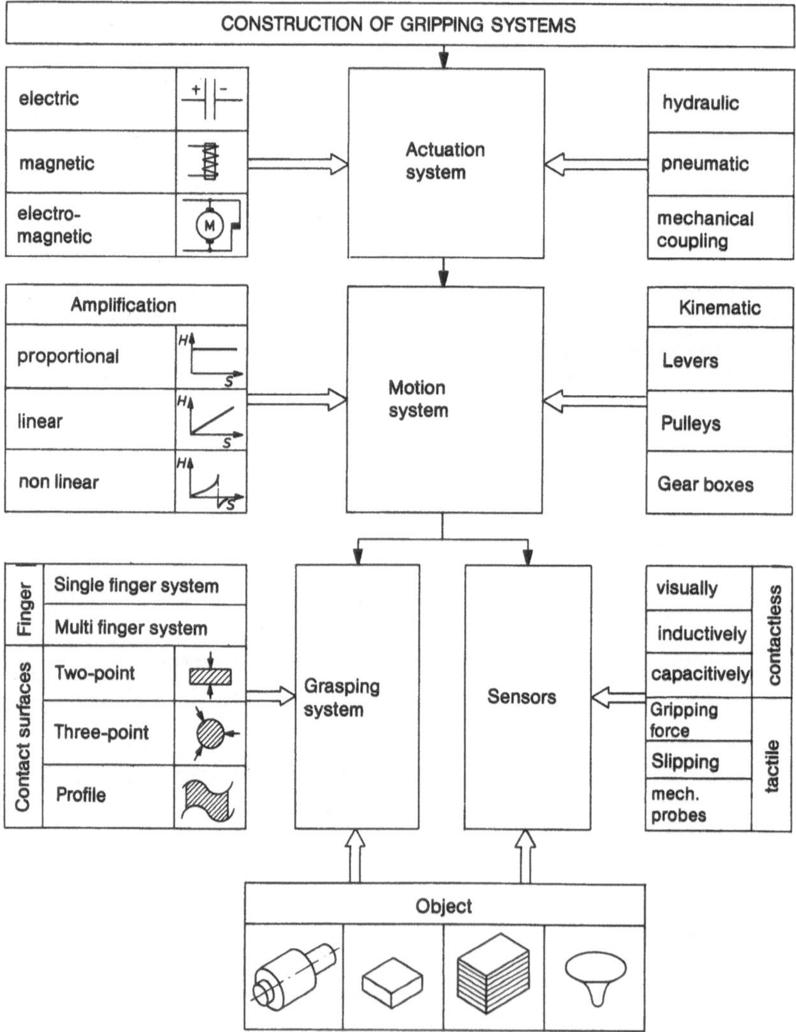

Fig. 4.34: Construction of gripper systems (AUER [4.4])

controlled by signal outputs. Programming of effector actions requires exact knowledge of the kinematics, drives and functionality, because a pneumatic gripper may only open and close, while an electric gripper can take arbitrary values in hand opening and may grasp with a defined force. A three-fingered gripper can hold round objects much more safely than one with only two jaws.

4.3.1 Simple Effector Instructions

Simple instructions cause only the opening or closing of the gripper, the operation of a tool, switching magnetic grippers on or off, etc.; they do not contain parameters like *jaw opening* or *gripping force*. Naturally, they assume simple effectors, otherwise the programming would not exploit all capabilities of effector systems.

SRL includes the general effector statement OPERATE, which acts upon effectors and their operations defined in the system specification part of the program.

SRL:

```
OPERATE screw_driver BY screw_in;
```

The logical numbers defined in the specification part are generated, such that the controller will cause the screwdriver device to tighten the screw.

The OPEN and CLOSE statements in SRL are designed for either an electrical or a pneumatic gripper.

SRL:

```
OPEN pneumatic_gripper;
CLOSE electric_gripper;
```

The pneumatic gripper is opened and the electric one closed, such that the jaws are moved together until an implementation-dependant, predefined gripping force is encountered.

PASRO:

```
gripopen;
gripclose;
```

It is possible to define a gripping force and a holding force, including a gripping time in PASRO (for the implementation on the Mitsubishi robot controller).

```
gripforce(6, 3, 1);
```

A high gripforce of 6 (7 is the highest) and a holding force of 3 is set up. The time for gripping and gripforce is 1 s.

The language AL does not recognize effector instructions without parameters; it needs at least a jaw opening demand (Sect. 4.3.2). This is tailored for an electric gripper, and an implementation with a pneumatic gripper demands a module from the interpreter or controller to decide on opening or closing the gripper. Unfortunately, actions cannot be decided simply by choosing OPEN for opening and CLOSE for closing. The instructions

```
OPEN HAND TO 5 * CM;
CLOSE HAND TO 5 * CM;
```

have both the same effect, i.e., opening the gripper if it was closed before.

In VAL, the instructions OPEN and CLOSE are designed for electric as well as pneumatic grippers. The jaw opening value is omitted for simple effectors.

OPEN and CLOSE are only executed during the following motion instruction, whereas OPENI and CLOSEI are obeyed immediately.

VAL:

```
OPEN
MOVE gripframe
CLOSEI
DEPART 80
```

The pneumatic gripper is opened during the approach to gripframe, in order to be closed before the robot departs along the z-direction.

The instruction RELAX disconnects the gripper from the air pressure.

There is no special effector instruction in HELP. However, the instructions SET, RESET and PULSE permit the control of a gripper or tool.

HELP:

```
SET(5);
DELAY(20);
MOVE(1,#1,500,#3,40);
RESET(5);
```

An electromagnet mounted on the robot is connected to channel 5. After the magnet is switched on, action is delayed by 0.4 s to allow the attraction to have full effect. Then the robot moves to the location x = 500 mm and z = 40 mm in order to drop the part there.

```
PULSE(1);                        !Close gripper
DELAY(effectortime);             !Wait for the closing
MOVE(1,#1,200,#3,250);           !Robot motion
PULSE(2);                        !Open gripper
DELAY(effectortime);             !Wait for the opening
```

PULSE outputs a pulse to the specified channel number.

SIGLA does not recognize gripper instructions either. The opening or closing of a gripper or slider is controlled by the AX instruction. A minus sign indicates the switching on, a positive value the switching off.

SIGLA:

```
AX/-1,-12;        Open gripper and slider before
MO/...;           the robot motion.
AX/1;             Close gripper
```

The effector to which the following instructions are referred can be defined in ROBEX (Fig. 4.35). The gripper or tool selected by a previously given channel num-

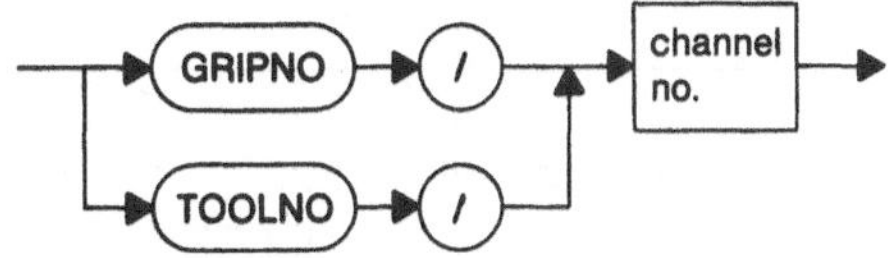

Fig. 4.35: Predefining the gripper/tools used in ROBEX

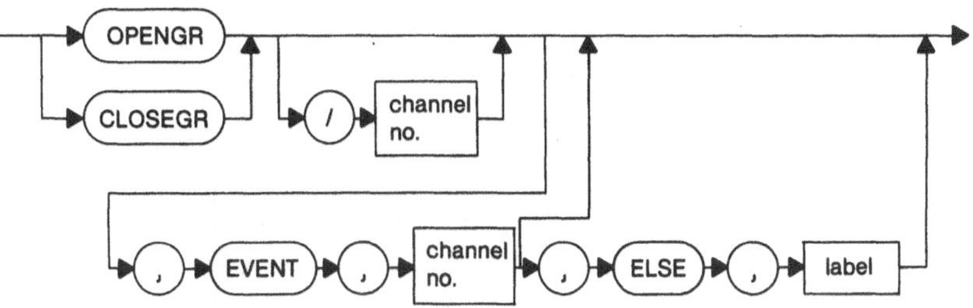

Fig. 4.36: Opening and closing of an effector in ROBEX

ber is controlled by the OPENGR or CLOSEGR instruction, if no channel number is given in the statement (Fig. 4.36).

Since the exact semantic definition is not available, the only statement possible is that a pneumatic gripper is opened by OPENGR and closed by CLOSEGR.

ROBEX:

```
GRIPNO/5
OPENGR
CLOSEGR/3
```

The gripper connected to output channel 5 is opened, the one at channel 3 closed.

4.3.2 Effector Instructions with Parameters

Just as in the case of motion instructions, the programmer may wish to specify further parameters to influence the effector control. These *parameters* concern

- Gripper orientation
- Duration of effector motion
- Speed of effector motion
- Gripping force
- Control tolerances
- Device specific instructions (such as welding current)

These are usually realized in the controller and not left to the programmer.

Several languages allow a specification of parameters for desired openings of two (or more) jaws (especially for grippers). Thus Fig. 4.37 shows how this may be specified in SRL and AL, using the TO keyword for absolute and BY for relative openings

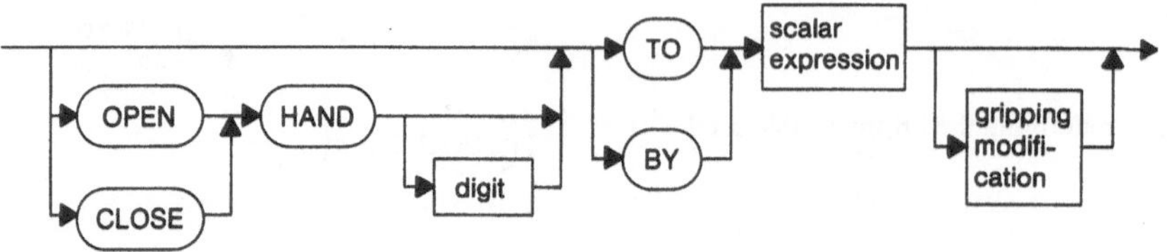

Fig. 4.37: End effector instruction in AL

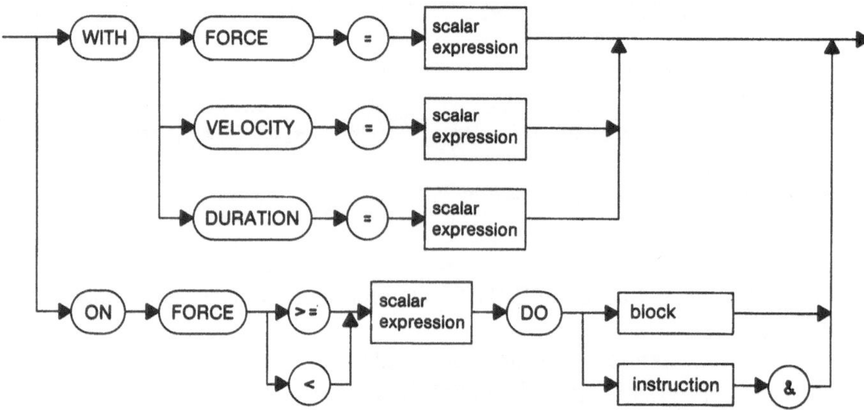

Fig. 4.38: Demanding and monitoring of gripping forces of an effector instruction in AL

for the electrical or hydraulic gripper. If OPEN or CLOSE HAND is missing, the keywords TO and BY refer to the last OPEN or CLOSE HAND statements.

SRL:

```
OPERATE cutter BY OPERATION(1) DURING MOVE;
```

The cutting is carried out during the next move of the robot associated with that tool.

```
OPEN gripper1 BY 3 DURING MOVE;
SMOVE puma TO targetframe;
CLOSE gripper1 TO 4
  WITH VELOCITY = 0.5;
```

During the next move, the gripper opens its jaws by an additional 3 cm. After the move, the gripper is closed to distance of 4 cm with a velocity of 0.5 cm/s. Note that the CLOSE command is monitored in SRL such that a closing of the gripper occurs (the equivalent is also true for OPEN).

```
GRIPWIDTH gripper2 TO 5;
```

The jaws of gripper2 are moved to a distance of 5 cm, irrespective of closing or opening actions.

PASRO:

```
gripwidth (5);
```

Same as in SRL.

AL:

```
OPEN HAND TO 8 * CM;          (* open to 8 cm              *)
MOVE ARM1 TO targetframe;
CLOSE HAND1 BY 5 * CM;        (* gripper opening = 3 cm *)
MOVE ARM1 TO @ + ZHAT;
TO 2 * CM;                    (* close hand 1 to 2 cm     *)
```

An effector instruction with a parameter for a gripper opening is included in VAL. The opening distance is given in mm.

VAL:

```
OPEN 45
```

Open gripper to 45 mm. Only absolute values are possible in VAL.

In addition to the opening- or closing-speed parameter in SRL and AL, an opening or closing duration is possible. (Parameters for gripping force are dealt with in the following sections.) The syntax shown in Fig. 4.38 shows an extension of the AL implementation at the University of Karlsruhe. The possibilities offered by any newly developed gripper controller are passed on to the programmer.

SRL:

```
ALWAYS WHEN max_current
  DURING
    OPERATE TOOL(1) BY OPERATION(3);
  END_DURING;
  DO START warning(17);
```

The tool with the logical number 1 is operated by the operation 3 (implemented by a special driver). Whenever an interrupt occurs indicating an excess of the maximum current, a task is called which prints a warning message.

```
CLOSE WITH MAXSPEED;
```

The gripper in use is closed with its maximum speed.

```
GRIPWIDTH gripcx TO box_open+1
  WITH V = 1;
```

The gripper opening is set to the result of the expression box_open+1 and the jaws are moved at a velocity of 1 cm per second.

```
WHEN start_time <= SYSTIME+close_duration
  MONITORED EVERY 50 MS
  DURING
    start_time := SYSTIME;
    CLOSE gripper2 with V = 0.5;
  END_DURING;
  DO STOP gripper2;
```

The time it takes to close the gripper is monitored every 50 ms.

AL:

```
OPEN HAND TO 5 * CM;
  WITH VELOCITY = 0.5 * CM;
```

Open gripper with speed parameter

```
BY 2 CM
  WITH DURATION = 5 * CM;
```

Opening/closing gripper by 2 cm in 5 s.

4.3.3 Effector Instructions with Sensory Monitoring

Some languages even allow *sensory* monitoring of effector instructions. Strictly speaking, moving to a desired jaw opening has to fall into this category (Sect. 4.3.2), since this usually incorporates some form of position sensors in the gripper which feed the current distance of jaws back to the controller.

In SRL, sensory monitoring of effector instructions can be done specifically for force monitoring in the gripper instruction itself, and generally with the WHEN statement.

SRL:

```
CLOSE gripper1 UPTO GRIPFORCE > gf1+0.3;
```

Gripper 1 is closed until a gripping force is encountered which exceeds the result of the expression gf1+0.3.

```
WHEN touch_sense = TRUE
  DURING
    CLOSE gripper2;
  END_DURING;
  DO STOP gripper2;
```

If a touch sensor mounted on the gripping jaws records the contact, further closing actions will stop.

Monitoring of force sensors can be specified in AL, as with motion instructions (Fig. 4.38). If the gripping force passes the indicated value, an action (program block or instruction) is initiated.

AL:

```
CLOSE HAND TO 2 * CM
  ON FORCE >= 100 * GM DO
    STOP HAND &;
```

If the force exceeds 100 g, the closing of the hand stops. The significance of the & symbols has been explained in Sect. 4.2.5.1.

Gripper motions in AML are monitored in the same way as ordinary robot moves (see Sect. 4.2.5).

VAL permits the monitoring of grasping an object by assuming a servo-controlled closing of the jaw. The GRASP instruction will move the jaws until a counterforce stops the motor. The opening is then measured and compared with the given value. If the jaw distance is smaller than the specified value, the program branches to the label.

VAL:

```
GRASP 14.8, 200
```

If the gripper does not hold an object with 15 mm diameter, the program branches to label 200.

The CENTER instruction in AL represents a special effector or motion command. It causes the gripper to move along the y-axis of the gripper frame until a touch sensor in the jaw responds. Now the gripper is moved back by half the distance travelled and the gripper closed by nearly that distance. This process is repeated at the other side etc. As soon as the effector movement stays within a certain tolerance, it grasps the object with both jaws. This CENTER instruction enables the grasping of an object (without shifting it) with a gripper, which can only move both jaws at once.

4.3.4 Effector Instructions with Sensor-Monitored Parameters

Only in very rare applications are effector instructions monitored by sensors, so that specific *parameters* may be observed, as may be necessary with an automatic screw-driver trying to exert a given torque.

The SRL and AL implementation in Karlsruhe allows the programmer to specify the opening/closing while keeping a particular force.

SRL:

```
CLOSE hand BY 5
  WITH FORCE = 50;
```

AL:

```
CLOSE HAND BY 5 * CM
  WITH FORCE = 50 * GM;
```

4.4 Stopping a Robot or Effector Movement

In the context of sensory monitoring, the stopping of robot or effector motions with program instructions - i.e., not by the operator - is sensible and necessary (see also Sects. 4.2.5 and 4.3.3). Such an instruction is only useful if the language permits *parallel execution* in some form or other, because the stop instruction has to be carried out during robot movements. Consequently, the position and orientation are undetermined and have to be read again.

The STOP statement in SRL can also have various parameters, for example, specifying an immediate termination of the robot or effector moves, a soft termination or stopping of all movers.

SRL:

```
STOP ALL;
```

All robots and effectors cease motions.

```
STOP hero1 SOFT;
```

The robot hero1 is stopped softly.

The STOP instruction in AL affects a robot, a gripper or a frame connected to the arm. A STOP directed to the gripper causes a termination of the open/close action of the jaws. If only STOP is specified, all robots and effectors in motion are stopped.

AL: `STOP box;`
The robot arm to which the frame box is attached, is stopped.

The STOPMOVE routine of AML terminates robot and gripper motions and returns the values of the aggregate representing the target frame. Therefore, the programmer can try again to reach the goal position. An example from the AML manual will illustrate this:

AML:

```
ATTN:      SUBR;
MOTPARMS:  NEW STOPMOVE;
           WAITMOVE;
           BREAK (EOL, 'ATTENTION REQUESTED');
           APPLY ('AMOVE', MOTPARMS);
           END;
```

The effects of calling the subroutine ATTN during the AMOVE execution are an immediate stoppage of the move and an assignment of the target frame to the variable MOTPARMS (done in the declaration of MOTPARMS). The WAITMOVE ensures that the motion has stopped before the program continues. The BREAK instruction prints the message "ATTENTION REQUESTED" on the terminal and interrupts program execution. Then the APPLY instruction performs a new AMOVE to the target frame of the previously stopped move.

Since there is no parallelism in VAL, there is no stop instruction.

HELP can stop only robots, not effectors. Since HELP has parallel processing, a signal may cause the stopping.

HELP: `HALT(2)`
Robot 2 is stopped.

SIGLA, too, has a stop instruction, and this causes a break in the program which can only be continued by the operator.

SIGLA: `HL`
The robot arm and the program are stopped. For a restart, the key "CYCLE START" has to be pressed.

4.5 Sensor Instructions

Sensors are necessary for safe and flexible robot employment in many areas, particularly for material handling and assembly. Currently, the eye-hand coordination is dominant, i.e., there is already a broad application of marketable **vision systems** (see also LIEBERMAN [4.5]). However, *switch contacts* which are triggered by touch, *cross-fire sensors* or *ultrasonic distance sensors*, all of which have a much simpler construction and are much cheaper than vision systems, can be counted among robotic sensors. Their information may be just as valuable for the program execution as complicated optical sensors. Since sensors have been used mainly in development systems and not on a broad industrial base, a general interface for physical, logical and program integration of sensors and their information is still missing. Most languages are content with instructions for binary and analogue input/output. This may have been sufficient up to now, but it is certainly not enough for the demands of structured and self-documenting programming with a high degree of compatibility. Therefore, *language extensions* have been written especially for vision systems, and apart from the VAL 11V (V for vision, VAL extension), instructions are introduced for a particular vision system containing the language RAIL by the company AUTOMATIX. The high-level language RAIL is PASCALor ALGOL-oriented and is designed for industrial processes, flexible manufacturing cells, and robots with vision systems. The vision instructions in RAIL span all functions necessary for the integration of sensors into a robot programming system:

- Supplying sensors with parameters which control their state or measurement accuracy
- Activating sensors
- Reading of specific sensor data
- Interrogating sensor status
- Disabling sensors

What is missing in RAIL is an interrupt capability permitting the sensor to run a routine for processing sensory information in parallel to the main program, initiated by an interrupt.

Again, this is possible in SRL. Although SRL does not contain any special vision commands, the appropriate data structures for the sensory data can be defined in the system specification part and the program then has access to it later on. Thus SRL can be adapted very flexibly to different sensor or vision systems without extensions to the language or changes in the actual program section.

However, program branches caused by sensory information, reading of sensor data and the programming organization of sensory monitoring are dealt with next.

4.5.1 Program Branching Subject to Sensory Information

Static testing of a line for binary signals or for signals exceeding or falling short of given thresholds is termed sensor-determined branching. If the tested condition is true, the program branches or a different segment is executed.

The *American AL programming system* has been oriented exclusively toward a force or torque sensor, which prompted a **general sensor instruction** to be added. Therefore, the IFSIG command was formulated in SRL for digital input sensors and the INPUT system routine was extended for general sensor values.

SRL:

```
IFSIG switch3 THEN
  PTPMOVE TO error_place
ELSE
  SMOVE TO targetframe;
```

If a high signal is detected on the channel specified in the symbol switch3, the PTP-move to a predetermined error point is executed, otherwise the robot is moved along a straight line towards the target frame.

```
INPUT (temperature);
IF temperature > temp_limit THEN
BEGIN
  SMOVE robot1 TO pickup;
  .
  .
  .
END;
```

First, the analogue value of a temperature sensor is read and stored in the variable temperature, then the value is compared with a limit and appropriate action taken.

Sensory information is read and evaluated by the SENSIO routine in AML.

AML:

```
IF SENSIO (LED, 0) THEN
  BREAK ('Part gripped');
```

If the light beam in the gripper is interrupted, the text shown is printed on the terminal.

Only binary signals can be tested in VAL. The test for branching permits a logical AND of up to four input signals (see Fig. 4.39). The necessary specification of commas in every case is unfamiliar for a language on that level. If the channel number has either a positive or no sign, a high potential is tested, and a minus sign tests for a low potential.

VAL:

```
IFSIG 1,-5,, THEN 150
```

If signal 1 is set and signal 5 reset, the program branches to label 150.

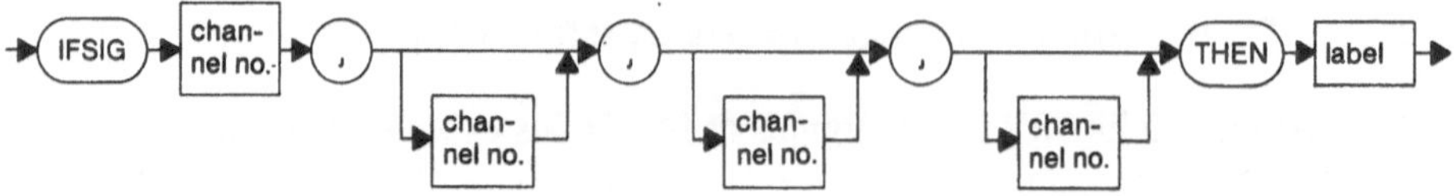

Fig. 4.39: Binary sensor interrogation with program branching in VAL

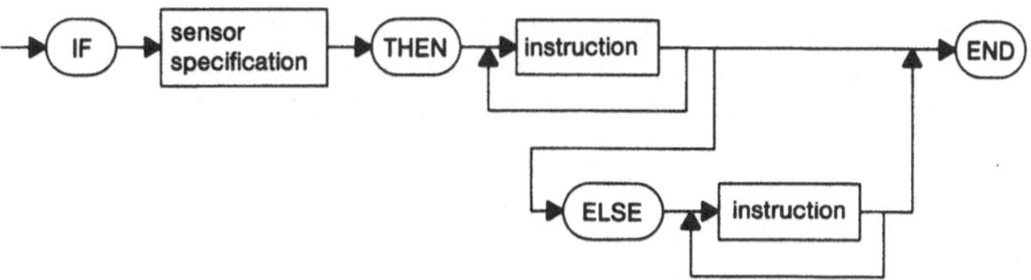

Sensor specification:

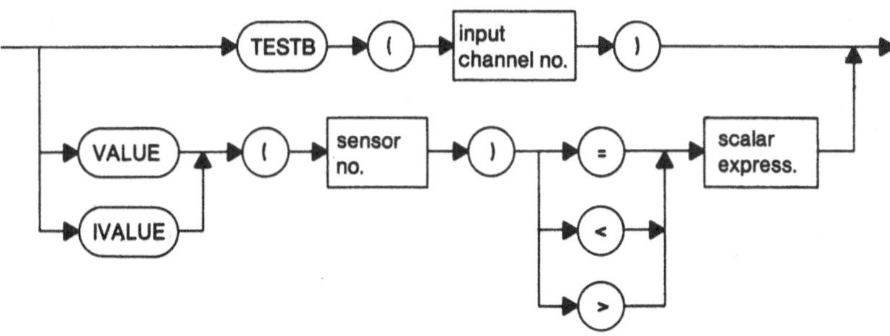

Fig. 4.40: Sensory branching in HELP

Sensor interrogation in HELP has a similar structure to that in AL (see Fig. 4.40). The command TESTB represents a logical function which delivers the value TRUE if the input is high, otherwise returns FALSE. VALUE and IVALUE are input instructions to read arithmetic values of a specified sensor. The value is entered by the run-time system at every clock interrupt. The function IVALUE delivers the last sensor value read. The function VALUE delivers a stabilized sensor value, reading the device until two successive values differ only in the least significant bit.

HELP:

```
IF TESTB(3) THEN
  DELAY(50)
END;
```

If sensor 3 is set, the program execution is delayed by 1 s.

```
IF VALUE(1) > 74.9 THEN
  MOVE (1,#1,140,#3,50);
  PULSE(2);
  MOVE (1,#1,120,#2,260,#150)
ELSE
  PRINT ("Too little paint");
  HOLD
END;
```

Sensor 1 measures the intensity of colours on objects; if it is sufficient, the parts are transported to a deposit. If not, the message "Too little paint" is printed out and an operator response is expected.

Although SIGLA is a simple language, sensory branches can be programmed according to the state of microswitches or previously entered sensor data. If the switch is closed and a positive switch number has been specified, a branch to the given label follows.

SIGLA:

```
PP/-1,12
```

If the microswitch is open, the program branches to label 12.

Alternatively, a sensor value can be entered and stored with the MT instruction. This value can then be used in the normal conditional branch instruction (see also Sect. 4.7.1.1).

SIGLA:

```
MT/1,M8,2;
BL/M8,75,3
   .
   .            Instructions if value >= 75.
   .
JU/4;
NU/3;
   .
   .            Instructions if value is too low.
   .
NU/4;
```

The value of sensor 1 is entered into counter M8. According to this value, the corresponding instruction sequence is executed.

In ROBEX, as in VAL, only a binary signal can be tested. If a signal is present at the input channel, the program continues execution at the label.

ROBEX:

```
ONSIG/EVENT,3,JMP,wait
```

If the signal at input 3 is set, the program branches to the label wait.

4.5.2 Input of Sensory Information

As was established earlier, *input of sensor data* is carried out by instructions to peripheral processes. However, even this possibility is not present in all robot languages.

In SRL, sensors and ports are specified in the system specification part of the program. The system procedure INPUT reads values from the specified sensor.

SRL:

```
SYSTEM_SPECIFICATION
  SENSOR: switch = CHANNEL(3);
          STRUCTURE switch = BOOLEAN;
          weight = CHANNEL(2);
```

```
            STRUCTURE weight = INTEGER;
END_SYSTEM_SPECIFICATION;
PROGRAM ...
    .
    .
    .
INPUT (switch);
IF switch THEN STOP robot3;
INPUT (weight);
IF weight > 75 THEN
  SYNMOVE ir600robot TO outplace;
    .
    .
    .
```

There are two sensors specified, a digital input from a switch and an analogous one from a weight sensor.

The functions TESTB, VALUE and IVALUE in HELP have been dealt with in the previous section. They may be used not only for sensor interrogation, but also to assign sensor values to scalar variables, thus making them available for future references.

HELP:

```
sens1 := VALUE(1);
```

The value of sensor 1 is stored in the scalar variable sens1.

As mentioned in the previous section, the value of a sensor can be stored in a counter or parameter in SIGLA with the MT instruction (see also Sect. 4.2.5.1). The specified delay refers to the addressing and reading of the hardware.

4.5.3 Sensory Monitoring

Sensor signals and the passing of sensor thresholds can invoke the *execution of an instruction sequence.* Here the time of the execution is unknown, so that the main program may be interrupted at any place and continued from there afterwards. Several languages offer special instructions for sensory monitoring (Sect. 4.2.5.1). However, instructions to assign sensor signals or thresholds to program sections, for instance in the context of a general task concept, are rarely available. Instead, certain reactions are realized which are all too often in the lower level of control, and hence are beyond the influence of the programmer.

SRL contains the sensor monitoring instruction ALWAYS WHEN, which is only valid within the block between DURING and END_DURING. If the given condition occurs, the interpreter interrupts the main program run, executes the instruction or task given after the DO keyword, provided the task has a higher priority, and then continues the main program.

SRL:

```
ALWAYS WHEN SENSOR(3) >= 5
  MONITORED EVERY 100MS
```

```
DURING
  SMOVE hds36robot TO handler;
   .
   .
   .
END_DURING
  DO START warning WITH PRIO=3;
```

As described in Sect. 4.2.5, in AML a monitor can be defined to react when a sensor value leaves a defined interval.

Although HELP has no specific sensor instructions, a cyclic sensor interrogation can be implemented with the help of the task concept. The monitor task consists of a sensor interrogation, the instructions to be executed if the condition is true, a delay instruction and a jump back to the interrogation (program section 4.7).

4.5.4 Sensor Instructions for Vision Systems

Most **vision systems** have three characteristics which aid their integration into a robot programming system:

1. Parts to be identified can be assigned symbolic names, represented by ASCII strings.
2. Position and orientation are evaluated in relation to a Cartesian coordinate system.
3. The system is "taught" those parts to be recognized by "showing" and specifying symbolic names for them.

Parts are identified by a *set of features*, which may contain area, number of holes, largest and smallest diameter or circumference. The user does not have to concern himself with the setting up of these parameters; the vision system automatically generates a set for the taught part and compares the stored values with the set of features of the current picture during program execution.

The extension of VAL, VAL 11V, contains several commands for a slave program which performs camera calibration, teaching and storage of feature sets of various parts under symbolic names. The user program can then communicate with the vision system by two instructions VPICTURE and VLOCATE. VPICTURE causes the cap-

```
suetask:  sens3 := VALUE (3);
          IF sens3 > 5 OR sens3 = 5 THEN
            PRINT ( 0, 'Machine started');
            m1on := 1
          END;
          DELAY (10);
          GOTO suetask;
```

Program section 4.7: Task suetask for sensory monitoring in HELP

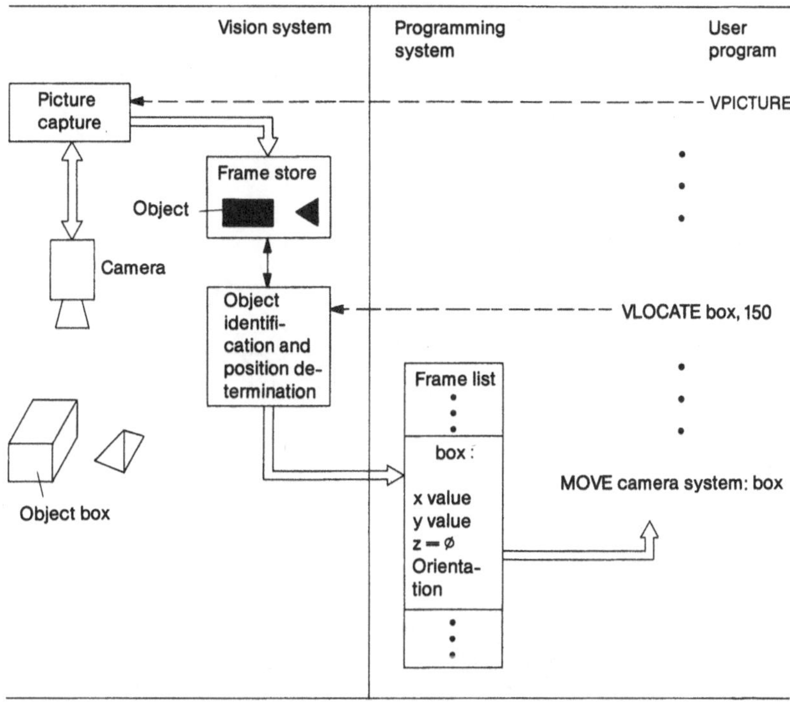

Fig. 4.41: Connection between the vision system and robot programming in VAL 11V

```
HERE calibrobot
MOVE restpoint
REMARK remove robot from field of view

VPICTURE
VLOCATE calibcamera, 150
INV invcalibcamera = calibcamera
SET camerasystem = calibrobot:invcalibcamera
           .
REMARK     .                    normal program flow
           .
VPICTURE
VLOCATE object, 150
MOVE camerasystem:object
           .
           .
           .
```

Program section 4.8: Calculation of the camera coordinate system within the robot coordinate system in VAL 11V

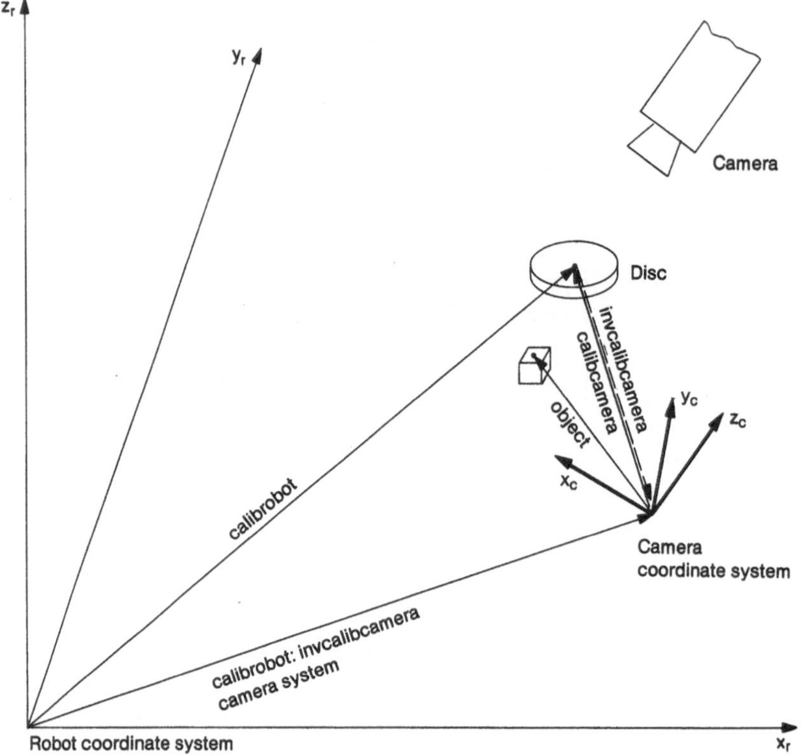

Fig. 4.42: Geometric representations of relative frames to calculate camera coordinates in the robot coordinate system

ture of a frame in the memory of the vision system. VLOCATE then activates a search of the stored picture for a specified object. If the search fails, the program branches to a label - if given - or the program stops and outputs an error message. However, if the search succeeds, the vision system enters the position and orientation under a *frame symbol*, which is the same as the *object name.*

The coordinate values of object frames refer to the *Cartesian coordinate system of the camera*, and not to the base coordinate system of the robot. Fig. 4.41 shows the connection between the vision system and the robot programming in VAL 11V schematically. With the aid of *relative frames* (Sect. 2.3.1), the user can easily calculate the coordinates of the camera system relative to the robot base coordinate system. In order to determine the relationships, he puts a disk- or ring-shaped object into the field of view of the camera, guides the robot to the center of that object via the teach-in procedure and defines the position and orientation as a frame `calibrobot`. After the robot has been removed from the field of view, the origin and orientation of the camera coordinate system within the robot coordinate system can be evaluated with the help of program section 4.8 (in a slightly simplified manner). Fig. 4.42 shows the geometric representations of relative frames necessary for the calculations. Only position vectors are considered, for the sake of clarification. Orientation is dealt with in a similar way.

The HERE instruction defines the center of the disk as the frame calibrobot, and VLOCATE defines the disk center as the frame calibcamera in the camera coordinate system, but obviously, both position vectors refer to the same point in space. The frame calibcamera is now inverted, hence the position vector of invcalibcamera points from the disk center to the origin of the camera coordinate system. The compound transformation calibrobot:invcalibcamera specifies the origin of the camera coordinate system within that of the robot as the vector of this frame pointing from the origin of the robot to the origin of the camera coordinate system. In order to simplify, the frame was assigned to another frame camerasystem. After this calculation, all objects found by VLOCATE can be used in the MOVE instruction relative to camerasystem. This example highlights the necessity of frame concepts as well as geometric operators with an application of vision systems.

The language RAIL from AUTOMATIX contains a wide spectrum of special instructions for communication and control of a vision system. It allows the user to program *object identification* by himself. For this, the function OBJ_FEAT delivers the values of various object features after a frame capture by the PICTURE command. For example, among a maximum of 45 features there are

- Position (OBJ_XMIN, OBJ_XMAX, OBJ_YMIN, OBJ_YMAX)
- Orientation (OBJ_ANGLE)
- Number of holes (OBJ_NHOLES)
- Largest/smallest diameter (OBJ_RMAX, OBJ_RMIN)
- Color (OBJ_COLOR)
- Area (OBJ_TOTALAREA)
- Center of gravity (OBJ_XCENT, OBJ_YCENT)

They may also be addressed by *predefined feature variables* - as shown in the examples.

RAIL:

```
IF OBJ_NHOLES == 3 THEN
  part = 1
ELSE
BEGIN
  magazine = 1
  part = 2
END
```

If the part currently looked at has three holes, it is part 1, otherwise it is part 2 and the variable magazine is set.

Usually, not all of the 45 features are necessary for positive identification. In particular, the evaluation of those 24 features which are determined at each frame capture can be suppressed. This is specified by a system switch containing the prefix REC instead of OBJ, followed by the feature name.

RAIL:

```
REC_NHOLES = OFF
```

The calculation of number of holes is suppressed.

The switch REC_FEAT is important because it disables all feature evaluations and enables special ones explained in the remainder of the section. However, before any features of an object can be investigated, the whole picture has to be analyzed. This is the purpose of the following instructions:

FIRSTPART	Take the first object within the current picture.
NEXTPART	Take the next object (always after a FIRSTPART instruction).
FIRSTBLOB	Take the first object or hole of the current picture.
NEXTBLOB	Take the next object or hole in the current picture (always after FIRSTBLOB).

In this context, the predefined variable VIS_NPART is important, since it contains the number of objects within the current picture. Now all objects within the picture can be analyzed systematically.

RAIL:

```
PICTURE
IF VIS_NPARTS > 0 THEN
BEGIN
  FOR numberofparts = 1 TO VIS_NPARTS DO
  BEGIN
    IF numberofparts == 1 THEN
      FIRSTPART
    ELSE
      NEXTPART
    IF OBJ_NHOLES == 3 THEN
      EXITLOOP
  END
END
```

After the picture has been captured, the objects within the field of view are examined sequentially for three holes. If they are present, the loop is exited and the part can be examined further.

Different vision systems or *user requirements* can be adapted by special system switches, parameters and functions. Since the system is geared to gray scale analysis, tables of thresholds achieve compatibility with the appropriate hardware. This variety of system switches should be set once by a specialist, and the user should only be bothered by them in exceptional circumstances.

RAIL:

```
VIS_DISPLAY = OFF
```

The picture captured in the frame store is not displayed on the monitor.

```
VIS_CAMERANO = 2
```

Use camera 2 for further exposures.

4.6 Block Structuring and Instruction Sequencing

The connections between life time and validity of variables and block structures have already been mentioned in Sect. 2.2.3. Thus, blocks are *instruction sequences* which may be bracketed by suitable symbols (such as BEGIN and END), within which variables can be defined. This convention is fully implemented in SRL and AL, as program section 4.9 shows.

Block 1 and 2 define new variables, while block 3 contains only a bracketed instruction sequence. The latter is called a **compound statement** and is needed for the program flow control, which is dealt with in the next section.

In SRL, there is a difference of syntax between a block, which is bracketed by the keywords BEGIN_BLOCK and END_BLOCK, and a compound statement delimited by BEGIN and END.

However, the same keywords BEGIN and END serve exclusively as brackets for compound statements in PASCAL or compound expression in AML. The main program, procedures and functions are blocks with respect to the management of variables (see also Chapt. 6). An example for this is given in program section 4.10 for PASRO and PASCAL.

The variables a, b and c are defined in mainblock of the main program, and i, x

```
BEGIN "block 1"
  SCALAR limit;
      .
      .
      .
  IF SENSOR (3) < limit THEN
  BEGIN "block 2"
    SCALAR b;
      .
      .
      .
  END "block 2"
  ELSE
  BEGIN "block 3"
      .
      .
      .
  END "block 3";
      .
      .
      .
END "block 1"
```

Program section 4.9: Blocks in AL

```
PROGRAM mainblock;
VAR a,b,c : INTEGER;

PROCEDURE thisisablock (x, y : REAL);
  VAR i : INTEGER;
  BEGIN (* sequence in thisisablock *)
    .
    .
    .
  END (* sequence in thisisablock *);

BEGIN (* sequence of mainblock *)
    .
    .
    .
END (* sequence of mainblock *).
```

Program section 4.10: Blocks and instruction sequences in PASCAL

and y are defined in the block of procedure `thisisablock`, where x and y are procedure parameters at the same time.

HELP has *dynamic*, *static* and *init blocks*. Instructions in a dynamic or init block are erased after their execution in contrast to static blocks. Variables defined in blocks remain only in the static and init blocks after their execution. Therefore, the init blocks serve only for the initialization of variables, as the name indicates. Static blocks are meant for the definition of subroutines; dynamic blocks for the segmenting of the program into individual parts, which are not needed after their execution and hence can be erased. For this, see program section 4.11. After the execution of the init block (which is always done first), only the variable v remains in memory. The subroutine sub1 can be called from all other blocks, since it is defined in a static

```
STATIC BLOCK
subr1:  .
        .
        .
        RETURN;
ENDBLOCK;

INIT BLOCK
  DEFINE v(3);
  v[1] := 1;
  v[2] := 3;
  v[3] := 0;
ENDBLOCK;

BLOCK              ! Program segment 1
        .
        .
        .
ENDBLOCK;

BLOCK              ! Program segment 2
        .
        .
        .
ENDBLOCK;

BLOCK              ! Program segment 3
        .
        .
        .
ENDBLOCK;
```

Program section 4.11: Blocks in HELP

block. The remainder of the program has been segmented into three parts which are executed sequentially and erased after their execution.

The purpose of this is to free memory from instructions and variables which are only needed once. Justification of this mechanism is, however, questionable, considering the constantly falling cost of memory and the complicated linguistic solution in HELP.

4.7 Program Flow Control

Much was written and argued about the advantages of **structured programming** and **top-down design** at the end of the 1960s and beginning of the 1970s. This will not be debated any further here; the interested reader may refer to the relevant literature, [4.6 - 4.10], for more details about the "argument about the GOTO statement". Unfortunately, no general definition has been found despite the series of publications about this subject which may be summarized by the expression "structured programming". [4.10 - 4.14]. One of the first publications was the article known as "Dijkstras GOTO letter" [4.15], where the author warned against the use of a *jump instruction* as it can be a source of frequent errors. Compare the two statements as an explanation

```
result := SIN (angle);
```

and

```
GOTO 500;
```

The effect of the first instruction is clear, whereas that of the second is not. The reason for this lies in the absence of an explanation as to why a branch follows and in what *structural context of the program flow control* this happens. An example of an unstructured flow diagram in Fig. 4.43 highlights the great difficulties with which all the possibilities of a dynamic program flow can be traced on a static flow diagram. This is exactly the reason why a frequent use of the jump instruction is error-prone.

Structured programming is concerned with *problem-oriented language constructs* introducing the basic pattern for dynamic flow and laying out the dynamics already in the static description. These basic patterns and their representations in **flow diagrams**, **structograms** and the languages concerned are described in the following sections. The representations of basic patterns as flow diagrams as well as structograms are given for the following reason: Flow diagrams are still more easily comprehended by the untrained reader than structograms, despite the disadvantages described above, whereas the latter is introduced for clarification of structured programming. *Structograms* are a means of representing flow structures graphically, developed by Nassi and Schneidermann [4.16]. The simplest elements are *structure blocks*. The combination of these to give new, more complex structure blocks follows the rules listed below (see [4.7 and 4.10]):

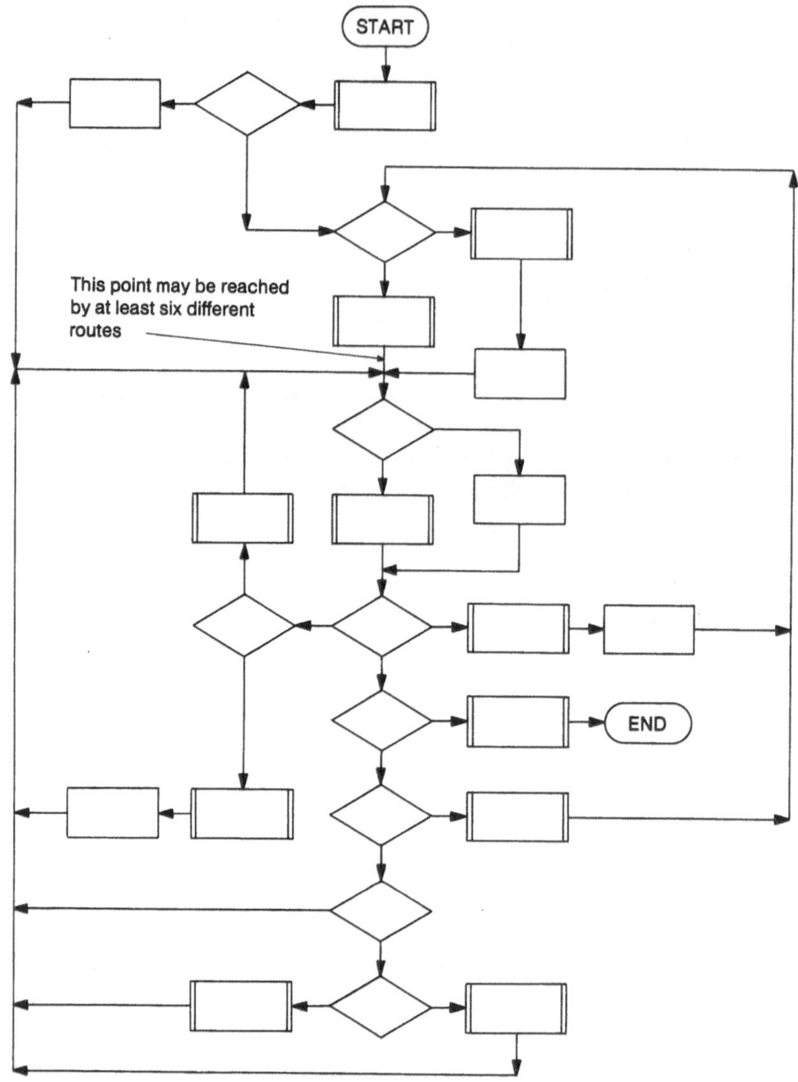

Fig. 4.43: Nonstructured flow diagram

- Structure blocks have *one* input and *one* output.
- The dynamic *flow control* is always from *top to bottom* through the structure block. This also holds for structograms combined to one structure block.
- A structure block is *one closed functional unit.*
- It defines clearly a *containment relation*; there are no overlaps.
- It *corresponds exclusively with its direct neighbors*: it receives its commands from its upper neighbor and passes them on to the lower neighbor.

However, let us consider first a robot specific problem of program flow control. Programming systems, like the American AL, which generate a simple environment model during compilation, run into problems with all kinds of branching. So, the American AL introduces a "plantime assignment", with which the program can communicate values to the simulator during compile time. However, even this does not solve the problem, since it is not in general possible to envisage the dynamics of a program from a static text. The problem is further complicated by the great importance of the analysis of sensory information to robots. Thus the version of AL developed at the University of Karlsruhe has omitted this kind of presimulation [6].

The following section often refers to <instruction>, which represents a general instruction, in particular including those for program flow control. A special instruction is the **empty instruction** present in SRL, PASRO or PASCAL, AL and AML. Empty instructions are useful for distinguishing cases which do not require any actions (see also Sect. 4.7.1.2). For example, see the following SRL statement:

```
CASE ch OF
  'i': writeln ('intput');
  'm': ;
  'o': writeln ('output');
  OTHERWISE
    writeln ('Error');
END_CASE;
```

If ch has the value m no action is performed. There is also a syntactical reason for the empty statement. In most languages a semicolon seperates statements. Therefore it is not possible to write

```
      .
      .
      .
  <statement>;
END;
```

But this notation would be useful for later program extentions, where statements might be inserted before the END. Therefore the spare between the semicolon and the END is interpreted as an empty statement.

As mentioned before, AML is an expression-oriented language. Thus every statement evaluates an expression and can deliver a result. The result of an empty statement is the null aggregate <>. Hence the semicolon assumes an important role in AML:

```
BEGIN      BEGIN
  .          .
  .          .
  A=5;       A=5;
  A+10;      A+10
END        END
```

The compound expression on the left-hand side delivers the null aggregate, because after the semicolon and before the END an empty expression is defined, whereas the one on the right returns the value of 15.

4.7.1 Program Branches

The simplest way of branching is that of the unconditional jump instruction. As mentioned earlier, the command itself does not allow any conclusions about the program structure. This may be slightly improved by the expressiveness of the label connected with the jump instruction. Hence there is a difference in clarity between

```
GOTO 100
```

and

```
GOTO loopstart
```

Only HELP and AML have enough room to maneuver and define text as labels, while PASRO and PASCAL, VAL and SIGLA allow only digits. The following text treats jump instructions in the various languages. In general, languages which offer structured instructions apart from jumps for flow control (like SRL, PASRO, AML and HELP) either exclude jump instructions into these structured instructions or their effects are compiler-dependent and hence should be omitted anyway.

The SRL language includes a special facility, the EXIT command. It terminates the execution of a procedure, task or loop at any place, hence it works just like an unconditional jump to the end of these program parts.

SRL:

```
REPEAT
  i := i + dt;
    .
    .
    .
  IF i > endval THEN
    EXIT_REPEAT;
    .
    .
    .
UNTIL scos > 0.9;
```

The normal termination condition for the repeat loop is a value in scos which is larger than 0.9. However, the loop is interrupted and terminated prematurely if i has reached the final value endval.

There is no other jump command in SRL. The programmer should use problem oriented-instructions for program flow control.

AL does not allow any jumps and instead offers problem-oriented instructions for program flow control.

In AML a label may consist of any number of alphanumeric characters starting with a letter. A labelled expression is of the form

```
<label>: <statement>
```

In PASRO a label must be declared in the program header and may consist of up to four digits. A labelled statement has the form

```
<label>: <statement>
```

A label may only consist of digits in VAL and has to be smaller than 32767. A labelled statement has the form

```
<label> <statement>
```

In HELP a label may consist of any number of alphanumeric characters, starting with a letter. Only the first six characters are significant. A labelled statement has the form

```
<label>: <statement>
```

There are considerable restrictions in HELP for the jump instruction because of the complicated block structure. HELP distinguishes between static and dynamic blocks, where the code generated in static blocks remains in memory, while dynamic blocks are erased after their usage (Sect. 4.6). Therefore, jumps to another block can only be done into a static block, and only back references are possible. Within a dynamic block, jumps may done in both directions. Jumps into dynamic blocks are generally forbidden.

A label in SIGLA may be an integer between 1 and 255, a parameter, a counter or an indirectly addressed counter, where the values have to fall within the limits above. The target is specified by

```
NU/<label>
```

where this code is placed before the instruction to be executed.

The shape of a label in ROBEX is not explained in detail in [14]. However, it can be concluded from the general specifications and the examples in [15] that a maximum of six alphanumeric characters is allowed, and always has to start with a letter. The jump target is of the form

```
<label>) <instruction>
```

Apart from its jump instruction executed at run-time, ROBEX also has a jump instruction for the compiler causing compilation to continue at the specified label.

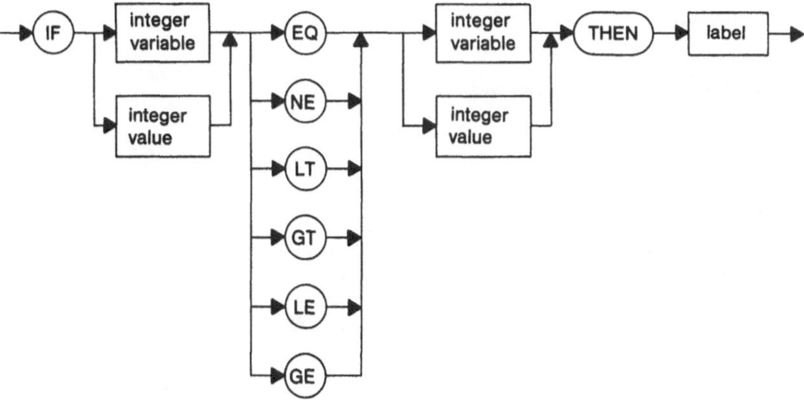

Fig. 4.44: Conditional jump instructions in VAL

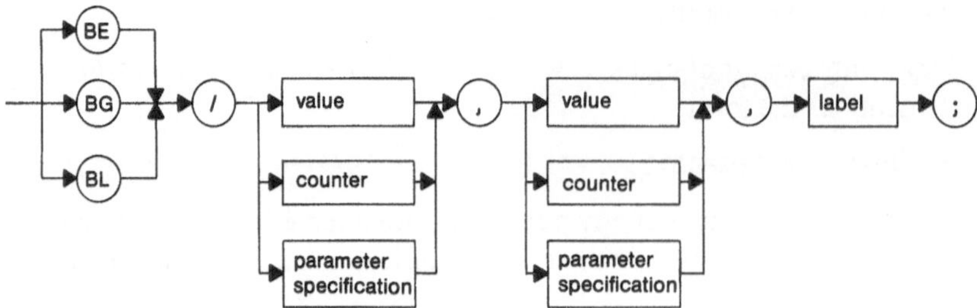

Fig. 4.45: Conditional jump instructions in SIGLA

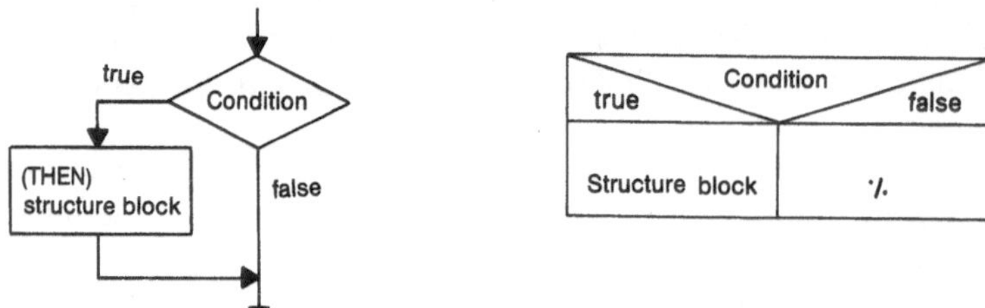

Fig. 4.46: Conditional instruction

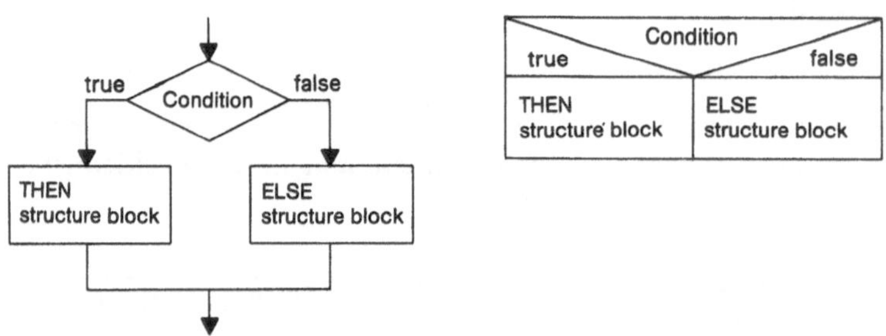

Fig. 4.47: Choice of two alternatives

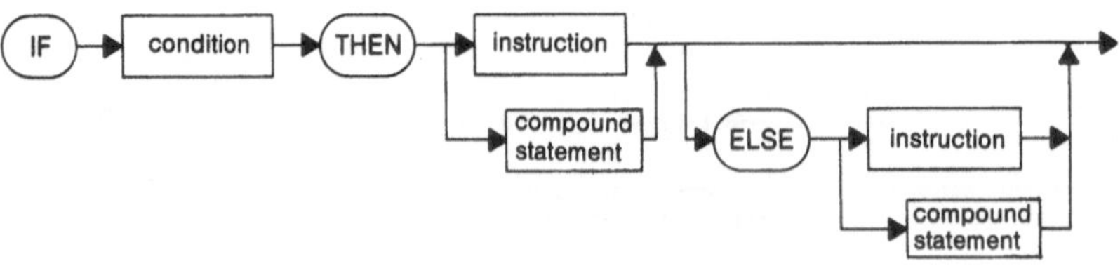

Fig. 4.48: IF instruction in SRL, PASRO, AL and AML

Since such constructs do not belong to the language itself (even if they are listed in appendix I), they will not be treated any further.

ROBEX does not offer any means of influencing the program flow, apart from the unconditional jump and the branching due to sensory information or subroutines. These standard types of flow control would have to be programmed explicitly in ROBEX as in VAL and SIGLA, since they have no direct linguistic means for conditional jumps, case distributions or loops, which will be considered later. In ROBEX, these can only be implemented in a very limited manner via sensory conditions.

4.7.1.1 Conditional Branches

First the conditional branches of VAL and SIGLA are introduced and afterwards the structured IF commands of SRL, PASRO, AL, AML and HELP.

In VAL (Fig. 4.44), the symbols are EQ for equal, NE for not equal, LT for less than, GE for greater or equal, LE for less than or equal and GE for greater than or equal. If

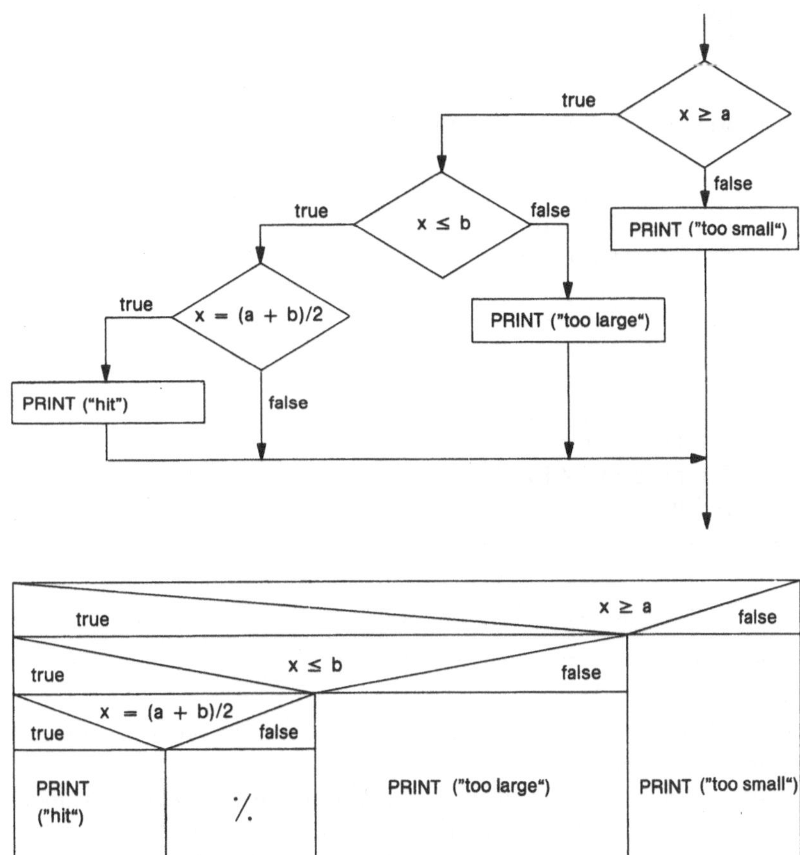

Fig. 4.49: Flow diagram and structogram for a simple example

```
IF  x >= a
THEN IF x <= b
     THEN
     BEGIN
       IF x = (a + b)/2
       THEN PRINT ("hit")
     END
     ELSE PRINT ("too large")
ELSE PRINT ("too small");
```

Program section 4.12: Program for Fig. 4.49 in AL and PASRO

```
IF (x > a) OR (x = a)
THEN IF (x < b) OR (x = b)
     THEN IF x = (a + b)/2
          THEN PRINT ('hit')
          END
     ELSE PRINT ('too large')
     END
ELSE PRINT ('too small')
END;
```

Program section 4.13: Program for Fig. 4.49 in HELP

```
    IF x GE a THEN 10
    TYPE too small
    GOTO 100
10  IF x LE b THEN 20
    TYPE too large
    GOTO 100
20  SETI temp = a + b
    SETI temp = temp/2
    IF x NE temp GOTO 100
    TYPE hit
100      .
         .
         .
```

Program section 4.14: Program for Fig. 4.49 in VAL

the condition is true, the program branches to <label>, otherwise the succeeding instruction is executed.

SIGLA knows about three types of jumps (Fig. 4.45). Here the symbols are BE for branch if equal, BG for branch if greater, and BL for branch if less. The program jumps to <label> if the condition is fulfilled, otherwise it continues with the next instruction.

```
                              Meaning:
BL/M1,M2,1;
BG/M1,M3,2;
SE/M4,M2;
IC/M4,M3;
SE/M5,0;                      \
NU/10;                        |
IC/M5,1;                      |
IC/M4, -2;                    > division by 2
BG/M4,0,10;                   |
BE/M4,0,11;                   |
IC/M5,-1;                     /
NU/11;
BE/M1,M5,3;
JU/100;                           Where:
NU/1;                             M1 contains x
NT/too small,M1;                  M2 contains a
JU/100;                           M3 contains b
NU/2;                             M4 contains a+b
NT/too large,M1;                  M5 contains result of
JU/100;                              integer division
NU/3;
NT/hit,M1;
NU/100;
```

Program section 4.15: Program for Fig. 4.49 in SIGLA

The structured IF command permits the formulation of **conditional instructions** and a choice of two alternatives. The general types are shown in Figs. 4.46 and 4.47.

For the languages SRL, PASRO, AL and AML the form of the language element is shown in Fig. 4.48, where <condition> is a Boolean expression.

Basically, HELP proceeds from an instruction sequence consisting of at least one instruction, and therefore the IF command always finishes with an END symbol.

Instructions or structure blocks may, of course, themselves contain an arbitrary number of program flow instructions and hence form a hierarchical structure of program branches. This may have the form shown in Fig. 4.49 for IF instructions. Program section 4.12 shows the corresponding program text for AL and PASRO, where in PASRO the PRINT instruction would have to be replaced by a WRITE.

The third test is bracketed by BEGIN and END. The reason for this is the missing ELSE with the third IF, and the rules for nested IF instructions are: A present ELSE part belongs to the next IF not having an ELSE. Hence, the third IF is not relevant for the ELSE PRINT ("Too large") because of the BEGIN-END and is assigned to the second IF.

This consideration is not relevant in SRL and HELP because each IF instruction is concluded with an END_IF resp. END. So the correspondence of IF and ELSE is unambiguous. Differences to AL and PASRO are minimal, as far as the program structuring is concerned, as shown in program section 4.13.

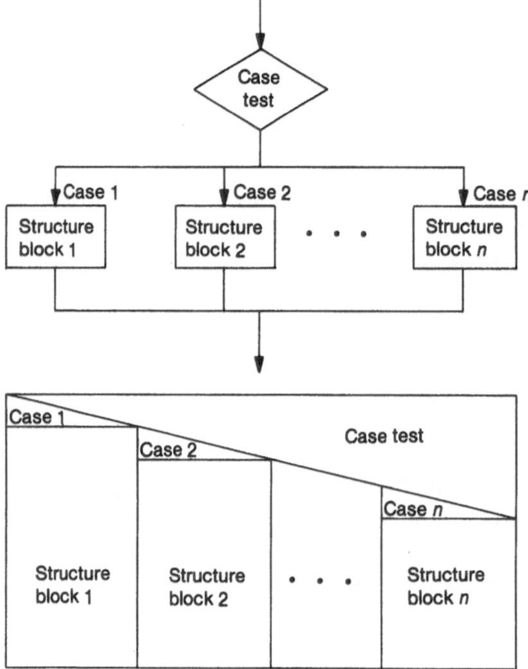

Fig. 4.50: Case distribution

Tests in the IF instructions have been chosen such that the structure of Fig. 4.49 has been maintained.

Clarity varies and diminishes more or less in the other languages, first in VAL and then in SIGLA (program sections 4.14 and 4.15).

Contrary to the structured commands of SRL, PASRO, AL, AML and HELP, neither VAL nor SIGLA allows the nesting of tests to be recognized from the program text itself, unless it is studied carefully. The program in SIGLA is significantly longer and more complicated, because the division has to be programmed explicitly (cf. program section 3.11). Although not all sections have been commented, the versions of SRL, PASRO, AL, AML and HELP may be easily read. This clarity diminishes in VAL, and SIGLA is obscure and practically illegible without comments and thus most prone to errors.

4.7.1.2 Case Statements

If there are more than two alternatives for a decision, one may use either a nested IF statement or a CASE statement. Figure 4.50 shows the flow diagram and the structogram.

Only SRL, PASRO, AL and HELP have suitable language elements available for this type of structure.

First SRL and PASRO: In Fig. 4.51, `<const>` has to be of the right type to accept the result from `<expression>`; for example `<expression>` may be of type `INTEGER` or `CHAR`. If no case label matches, the situation is not defined in standard PASCAL. But most PASCAL implementations offer an `OTHERS` exit, just as SRL does.

There are two versions of the CASE statement in AL, one with labels and one without. The <scalar_expression> is evaluated, and the integer part specifies the section to be executed, where count starts at zero. The second version does not allow negative values. If no constant according to the scalar expression can be found or the i-th statement does not exist, an error is generated in AL, unless an ELSE part is present in the second version.

Although AML does not offer a CASE construction, it is possible to construct an easily understandable one by using the BRANCH instruction, as the following example shows:

AML:

```
BRANCH (IF N GE 1 AND N LE 4 THEN
    <CASE1, CASE2, CASE3, CASE4> (N)
    ELSE ERROR_CASE );
CASE1:
    .
    .
    .
  BRANCH (CASE_END);
CASE2:
    .
    .
    .
  BRANCH (CASE_END);
CASE3:
    .
    .
    .
  BRANCH (CASE_END);
CASE4:
    .
    .
    .
  BRANCH (CASE_END);
ERROR_CASE:
    .
    .
    .
CASE_END:
```

The selector variable N is first tested for a valid range and is then used as an index for the aggregate.

HELP does not have a CASE construction as such, but only the **computed GOTO**, as it is known from FORTRAN. A case statement can thus be implemented easily. The syntax is shown in Fig. 4.52. <expression> is a scalar expression, which is rounded. Its value selects the i-th <label> as the jump target, where the count starts at 1. If the value exceeds the number of labels, no jump instruction is executed and the program continues with the next statement. The other case, if a value less than

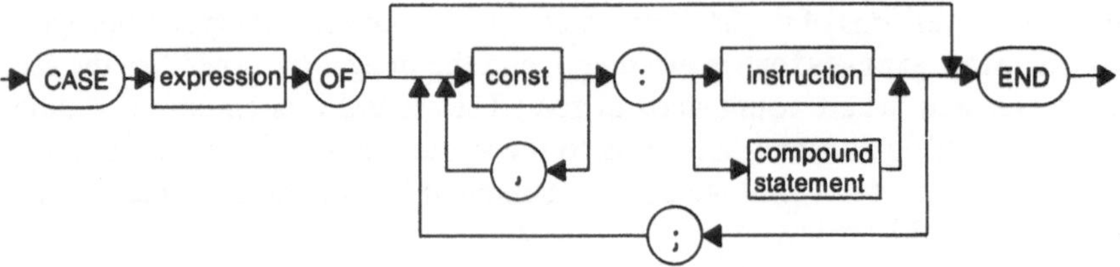

Fig. 4.51: CASE instruction in PASRO or PASCAL

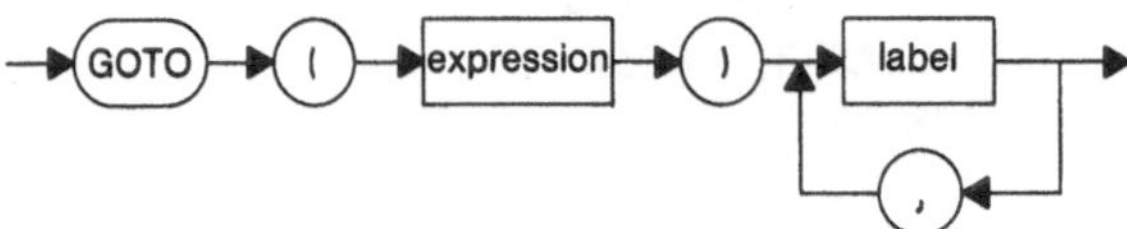

Fig. 4.52: Computed GOTO in HELP

```
        IF case < 0.5 THEN GOTO caseend END;
        GOTO (case) case1, case2, ... , casen;
        GOTO caseend   ! do nothing if case is out of range
case1:  <instructions>
        GOTO caseend;
case2:  <instructions>
        GOTO caseend;
            .
            .
            .
casen:  <instructions>
caseend:    .
            .
            .
```

Program section 4.16: Case statement with computed GOTO in HELP

0.5 occurs, is undefined in HELP. The same restrictions hold for the computed GOTO as for the ordinary jump instruction (Sect. 4.7.1). Hence a case statement may be constructed in HELP - as program section 4.16 illustrates.

As has been mentioned at the beginning, case statements may be built with nested IFs. This solution is to be preferred, if

- the probabilities of the individual cases differ drastically and the distribution may be estimated at the time of writing and
- testing cannot be reduced to a calculation of index values.

Case distributors have to be programmed explictly with the GOTO statement in VAL and SIGLA, and in ROBEX they are only conceivable via sensor values.

4.7.2 Loops

The last basic structure for program flow control listed is that of loops, in which an instruction or compound statement is executed repeatedly until a certain condition becomes TRUE. Loops are distinguished by the type of condition and the place of testing for continuation. Corresponding language elements are only contained in SRL, PASRO, AL, AML and HELP; they have to be programmed explicitly in VAL and SIGLA via the GOTO instruction. Only sensor-dependent loops may be used in ROBEX.

It should be pointed out that a jump into a loop may lead to unpredictable results and is even forbidden by some languages or implementations.

4.7.2.1 Counting Loops

Counting loops are executed until a count variable reaches or passes a predetermined value by up or down counting. The FOR instruction is intended for this in SRL, PASRO, AL and HELP. SRL, AL and PASRO test the condition on entry of the loop (and hence before the first execution), while HELP, unfortunately, does not specify the test point. Figure 4.53 shows the general form of counting loops for up or down counting with test on entry as a flow diagram and structogram.

The form given also executes the block for the `<finalvalue>`. If this is to be excluded, the comparison has to be changed from "larger or equal" to just "equal".

Only the step value of 1 is possible in SRL and PASRO (Fig. 4.54). Both expressions for start and final value are calculated on loop entry. Changes in the count variable are not allowed; its value is undefined after the end of the condition evaluation. The loop variable has to be of type INTEGER, CHAR or any other (by the user) linearly ordered data type.

The AL version of the FOR loop offers more possibilities. According to the sign of `<scalarexpression>` of the step value, the counter is incremented or decremented until the final value is passed. Here, too, the expressions are only evaluated once at the beginning of the loop. The value of the loop variable is still defined after the iteration contrary to PASRO and PASCAL. The effect of changing the loop variable

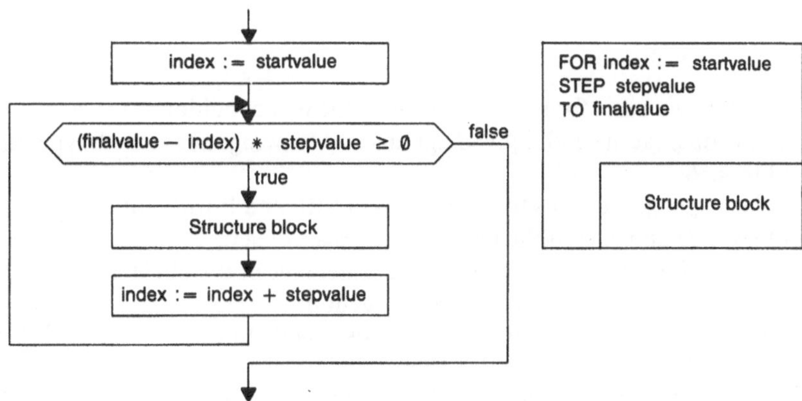

Fig. 4.53: Counting loop

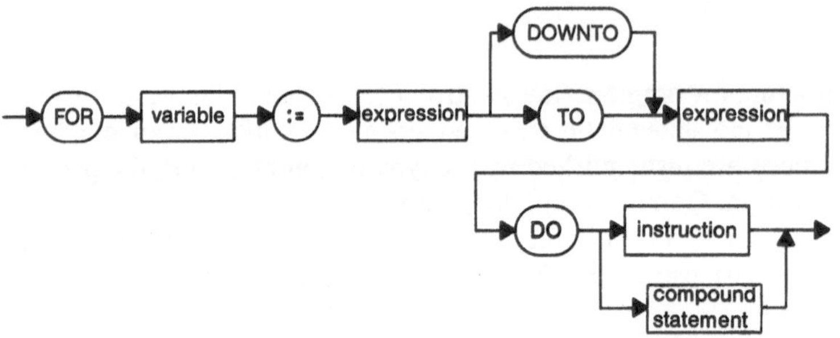

Fig. 4.54: Counting loop in SRL and PASRO

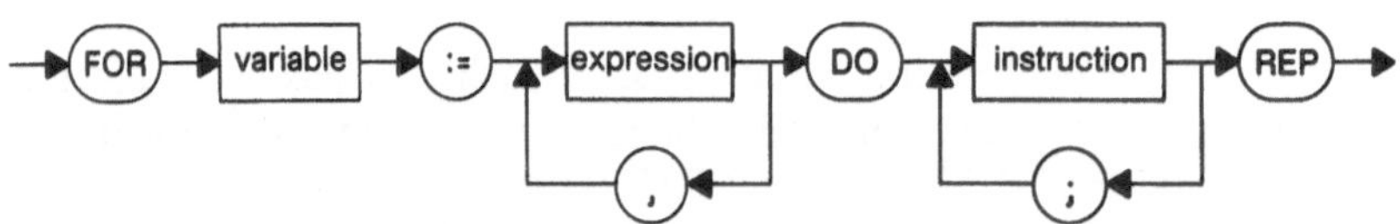

Fig. 4.55: Counting loop in HELP

within the loop is not defined. The loop variable does not have to be declared, because of possible implicit declarations.

HELP has two different forms of the FOR instruction. The first corresponds to the ones shown previously.

The loop variable can only be incremented, and if the step value is omitted, it is incremented by 1. Here, too, the values are calculated before loop execution. In contrast to the loops treated so far, HELP allows a value to be specifically assigned to the loop variable, in order to shorten the loop and exit it prematurely. The last value of the loop variable is still available after loop exit.

The second form is shown in Fig. 4.55. The loop variable takes all values of those n expressions and executes the loop body - consisting of one or more instructions - sequentially.

4.7.2.2 Conditional Loops

There are two basic forms here: One tests the exit condition at the start of the loop, and the other at the end. Flow diagrams and structograms of both types are shown in Fig. 4.56.

The language elements for both loops in SRL and PASRO or PASCAL are given in Figs. 4.57 and 4.58. If the loop body consists of several instructions, they have to be grouped together into a compound statement in the WHILE statement, whereas this is not necessary for the REPEAT instruction.

In AL the WHILE command corresponds exactly to the one in PASRO, while the REPEAT only differs in the syntax (Fig. 4.59).

The WHILE in AML command corresponds to the one in SRL and PASRO, while the REPEAT is similar to the DO instruction in AL (see Fig. 4.59). The differences arise

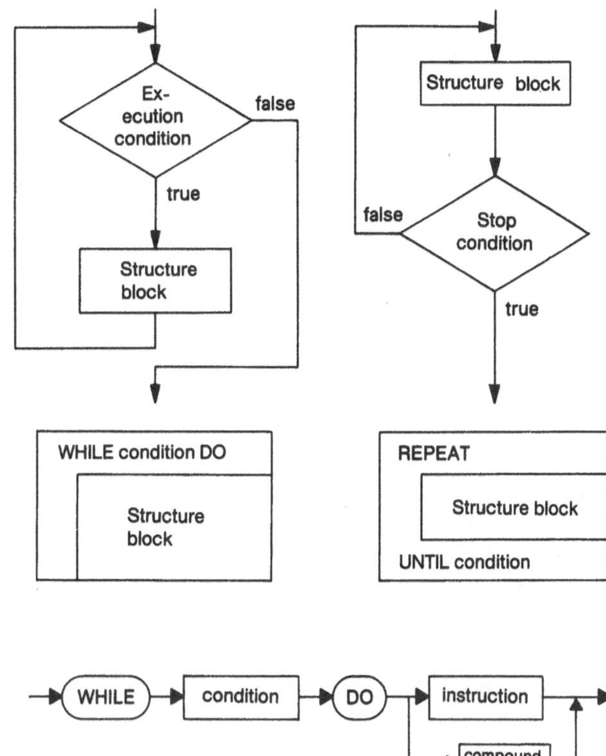

Fig. 4.56: Conditional loops

Fig. 4.57: `WHILE` loop in SRL, PASRO, AL and AML

from the point that it is a reserved word and a compound expression instead of a compound statement.

HELP contains a `WHILE` instruction.

4.7.3 Synchronization Commands

In order to synchronize parallel processes, as described in Sect. 2.2.6, special variables - **semaphores** - are made available. These may be set or reset in the same way as a Boolean variable, or incremented or decremented as a counter. In the first case, a process may be activated or deactivated, whereas the second case also administers the number of lockings and unlockings which have occurred (see also Sect. 2.2.7). Details and organisation of these parallel processes are described in Chap. 7.

Variables of type `EVENT` are used for synchronization in SRL and AL. The statement

```
SIGNAL <event-variable>
```

increments the value of the semaphore `<event-variable>`. If the value is larger or equal to zero, a waiting process is released. The statement

```
WAIT <event-variable>
```

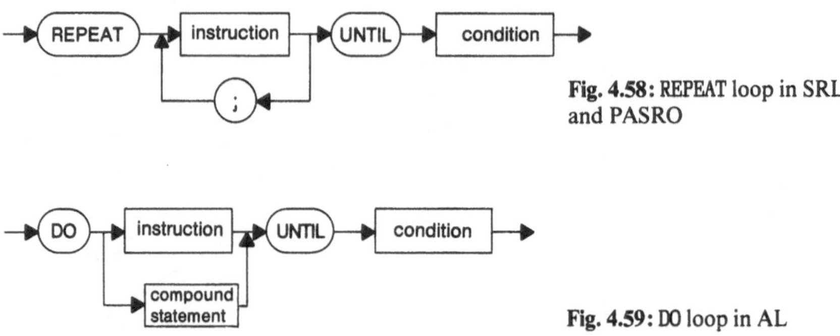

Fig. 4.58: REPEAT loop in SRL and PASRO

Fig. 4.59: DO loop in AL

decrements the value and the process within which the instruction is contained is stopped; it waits. However, if the semaphore has a positive sign, the process continues. Thus, two or more processes may be synchronized.

HELP only contains Boolean semaphores. The instructions have a similar form to SRL or AL:

```
SIGNAL (<n>)
WAIT (<n>)
```

where <n> is the number of the semaphore. The number of semaphores is implementation-dependent.

WAIT halts until the specified semaphore is set by SIGNAL and then resets it. Should the semaphore be set already, the behavior of WAIT is not defined (presumably there is no waiting and the semaphore is reset).

The function

```
TEST (<n>)
```

examines the value of a semaphore. It delivers TRUE if it is set.

In SIGLA the statement

```
ES/<variable>;
```

may set a counter, parameter or indirectly addressed counter. When the command

```
EW/<variable>,...,<variable>;
```

is reached, the program stops until all the specified variables are set.

PASRO, AML, VAL and ROBEX do not offer any language elements for semaphore synchronization. If, nevertheless, one wanted to synchronize several programs, one may succeed by using a trick and the instructions for binary signals which are intended for external systems. If the corresponding input and output signals are connected, the instructions WAIT and SIGNAL may be used in VAL, and SWITCH and WAIT in ROBEX.

The dynamic behaviour of synchronized processes has to be analyzed very carefully if more than two processes are involved or the programs themselves are complicated. There may be situations in which a **deadlock** may block the whole system. Such a case is illustrated in program section 4.17 in SRL. If the robot of process 2

```
Process 1:                Process 2 :
----------                -----------

        .                    .
        .                    .
        .                 ok := false;
WAIT gripped;             SYNMOVE arm TO grip_position;
        .                 CLOSE hand TO 5 UPTO MAXFORCE;
        .                 INPUT (force_sensor, force_value);
        .                 IF force_value > 200 THEN
        .                   SIGNAL gripped;
SIGNAL continue;            ok := true;
        .                 END_IF;
        .                    .
        .                    .
        .                 WAIT continue;
        .                 IF NOT ok THEN
        .                   (* error handling *)
        .                    .
        .                    .
        .                    .
        .                   SIGNAL gripped;
        .                    .
        .                    .
        .                    .
        .                 END_IF (* error handling *);
```

Program section 4.17: Example for a deadlock situation in SRL

does not grip anything, the system halts because the WAIT instruction has mistakenly been put before rather than after the error situation handling.

4.7.4 Wait Instructions

All robot languages contain instructions which halt the system until a particular condition occurs. These conditions may be distinguished as:

- Timing conditions
- Terminal response
- Pressing of special keys
- External signal events (see also Sect. 4.5)

Table 4.2 contains a summary. The FREEZE routine of AML can prohibit all robot motions, and the SHUTDOWN instruction stops all current robot movements, turns off the hydraulics and enters the idle mode.

Table 4.2: Wait instructions

Wait for	SRL	PASRO	AL	AML	HELP	VAL	SIGLA	ROBEX
Timing condition	SUSPEND <r> MS	./.	PAUSE <s> in seconds	DELAY (<r>) in 20 ms	DELAY (<s>) in 20 ms	DELAY <r> in seconds	WA/<A> in 20 ms	DELAY/<n> in seconds
Terminal response answer	user programmable	user programmable	PROMPT (<print>) P	BREAK (<text>) any key	./.	PAUSE <text> PROCEDD	./.	./.
Activate special keys	ALWAYS WHEN INTERRUPT <name> (programmed)	./.	./.	FREEZE SHUTDOWN	HOLD MAYRUN	./.	HL	CSTOP
External signal	ALWAYS WHEN INTERRUPT <name>	./.	WAIT <e>	./.	HIGH (<k>, <s>), <s> in 20 ms	WAIT <k>	./.	WAIT/ EVENT, <k>

<k> is a channel number.
<s> is a scalar.
<r> is a real.
<a> is a counter, parameter or indirectly addressed counter.
<print> is a <printlist> of AL.
<e> is a predefined event variable.

In HELP, the instruction HIGH waits for <s> `* 20 ms` for the arrival of a rising edge on channel <k>. If no time is specified, the wait time is set to be 30 s. Should no signal change occur during that time, a `HOLD` is executed and the program waits until the key "RUN" is pressed. Also, a message is sent to the terminal if it is connected.

4.8 System Switches and Status Report

Some programming systems allow the setting or resetting of **system switches** by an instruction in a user program to activate or deactivate special system functions. This may be useful because of safety or time-saving considerations. Instructions for robot *calibration* and the definition of new origins for all axes are particularly important.

There is a system switch present in AML. It determines the specification of robot motions in either inches (default) or mm. The routines `FREEZE` and `SHUTDOWN` may put the system into a mode in which no robot motions are allowed or in which even the hydraulics are switched off.

The AML user can interrogate the status of the system through many routines. They include

QCHAN	attributes of the device opened in the specified channel
QGOAL	current position as generated by the control algorithm
QIODEF	attributes of the specified sensor
QMANIP	various robot (manipulator) state indicators
QMONDEF	attributes of the specified monitor
QMONITOR	informs about whether or not the monitor has been triggered
QPOSITION	actual robot position and orientation (as read by the the servo feedback)
QSTORAGE	available memory
QVOLS	names of all available volumes

Thus the user can clarify the current situation of the robot, sensors and other devices at any place in the program. This enables a proper reaction to any external environment.

VAL contains the instructions ENABLE and DISABLE for setting or resetting system switches. The particular switches are

CP	Movements with intermediate frames (Sect. 4.2)
CRT	For terminals with a RUBOUT key
EHAND	System contains an electric gripper (Sects. 4.3.1 and 4.3.2) or a pneumatic one, respectively
MESSAGES	Control of terminal printout with TYPE and TYPEI instructions of Sect. 3.4
SRV.ERR	Enable/disable automatic monitoring of special hardware errors
VISION	Initialize the vision system and establish communications

VAL:

```
ENABLE EHAND
OPEN 50
```

The system is informed about the control of an electric gripper, and the OPEN instruction causes the gripper to separate its jaws by 50 mm.

4.9 Treatment of Exceptional Situations

Exceptional situations are defined here as interrupts in the normal program flow by **exception conditions.** Such a situation may arise from data-dependent errors like a division by zero or a number which is too small, or by hardware or time-out conditions like wrongly adjusted data rates or conflicts in synchronizations. Conventionally, the run-time system reacts to such exceptions *automatically* by removing the causes or stopping the program. The programmer himself usually has no or only limited influence upon *system reactions.* Recently designed languages - particularly in the area of process data manipulation - contain constructs to deal with exception situations, e.g., languages like Industrial-Real-Time BASIC, PL/1, MicroPower/

Pascal or ADA. Currently, no robot languages are known which contain *robot-specific elements for exceptions.* A run-time error must not halt or abort the user program, particularly in cyclic manufacturing tasks. First, a point for normal program continuation should be found, or the program should come to a defined halt.

There are two kinds of *exceptions*:

1. Conditions given by the system which are well defined and specified by the programmer by fixed names or characteristics.
2. The programmer defines the conditions himself by formulating logical expressions like DURATION > 5.

AML conditions can be reported by the system (case 1) and can be reacted upon by a user program through a handler. A specification of exceptions by the programmer in case 2 is possible in SRL in general, and in AL and HELP, limited to motion instructions (Sects. 4.2.5.1 and 4.2.7).

The *handler*, i.e., the section of program executed after the exception condition has occured, may consist of one instruction, a block or a special subroutine. Handlers may be assigned to exception conditions dynamically during the program run, or statically during the definition of the handler.

Usually, the occurrence of an exception is registered by the system when it happens. Reading past the end-of-tape flag for example causes the setting of the exception condition ENDOFTAPE. An error handler, which can react to exceptions or program errors, can be declared in AML by the ERRTRAP instruction. The user can display the error number and data as well as react.

```
AML:  TRAP:  SUBR (NUM,DATA);
               DISPLAY ('ERRORNUMBER=', NUM, EOL);
               IF ?DATA THEN
                 DISPLAY ('ERRORDATA=', DATA, EOL);
               BRANCH (RETRY);
             END;
      ERRTRAP('TRAP');
```

The error handler TRAP displays the number and the error information, if there is any data, and then branches to the label RETRY.

New exceptions specifically for robotic systems may be defined apart from those introduced so far. An example for this is the GRASP instruction in VAL (Sect. 4.3.3). The exception here is "part not grasped", and the program continues at a given label. The exceptional situation is recognized in this case by position sensors measuring the opening of the gripper after it has been closed. It is also possible to use a vision system to monitor the position of the parts in the gripper. Exceptional situations in robot programming are as follows:

- Robot switched off
- Part to be gripped not present or wrongly positioned
- Programmed robot move may lead to a collision or the endpoint of the motion is outside workspace of the robot

- An attempt made to pick up a part which is either too heavy or fixed to the work surface
- Wait time for motion end or a signal has been exceeded
- The borehole cannot be found during an insertion
- Handled parts are suddenly heavier/lighter, larger/smaller (scrap)
- A magazine is full/empty
- A sudden force experienced by the effector or the robot

5 Integration of a Teach-In Procedure

Currently, most industrial robots are not programmed via textual systems but by **teach-in** procedures (Sect. 2.1). The reasons for this can be found in low complexity of tasks, no interaction with sensors, nonexistence of databases describing the environment, and the inability of programmers to describe a frame of motion abstractly, either in Cartesian or cylindrical coordinates. While it is possible to solve the former problems with the help of a robot language, the last point can at best be improved with a good simulation system, since a "weak point" of *textual programming* is the definition of frames of motion. The programmer has to measure coordinates explicitly, since humans have a very limited ability to estimate positions visually and especially to program them exactly according to an impression of space. Hence nearly all textual programming systems use a **teach-in procedure** for this particular part of frame definitions.

The teach-in procedure defines *motion points* - or *frames* - by positioning the robot or, to be more precise, robot-mounted effector with its tool center point TCP (Sect. 2.3.2), the effector taking up a certain position and orientation. For this purpose, the programmer has to direct the industrial robot in its motions. This is done by pressing specific keys, and as long as a key is depressed the robot executes a motion, which is shown as a symbol on that particular key. The programmer observes the robot and through continuous visual checking determines position and orientation. It is interesting to note in this context that the symbols are line drawings which represent movements of the robot in space (for example, a semicircle with an arrowhead signifies a turning motion in that particular direction). The symbol suits the user's spatial impressions and hence is more striking, particularly since information can be absorbed in a very short time by humans. This fact also highlights the limitations of explicit motion programming by human beings. Advantages of teach-in are constant checking and instant correction of frame definitions by the programmer. This form of **feedback** is not possible during offline programming away from robots, and has only limited use during simulation. An additional disadvantage is the dead time of robots during manufacturing. Hence the online part of the programming is limited to frame definitions, which are tedious to program in textual systems, and the remainder is programmed *offline*, without the robot. The *total programming procedure* consists of two parts:

1. During the **offline part** of programming, the operator constructs a **skeleton program** containing all instructions (including the ones for motion) and explicitly given data (constants), but excluding frame coordinates, which cannot be given textually.

2. The operator defines the missing frames during the **online part** of the system (the **interactive component**) with the help of the robot.

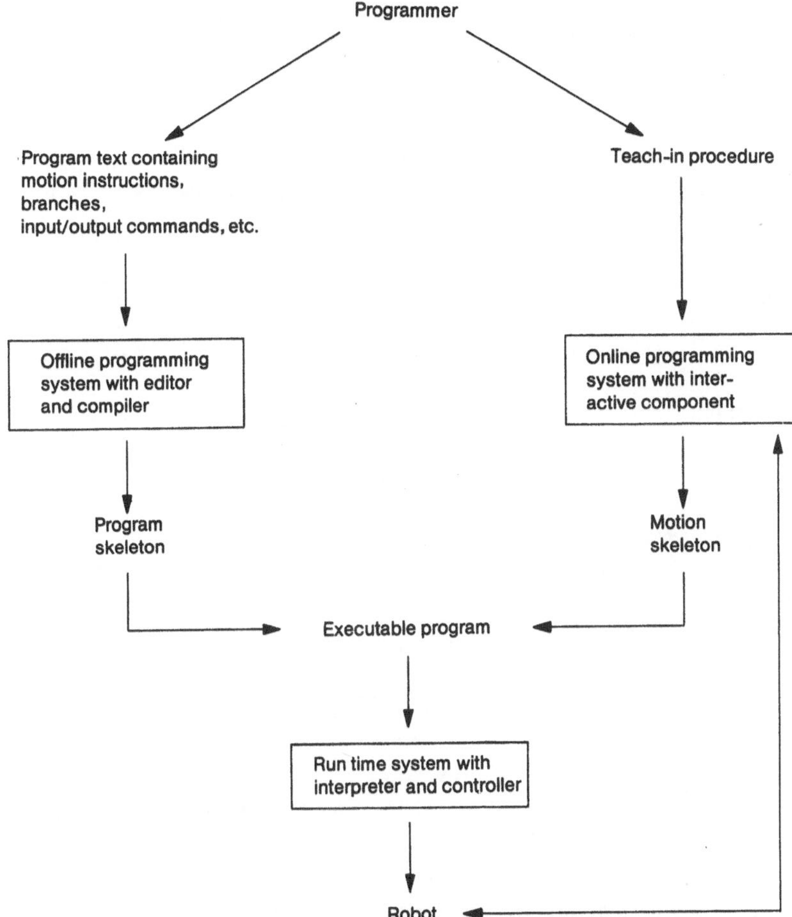

Fig. 5.1: Combined programming of industrial robots via online and offline programming

The **run-time system**, which realizes the program execution, constructs an **executable program** from the frames and the "skeleton" (cf. Fig. 5.1). *Assignment of frame coordinate values* to motion instructions in the program text can be carried out by either of two principles:

1. Each undefined frame or corresponding move command is *marked in the program code*, and storage space is reserved for coordinate values. During the first program execution, the run-time system halts the program at the marked places and outputs a message. Now the programmer has to define the frame via the *teach-in procedure.* Once coordinate values have been confirmed by the operator, they are stored and program execution continues (cf. Fig. 5.2). The disadvantage of this method is the necessity of executing all commands with teach-in markings during the first run, which is difficult to realize with branching and conditional sensory subroutines. For example, if the condition in an `IF <logical condition> THEN statement` was false, but the THEN part contained a frame defini-

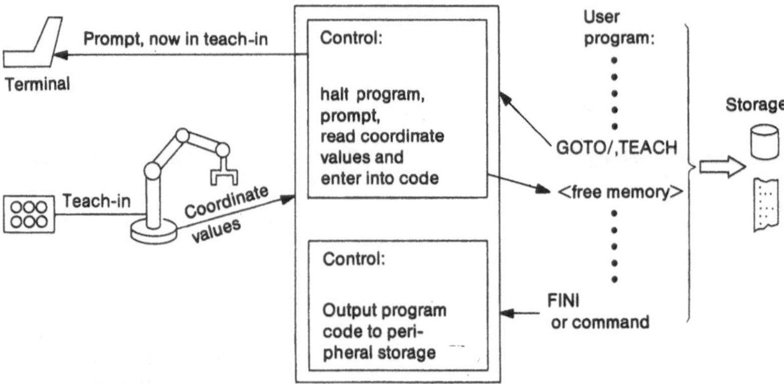

Fig. 5.2: Integration of the teach-in procedure by labelled instructions in ROBEX

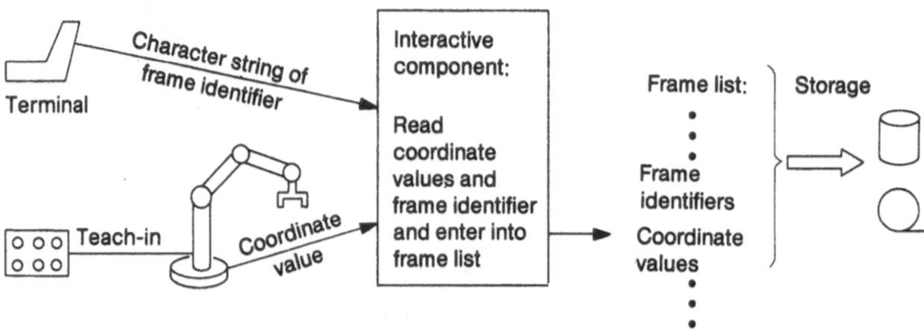

Fig. 5.3: Integration of the teach-in procedure by frame lists in AL and VAL

tion, then no teaching of that position would be done with teach-in procedures. However, when the logical condition is fulfilled later on by variables or sensors, for example the 40th program cycle, the run-time system or the robot controller will stop execution right in the middle of the production run.

Another approach to that way of teach-in integration is not to execute the program but to search for the undefined frames. This will result in a sequence of motion points to be taught, which will differ from that during normal program execution. Especially with assembly applications this may be very hard to do, if not impossible.

Furthermore, all the program code has to be stored again on peripheral memory, unless the teach-in procedure is repeated after each reloading of move instructions. A feature in favor of this method is the lack of overheads during normal program execution.

2. Definitions of frames take place *completely independently* from offline programming and also run-time systems. Values of frames of motions, which were entered during the teach-in phase, are entered into a **list of frames** with specified frame names. During program execution, the run-time system or interpreter looks up the values from this list for all move commands without values. The

frames are identified by their names (Fig. 5.3). The advantage of this method is the complete independence from program assembly (teach-in can be done before or after), as well as an easy change in already defined frames and a clear interface to CAD data. A disadvantage is the slightly higher overheads during program execution, since values have to be looked up in the list after every program start and entered into frame variables.

Flexible and powerful data concepts in SRL and PASRO allow an easy extension of the frame list by the user. Thus it is possible to define a frame file (which is predeclared in PASRO):

```
TYPE
  gripstatustyp = (gopen, gclose);
  ffilename     = PACKED ARRAY [1..ffmax] OF CHAR;
  framename     = PACKED ARRAY [1..ffmax] OF CHAR;
  ffilerecord   = RECORD
                    name        : framename;
                    value       : frame;
                    gripvalue   : REAL;
                    grips       : gripstatustyp
                  END;
  ffiltype      = FILE OF ffilerecord;
```

The programmer can now use a program for the teach-in procedure which has been written in SRL or PASRO itself. (The system program TEACHIN for teaching by showing is part of the system kit in PASRO.) If an extension for the definition of robot angles in the frame file is needed at a later date, the programmer can change the definition of the frame file above by simply adding the type

```
robotangletyp = ARRAY [0..5] OF REAL;
```

and into `ffilerecord` the variable

```
robotangles : robotangletyp;
```

Robot angles can now be read from and written into the frame file during the teach-in program.

Such flexibility in defining and structuring data allows the extension of frame files to include world models. The programmer or system programmer can describe the attributes of objects which are present in the physical world - such as geometry, color or weight - with the help of structured data-type definitions. Thus SRL and PASRO can be extended easily into an implicit programming system.

There are five file-handling procedures in PASRO, of which three can be used to integrate frames defined by the teach-in into user written programs. These are:

`openframefile (filename: ffilename; VAR status: INTEGER);`

Opens the existing frame file called `framefile` for the purpose of reading or writing and returns a status value.

```
closeframefile;
```

Closes the frame file.

```
frameinitalize (VAR fout:   frame;
                VAR gvout:  REAL;
                VAR gsout:  gripstatustyp;
                    fname:  framename;
                VAR status: INTEGER);
```

The frame called `fname` is looked up in the previously opened frame file and its value assigned to the parameters `fout` (which stands for the user program variable), `gvout` and `gsout`.

The following procedures are useful for creating or changing frame files. This task is also supported by the system program TEACHIN, or the user may want to write his own tools for manipulating frame files.

```
createframefile (filename:ffilename; VAR status:INTEGER);
```

Creates a frame file called `filename` and returns the resulting status. If the operation is successful, the file can be read from or written to.

```
framewrite (    fname:  framename;
                fin:    frame;
                gvin:   REAL;
                gsin:   gripstatustyp;
            VAR status: INTEGER);
```

The frame `fin`, the gripping value `gvin` and the grip status `gsin` are written to the previously opened frame file together with the frame name `fname`.

SRL, PASRO, the AL implementation of the University of Karlsruhe and VAL integrate a *teach-in procedure via a frame list.* While AL stores only Cartesian coordinates, VAL offers a distinction between frames with Cartesian coordinates and precision frames with robot coordinates. However, SRL, PASRO and AL allow definition of frames with explicit values during program execution, so that they no longer have to be defined with the teach-in procedure. The list of frames in VAL is not stored in a separate area as in the other languages, but is embedded as a linked list in the code of move commands, which could have been entered in between frame definitions.

In addition, VAL provides a very convenient method which allows motion programming via teach-in, resulting nevertheless in *readable and correctable* text for individual move commands. For this purpose, the *editor* contains a special mode invoked by the T or TS command, followed by a frame name, e.g., `store`. The programmer need not input any text from now on, but moves the robot through a succession of frames via the teach-in procedure. The "RECORD" button on the teach box is pressed every time a desired position is reached, and VAL enters a move command to frames `store1`, `store2`, etc., including current gripper position, into the program. Upon input of carriage return, this mode is exited and the program can be edited normally. The T command causes *joint interpolation* moves (MOVET), and the TS command moves with *linear Cartesian interpolation* (compare Sect. 4.2.4).

VAL:	`T store`	(Start of teach mode during edit)
		(Execute "teach-in" of first frame) (Press RECORD button)
	`MOVET store1, 0.0`	(Line entered, gripper was closed)
		(Execute "teach-in" of second frame) (Press RECORD button)
	`MOVET store2, 40.7`	(Line entered, gripper was opened at 40.7 mm)

This way, a complete sequence of motion can be programmed without explicit text input.

AML/E offers a convenient procedure to teach frames using an IBM PC, which is connected to the robot control. The system guides the user by integrating textual descriptions of frame values, as well as their names, into the source code. As the AML/E compiler is able to include files, an integration of textual frame lists can be performed.

HELP implements a modified method with flagged or special commands for teach-in integration. The programmer writes a so-called *"Self-Teach" program* where the MANUAL command enables teach-in. Coordinates are read in by the program and used for assignment of variables. Instead of storing variables themselves, instructions are output to a peripheral store (The American AL solves the problem in a similar manner). Commands described in Sect. 3.1.5.3 are used to write to a file.

HELP:

```
MANUAL(1);              ! Enable teach-in for arm 1
CREATE(<teachprog>);    ! Create a file called
                        ! "teachprog"
PRINT('teach frame on'); ! Message to terminal
COORD(1);               ! Wait for key press on joy-
                        ! stick and reading of coordi-
                        ! nate values of actual posi-
                        ! tion.
         ! Write commands  to file "teachprog":
RECORD(#0,'strt:= ', AX(1),';');
RECORD(#0,'strty:= ',AY(1),';');
 ...
RECORD(#0,'MOVE(1,#1,strtx,#2,strty,
   ... ,#6,strtyw)');! further instructions and
                        ! teach-in executions
 ...
RECORD(#0,'STOP');      ! Indication of file end
```

```
CLOSE                    ! Close file "teachprog"
STOP                     ! Program end
```

After enabling teach-in for robot arm 1, a file for instructions is opened, a prompt issued to the programmer and the robot moved to frame `strt` via teach-in. This may be done with the help of a joystick. The COORD system routine holds program execution for that period until a specific key on the joystick is pressed and coordinate values are read in. These values are then substituted into the following instructions and stored in the file `teachprog`. This procedure is repeatable for all other frames.

In SIGLA a special system mode for the integration of teach-in is provided.

ROBEX integrates teach-in through the command TEACH (Fig. 5.2). The complete program code is output to an external store after the first program run.

6 Subroutines, Procedures and Functions

The general concepts of subroutines, procedures and functions as well as recursions have been explained in Sects. 2.2.4 and 2.2.5, the respective language elements are introduced in this chapter, and some general problems of their practical usage will be discussed.

Not only does a well-structured program avoid the use of jump instructions, it also distinguishes itself by segmenting programming tasks into subtasks and their refinement into the simplest elements. These elementary tasks, or at least some raw subtasks, should be programmed, so that any changes "within" them do not influence the environment. In order to exert an influence, *interfaces* are defined specifying the data area which subtasks may access or alter. Procedures and functions - if they are provided by a language - are suitable linguistic means of interface definitions because of their parameter passing mechanisms. The task itself is described in a program section in the *procedure body.* Obviously, a procedure may be divided further into local tasks which are defined within the procedure. Such local tasks are not visible to the outside and thus cannot be part of another task. Those elementary tasks which are common to several tasks have to be defined globally.

A clean interface design in programs is extremely important with respect to **program security**, and every high-level language should provide the following support for this:

- Specification of parameters in procedure headers
- **Call-by-value** passing mechanism for input parameters
- **Call-by-reference** passing for parameters changed by procedures (to make side effects apparent)
- Exit from a procedure only at a single well-defined place

Out of the languages considered here, only SRL, PASRO and AL fulfil these requirements completely, while AML fulfils not all aspects.

If all incoming and outgoing data is treated as described above, a good interface can be achieved, in contrast to one which exploits **side effects**. Side effects are direct changes (that is, without insertion of parameters) of global data in the procedure body. Programs containing side effects reveal a bad programming style or an uncertain structuring of the tasks in question and are all too often the cause for logical errors. Side effects are justified only in exceptional circumstances; for example, in routines which manipulate data exclusively. These are known as memory functions.

Robot programming does not only change internal data, its main objective is to move real objects. Thus motion instructions in procedures may also lead to side effects, because an arm may be left in a different state. Hence procedure environments should be made aware of this new state via the interface. The state should consist of at least position and orientation components, as well as a possible gripper or tool state.

6.1 Subroutines

It can be seen in Table 3.3 that all languages apart from ROBEX contain possibilities to define subroutines. First, consider VAL and HELP:

The VAL instruction illustrated in Fig. 6.1 executes a program starting with the first statement, until a RETURN command (Fig. 6.2) is encountered.

If no <index> was specified or if it is less than or equal to zero, the program continues with the instruction following the subroutine call. A positive <index> is added to the address of the next instruction. Thus an <index> of 2 carries on at the third statement after the GOSUB command. During the execution of the main program, a RETURN instruction is regarded as a program terminator. Subroutines may also be called with the instructions REACT and REACTI, as already mentioned in Sect. 4.2.6, and now a RETURN statement refers to the same instruction in REACTI, whereas in REACT, the <index> starts at the next statement.

Up to ten nested levels of subroutines may be called in VAL.

HELP permits a similar construction to that in VAL, but the RETURN command refers always to the succeeding statement in the main section, and subroutines are defined by labels in the main program itself. The ways of calling subroutines are illustrated in Fig. 6.3.

The second form, which represents a case distributor, behaves exactly the same as the computed GOTO instruction discussed in Sect. 4.7.1.2. Remarks made in Sect. 4.7.1 regarding jump targets are also valid for subroutine jumps.

The description of the language mentions neither nesting levels nor possibilities

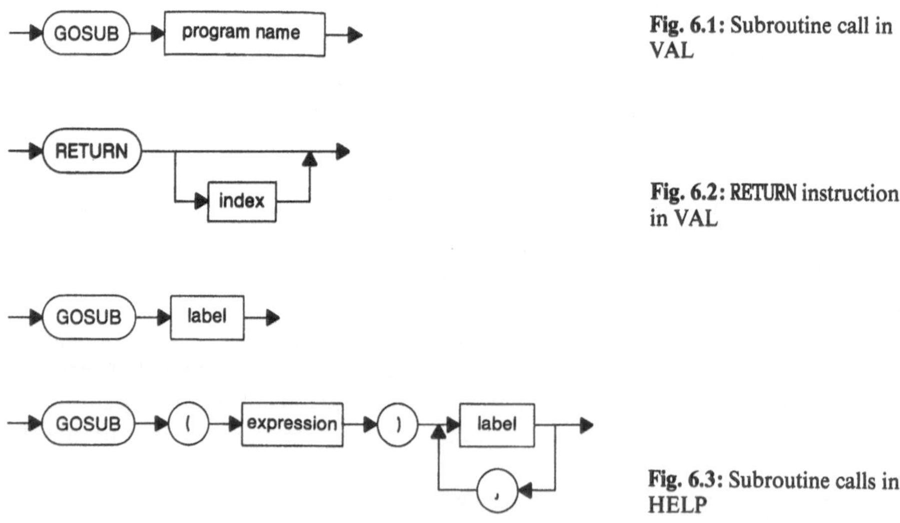

Fig. 6.1: Subroutine call in VAL

Fig. 6.2: RETURN instruction in VAL

Fig. 6.3: Subroutine calls in HELP

```
L10:        .
            .
            .
        RETURN;
L20:        .
            .
            .
        IF a < b THEN GOTO L30 END;
        RETURN;
L30:        .
            .
            .
        RETURN;
```

Program section 6.1: Several subroutines in HELP

```
L10:        .
            .
            .
        RETURN;
L20:        .
            .
            .
        IF (a>b) OR (a=b) THEN RETURN END;
            .
            .
            .
        RETURN;
```

Program section 6.2: Two subroutines in HELP

for recursion. This is probably implementation dependent. The behavior of main programs executing a RETURN command is also open. It is quite possible for a main program section to come across a RETURN of a subroutine, if it branched into the subroutine via a jump instruction.

Subroutines in HELP are not very obvious within the program text, particularly if ordinary jumps and jump targets are mixed with subroutine labels. This nuisance may be remedied to a certain extent by comprehensive comments. However, it is desirable for the language itself to contribute, for instance starting a subroutine not just with a label, but with the declaration

```
SUBROUTINE <label>;
```

This difficulty will be illustrated in the following three program sections 6.1 to 6.3. How many subroutines have been defined in program section 6.1 two or three? The section starting at label L30 up to the last RETURN may be regarded as a subroutine in its own right. If the programmer had intended only two subroutines, it would have been better to write program section 6.2.

However, if he wanted three subroutines, the structure in program section 6.3 would have been more suitable.

```
L10:       .
           .
           .
          RETURN;
L20:       .
           .
           .
          IF a < b THEN GOSUB L30 END;
          RETURN;
L30:       .
           .
           .
          RETURN;
```

Program section 6.3: Three subroutines in HELP

```
L10:       .
           .
           .
          IF a < b THEN GOTO L11 END;
L12:       .
           .
           .
L11:       .
           .
          IF b > 5 THEN GOTO L12 END;
           .
           .
           .
          RETURN;
```

Program section 6.4: Subroutine with several entries in HELP

This problem could also be solved if the subroutine exit had been permitted only at a single place.

HELP allows more than one entry point into a subroutine, possibly providing further problems, as shown in program section 6.4. How many subroutines are in this segment? This question may only be answered by analyzing the remainder of the text to see if the labels L11 and L12 are used in a GOSUB statement. It would be useful, however, if one could recognize this directly from the text.

The conclusion may be drawn, that the concept of subroutines in HELP is not consistent with a structured programming approach and that it tempts programmers into writing unreadable and thus error prone code. The concept in VAL is not structured, either. However, since it distinguishes between jump targets and subroutine names and allows subroutine entry only at one point, it forces a clearer programming style.

The other languages - except ROBEX - offer procedures which include the subroutine concept as an exception, that is, a procedure without local data and parameters can be regarded as a subroutine.

6.2 Procedures

In SRL, PASRO, AL, AML and SIGLA, procedure declarations and calls have the forms explained in the following sections.

The procedure definition of SRL and PASRO is illustrated in Fig. 6.4. Like any other SRL or PASRO program, the <block> consists of a declaration part and a set of instructions, the **procedure body**, which is bracketed by BEGIN and END in PASRO and PASCAL or BEGIN_PROCEDURE and END_PROCEDURE in SRL. Labels (only in PASRO), types, variables, procedures and functions which are local to this procedure may be defined in the declaration part (see also Sect. 4.6). Parameters containing the addition VAR are passed by **call-by-reference**, otherwise values are transmitted. Sections (only in SRL), procedures and functions may also be passed as parameters.

There is a FORWARD declaration, since all procedures have to be defined before they can be used. This enables constructions like procedure "P" calls "Q" and "Q" calls "P" (indirect recursion). The block in this declaration is replaced by the symbol FORWARD, and the procedure itself is declared at a later stage, without a repetition of the parameter list.

The procedure finishes when the end of the block is reached. Thus the procedure concept of SRL and PASRO or PASCAL fulfils the requirements stated in the beginning of this chapter. The method of calling a procedure is as follows. The correspondence of data types and number of parameters is checked against those of the actual parameters, and it is confirmed that parameters defined as call-by-reference actually pass variables, not values.

There is one major difference between PASRO and SRL: Local variables within a procedure can be declared in two ways in SRL. If is it done in a VAR declaration, they are treated the same as in PASRO, that is, they live as long as the procedure is

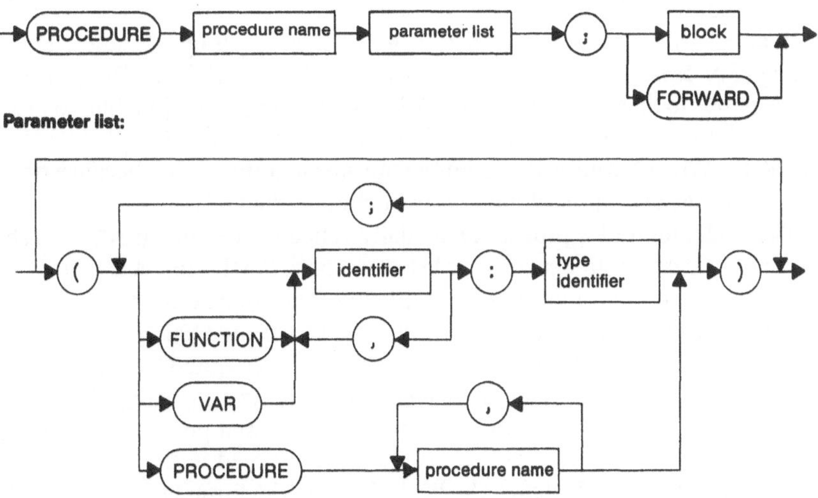

Fig. 6.4: Procedure declarations and calls in SRL and PASRO

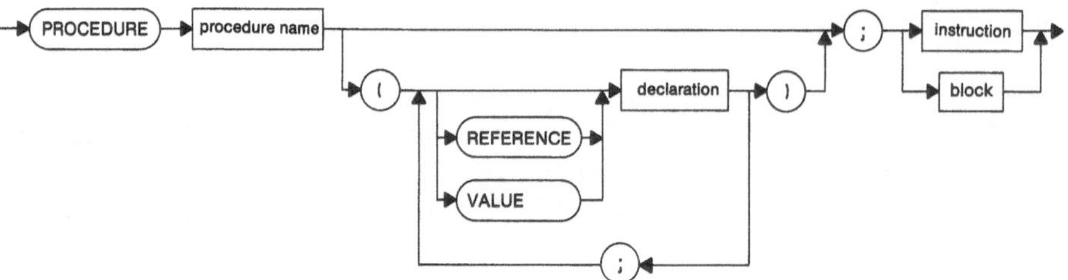

Fig. 6.5: Procedure declaration in AL

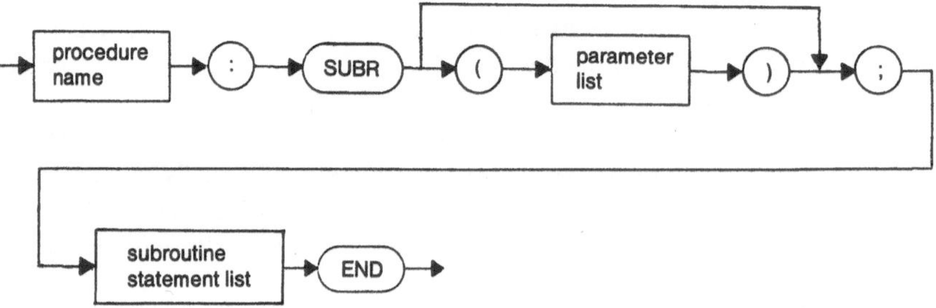

Fig. 6.6: Procedure declaration in AML

executing and any occupied memory space is released upon procedure exit. If a procedure call should work with the value of the variable, which has been evaluated in the previous call, the variable must be declared in an OWN_VAR declaration. The life of these variables does not depend on the procedure execution, however, they are only accessible by the procedure. This separates them from global variables and the appropriate use thereof increases program security.

The procedure declaration in AL is similar to that in PASRO and PASCAL (Fig. 6.5). Here `<declaration>` contains the usual definition of variables and arrays, and the procedure body may consist of a block or a single instruction. In contrast to SRL and PASRO, the standard parameter passing is call-by-reference. This may be changed or emphasized by the keywords VALUE or REFERENCE. The procedure call corresponds to that in PASRO and PASCAL, with the exception that no procedures can be passed as parameters. FORWARD declarations as in SRL and PASRO are not necessary. AL, too, fulfils the requirements given at the outset, because procedures finish with the execution of the last instruction in the body.

The declaration of a procedure, actually called subroutine in AML, is shown in Fig. 6.6. The subroutine statement list consists of AML language statements. and the last one is not terminated by a semicolon. The context determines whether the whole of the declaration has to be terminated by a semicolon or not.

The parameter list consists of formal parameters separated by commas. No types can be declared with formal parameters, nor is any type checking done by a procedure call. As in SRL or PASRO, parameters are passed positionally. AML knows about optional parameters, which are not actually used in the procedure body or have a default value. The latter can be declared in the formal parameter list by writing

```
<formal parameter> DEFAULT <expression>.
```

Optional parameters need not be supplied in a procedure call. They can be substituted by a null parameter. For example, the procedure A with three parameters, of which the second and the third parameter are optional, can be called in the following ways

```
A(1), A(1,) or A(1,,)
```

all of which have the same result. If only the second parameter is to be a null parameter, one would have to write

```
A(1,,3)
```

The system subroutine PARMS is useful in order to identify the actual parameters provided in a specific call. It returns an aggregate which contains the actually supplied parameters, excluding any default types. As there is no type checking, a token-type operator evaluates the type of a specific variable. Thus type checking can be programmed by the user. Another feature conflicting with the objectives for safe programming is the possibility to provide more parameters in a procedure call than are actually necessary according to the procedure declaration. However, any superfluous parameters are ignored and no error occurs.

AML offers reference and value parameters. The call-by-value is the default, as it is in SRL and PASRO. The definition of a reference parameter can be written as

```
!<parameter>
```

in the formal parameter list. If a variable is passed to the procedure, it may be changed after the call (except for AML system variables, which can only be passed by value). If a pointer value is passed, dereferencing is performed. Passed expressions are treated as call-by-value and no errors occur (!). In fact, if a pointer is passed to a formal value parameter, the pointer itself is passed.

Just as SRL, AML offers two types of local variables. Variables declared in a NEW declaration exist only during procedure execution, but those declared by a STATIC declaration can be used in any later call.

Procedures cannot be declared in SIGLA in contrast to the above mentioned languages. In SIGLA, one program may be called from another with passing of parameters. Such programs can be executed by themselves, as in VAL, and thus distinguish themselves from procedures explained above. Parameters are passed in up to sixteen memory locations identified by the character P and a subsequent number. These parameters are the only entities which are local to programs - and thus to procedures. Memory locations identified by an M and a subsequent number are global. When programs are called as procedures, the values of parameters are copied, thus they are passed by call-by-value. A call-by-reference mechanism may be simulated by skillful exploitation of the indirect addressing (program section 6.5). Hence, for a language at assembly level, SIGLA offers the remarkable luxury of having that calling mechanism to programs. The program call is shown in Fig. 6.7. Here, <fileno> is the number of the called program. Constants, parameters, and directly and indirectly addressed counters are admitted for <parameter>. Finally, here is an example for call-by-reference parameter passing in SIGLA.

```
Program 2:        SE/M1,P1
                  IC/M1,P2
                  SE/I3,M1

Main program:     SE/M10,5
                  SE/M20,17
                  EX/2,M10,M20,100
```

Program section 6.5: Passing of parameters by call-by-reference in SIGLA

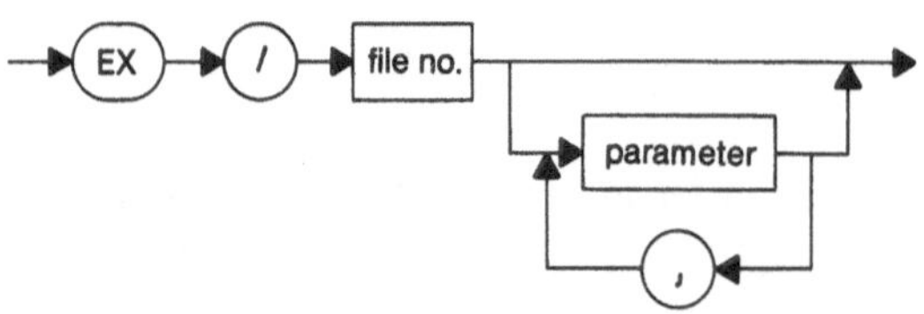

Fig. 6.7: Procedure call in SIGLA

The main program calls program 2 with the parameters
P1 = content of M10, i.e., "5",
P2 = content of M20, i.e., "17",
P3 = "100".

In program 2, P1 and P2 in M1 are added and the result is stored in the counter identified by I3. It contains the number specified in P3 at the time of the call in the main program, namely, "100". Thus the counter M100 receives the result, which is "23", and hence M100 has been passed to the subroutine as a reference variable.

6.3 Functions

Only SRL, PASRO, AL and AML contain functions. They deliver results which may be used directly in equations. This is the main difference from procedures which may deliver values (and not just a single value) as well, but only via reference parameters or side effects.

In order to extend procedures to functions, two additions have to be implemented in SRL, PASRO and AL: Firstly, the definition has to include the type of the result, and secondly, the result has to be assigned to the function.

The first demand is carried out in SRL and PASRO by writing FUNCTION instead of PROCEDURE and inserting the type of the function result after the parameter list. AL precedes the symbol PROCEDURE by the type of the result.

The result itself is passed out of the function by assigning it to the function name in SRL and PASRO, and by the RETURN instruction in AL. The RETURN instruction of AL does not terminate the execution of the function, and the value assigned to the last RETURN is the result of the function call upon exit.

All data types are permitted in AL, except EVENT, and in SRL and PASRO INTEGER, CHAR, BOOLEAN, REAL, pointers and certain user-defined linear data types are possible functions results.

```
SCALAR PROCEDURE vectorsum (VALUE VECTOR v);
  RETURN  (v.VECTOR (1, 1, 1) );
```

Program section 6.6: Function vectorsum in AL

```
FUNCTION vectorsum (v: vector): REAL;
  BEGIN (* vectorsum *)
    vectorsum := v.x + v.y + v.z;
  END (* of vectorsum *);
```

Program section 6.7: Function vectorsum in PASRO

As AML is an expression-oriented language, every procedure call evaluates a result. This will be the null aggregate <> if there is no call to the RETURN system subroutine in the body of the procedure. RETURN has a parameter, and the value of this is the result of the procedure or function call. AML's RETURN also terminates execution of the procedure.

Now the function vectorsum, which was formulated in a hypothetical language in Sect. 2.2.4.2, can be defined in PASRO and AL. This assumes the vector declarations of program section 3.1 for the PASRO case (compare with program sections 6.6 and 6.7). It is obvious that a definition of a special procedure in AL is superfluous, since the dot product is adequately represented by an operator.

The use of a function is shown in the next example:

```
basevalue <- 5 + vectorsum (targetvector);
```

where basevalue is a scalar variable (or REAL in PASRO; the assignment operator, too, would have to be changed in PASRO).

Calling functions as instructions, as in the case of procedures, is forbidden in SRL and PASRO, but not in AL and AML. Hence it would be possible to write in AL

```
vectorsum (targetvector);
```

Any results are lost, however, and thus the example above is meaningless. Nevertheless, it may sometimes be valid to execute a function without having any further interest in the result.

6.4 Recursive Procedures and Functions

Recursion has already been discussed in detail in Sect. 2.2.5 and a robot-specific example has been considered. Out of the languages investigated here, SRL, PASRO, AL and AML offer *recursive procedures and functions.* These do not need any special identification.

An AL example for administration of world models (cf. Sect. 4.1) is introduced here to show once again the usefulness of recursions in robot languages. This example shows how *hierarchical data structures of world models* may be managed efficiently and clearly by recursive procedures.

```
BEGIN "background block for example"
  SCALAR frameno;
       .
       .
  BEGIN "example"
    FRAME   ARRAY framearray [1:frameno];
    SCALAR  ARRAY affixarray [1:frameno];
    SCALAR  searchframe, affixframepointer, targetframepointer,
            searchframepointer, anchor, failureindicator;

    {***************************  search  ***************************}
    PROCEDURE search ( VALUE SCALAR pointer);
    BEGIN "search"
      IF affixarray [pointer,1] = searchframe THEN  { Frame found ?     }
      BEGIN "found"
        IF searchframepointer > 0 THEN  { Frame duplicated in tree      }
                                        { structure ?                   }
          failureindicator <- 1;        { Set it if duplicated          }
        searchframepointer <- pointer;  { Set pointer to found frame    }
      END "found";
      IF affixarray[pointer,2] > 0 THEN { Branch end downwards ?        }
        search (affixarray[pointer,2]); { Search across                 }
    END "search";

    {***********************  affixment  ****************************}
    PROCEDURE affixment (VALUE SCALAR affixframe, targetframe);
    BEGIN "affixment"
    SCALAR searchpointer;
    searchpointer <- anchor;          { anchor starts at linked list }
    affixframepointer <- 0;
    DO        { Search loop whether affixframe is attached to WORLD  }
      IF affixarray[searchpointer,1] = affixarray THEN
        affixframepointer <- searchpointer;   { Frame found          }
      IF affixarray[searchpointer,3] <> 0 THEN { Another frame here? }
        searchpointer <- affixarray[searchpointer,3] { Attach next    }
                                      { frame appended to WORLD      }
      ELSE
        searchpointer <- 0;           { Terminate search             }
    UNTIL searchpointer = 0 OR affixframepointer <> 0;
    searchframe <- targetframe;
    search (anchor);                  { Look for reference frame     }
    targetframepointer <- searchframepointer; { Label reference frame}
    IF affixframepointer = 0 THEN
    BEGIN "frame to be appended not attached to WOLRD"
      searchframe <- affixframe;
      searchframepointer <- 0;
      search (anchor);        { Test whether frame is already affixed}
```

```
      IF searchframepointer > 0 THEN
        failureindicator <- 2 { Indicate frame already affixed        }
      ELSE
        affix;                  { Procedure not defined here, it links }
                                { frame to list                        }
    END "frame to be appended not attached to WORLD"
    ELSE
    BEGIN "frame to be appended is attached to WORLD"
      searchframe <- targetframe;
      searchframepointer <- 0;
      search (affixframepointer);       { Check whether reference frame}
                                        { already affixed - i.e. for a }
                                        { vicious circle               }
      IF searchframepointer > 0 THEN
        failureindicator <- 3 { Error: vicious circle present         }
      ELSE
        affix;                  { Procedure not defined here, it links }
                                { frame to list                        }
    END "frame to be appended is attached to WORLD";
  END "affixment";

  {************************** Example *****************************}
  anchor <- 1;                  { Initialise list start                }
  FOR i <- 1 STEP 1 UNTIL frameno DO
  BEGIN "initialise affixarray"
    affixarray[i,1] <- 0;
    affixarray[i,2] <- 0;
    affixarray[i,3] <- 0;
  END "initialising affixarray";
        .
        .
        .
  affixment (14,6);             { Affix frame 14 to frame 6            }
  IF failureindicator > 0 THEN
  BEGIN "failure treatment"
        .
        .
        .
  END "failure treatment";
        .
        .
        .
 END "example";
END "background block for example"
```

Program section 6.8: Administration of affixments in AL with the recursive procedure search (AL)

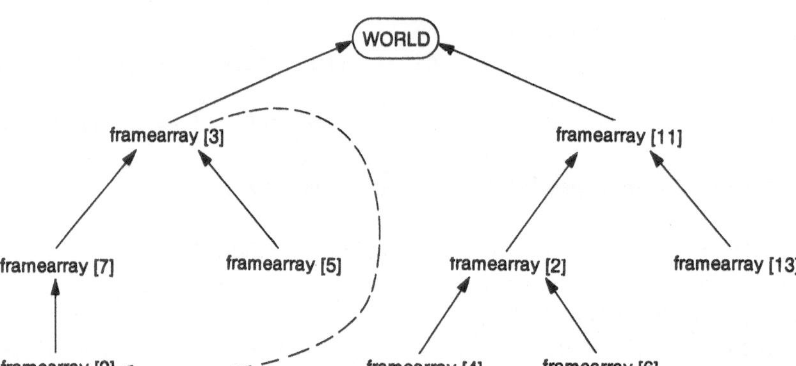

Fig. 6.8: Tree structure of affixments

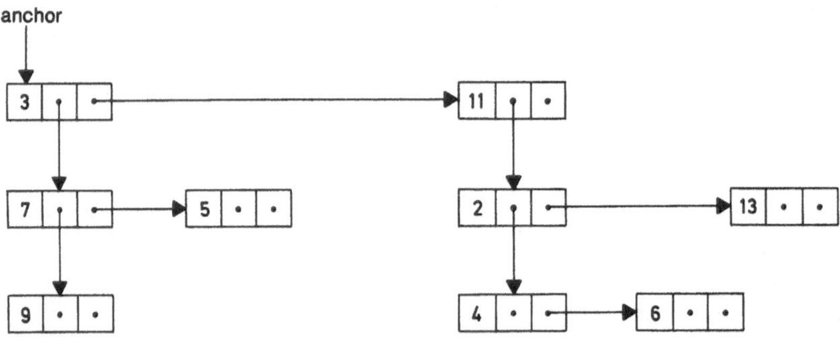

Fig. 6.9: Representation of affixment structures as a linked list

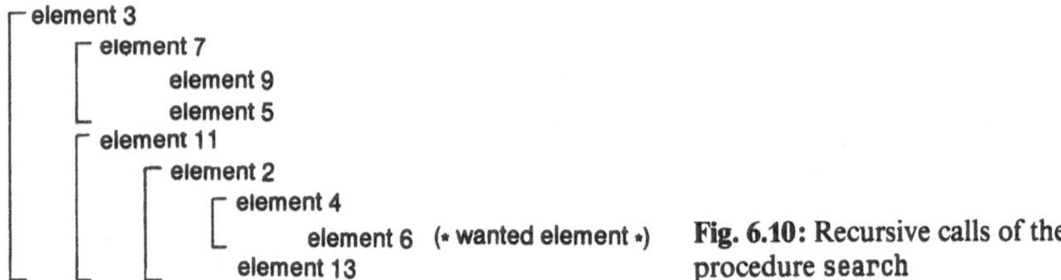

Fig. 6.10: Recursive calls of the procedure `search`

A programmer has generated a `framearray` for all his frames; its number is specified by `frameno`. At the same time he generates an integer or scalar array `affixarray`, which represents all his affixments. It, too, consists of `frameno` units, each of which in turn consists of three components:

1. A value, which is the index into `framearray`
2. A pointer to the first frame which is appended
3. A pointer to the next frame, which has been attached to the same frame indicated by this particular unit

Figure 6.9 shows the linked list of the units, which reflects the tree structure of the affixments in Fig. 6.8. However, note that the "appending" relation has been inverted, hence the link follows from reference frames to concatenated frames.

The procedure of interest here is called `search`. Its parameter is a pointer to units in the list `affixarray`. It searches `affixarray` for the presence of the given `searchframe`. If it is found, the `searchframepointer` is set to point to this frame. In order to cover all branches in the linked list, a recursive call of `search` is necessary from within the procedure itself. Otherwise, an extravagant administration for the return points would be required, such that the next downward link may be searched after the horizontal one has finished. First, the procedure `search` tests whether the wanted frame is the list element addressed by `pointer`. If this is the case, a further test is undertaken to see whether the same element has been entered in the list before and `failureindicator` is set to 1, correspondingly. Then another recursive call of the procedure `search` is executed following the link downwards. As soon as the end of the pointer chain is reached, linkages across it are investigated. Any further downwards references are processed first, and so on. For example, if the procedure `search` is called from the procedure `affixment` with the two parameters link start pointer `anchor` and `searchframe`, which is identified by index 6, the elements are processed as illustrated in Fig. 6.10, following the structure shown in Fig. 6.9.

The procedure `search` is needed in `affixment`, which enters an affixment into `affixarray`. A check is performed at the beginning of the procedure `affixment` to see whether the frame to be appended already exists as a reference frame with other attached frames. Such a frame may only be joined to the `WORLD` identification (Fig. 6.8) or to the horizontal linkage which starts at `anchor` (Fig. 6.9). If the frame exists, it is affixed to the reference frame with the whole of its own "subtree". This may lead to the following error: If the newly referenced frame already exists in the subtree of the appended frame, a *vicious circle* is created. The correction of an appended frame caused by changes in another frame in the vicious circle leads the interpreter into an endless loop. The dotted line in Fig. 6.8 indicates such a circle by the affixment of `framearray[3]` to `framearray[9]`. After the search for `targetframe` in the list in program section 6.8, it is checked whether the frame, which is to be affixed, already exists together with a subtree. If this is the case in which `affixmentpointer` has been set in the `WHILE` loop, `targetframe` is given as `searchframe`. Thus the procedure `search` looks through the subtree to see whether it already exists in there. This is initiated by the call

```
search (affixframepointer);
```

If the frame `targetframe` has not been entered in the list with the subtree, a vicious circle does not exist. Additionally, it is checked whether the frame to be appended appears at all in the list `affixarray`. If this is the case, `failureindicator` is set to 2, because no frame is allowed to be appended twice. The actual concatenation of frames is carried out by the procedure `affix`, which has not been shown for reasons of clarity. The affixment is then requested by the main program with the statement

```
affixment (14,6);
```

Any errors detected during entry into `affixarray` may be determined by checking `failureindicator` in the main program section.

7 Multitasking and Synchronization

Industrial applications of robots usually involve the coordination of one or more machines and/or robots. Thus the problem of timing manipulation sequences arises regardless of whether the units are programmable or not. Their programs have to be synchronized if they are programmable. Otherwise certain conditions such as "part ready" have to be communicated to the program of the serving robot or to the more global supervisory program. Hence timing problems may be reduced to mastering techniques of parallel processes, and any technological problems of suitable sensors and other signalling systems will not be considered here.

Computer science has developed a wealth of mechanisms to control parallel and asynchronous processes. **Task concepts** and their synchronization by **semaphores** (as discussed in Sect. 4.7.3) belong to those standards which have proven themselves. At this point, the danger of system deadlocks is emphasized again, as they are illustrated in program section 4.17. Such situations may result from programming as well as conceptual errors. Practical experience shows that hardly any software project is spared and that deadlocks usually appear when programs have to deal with exceptions or error situations correctly.

7.1 Parallel Blocks

Parallel sequences can be described by simple means with the help of parallel blocks. Of those languages considered here, only SRL and AL contain such a concept. Contrary to ordinary blocks, all instructions contained in parallel blocks are marked in AL by `COBEGIN` and `COEND` (hence the name *CO-block*) and in SRL by `BEGIN_PARALLEL` and `END_PARALLEL`. They are executed in parallel or quasi simultaneously. Quasi, because the execution of n parallel processes would require n processors. Usually, only one processor is available for program interpretation (under certain circumstances it may delegate partial tasks such as geometric calculations to subprocessors), and several processes are run by executing small parts sequentially during short time intervals. The processes are switched continuously, so that a true parallel execution appears externally. The problems of switching and distribution of processor time for the tasks will not be discussed any further here as it enters the realms of operating systems and various concepts are known from mainframe computers (keyword: Multi User Systems).

Synchronization between program parts which are executed in parallel is achieved by the AL instructions `SIGNAL` and `WAIT`, which have been introduced in Sect. 4.7.3. Program section 7.1 illustrates their application in the hand-over example of a box from `arm1` to `arm2`.

```
    BEGIN
      EVENT handover, gripped, released;
      FRAME box, handoverpos, grippingpos, target;

      COBEGIN
        BEGIN "arm1"
(         MOVE arm1 TO box;                (* arm1 takes box          *)
(         CENTER arm1;
(         AFFIX box TO arm1;
1                                          (* and moves it to         *)
(                                          (* handover position       *)
(         MOVE box TO handoverpos
(              WITH APPROACH = -6*cm;
(         SIGNAL handover;                 (* arm1 is ready           *)
          WAIT gripped;                    (* arm1 waits for arm2     *)
(         OPEN hand1 TO 6*cm;              (* arm1 releases box       *)
2         UNFIX box FROM arm1;
(         SIGNAL released;                 (* and flags it            *)
        END "arm1";
1
        BEGIN "arm2"
(         OPEN hand2 TO 6*cm;              (* Moves opened hand2 to   *)
A         MOVE arm2 TO grippingpos         (* gripping position       *)
(           WITH APPROACH = -6*cm;
          WAIT handover;                   (* and waits for arm1      *)
(         CENTER arm2;                     (* arm2 grips the box      *)
B         AFFIX box TO arm2;
(         SIGNAL gripped;                  (* and flags it            *)
          WAIT released;                   (* arm2 waits for arm1     *)
(         MOVE arm2 TO target              (* arm2 transports the box *)
C              WITH DEPARTURE = -6*cm;
        END "arm2";
      COEND;
    END;
```

Program section 7.1: CO-blocks in AL

Both parallel processes are divided into five parts (1, 2, A, B, C), of which only the first two (1 and A) are executed in quasi parallel. Further actions are synchronized by semaphores and are executed in the following sequence: B, 2, C. The movements of both robots have to avoid collision at the hand-over point. Figure 7.1 demonstrates this. The motion instructions have to make sure - with the help of an approach point (Sect. 4.2.4) - that both arms approach each other on an imaginary line through both effector axes. In the example, the CO-block consists of two instructions executed in parallel, the blocks arm1 and arm2. As mentioned earlier, several instructions (or blocks) for the control of further arms and other machines are also feasible.

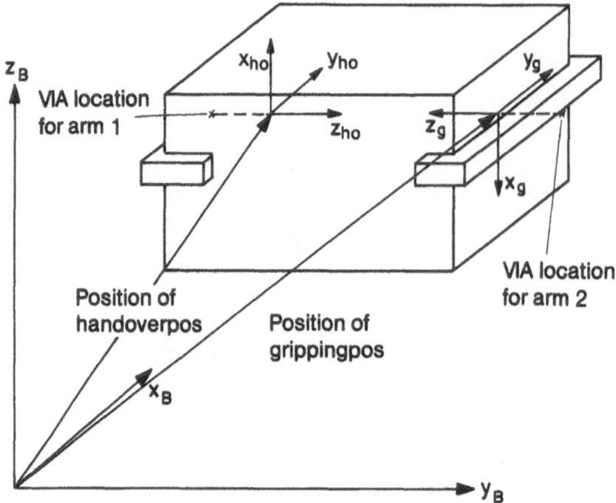

Fig. 7.1: Hand-over position for two grippers

After termination of all quasi parallel instructions within the CO-block, the program is continued with the instruction following COEND. Since all data is global to the CO-block, all instructions for all parallel processes have access to all data (except for any locally defined data of a block, which has been defined within a CO-block) and thus may alter it. The latter increases program complexity and conceals considerable danger for secure synchronization.

The example program does not contain any checks whether a box has actually been handed over or gripped correctly. This would be essential for any practical programs, and the apparently simple program structure of the example should not give rise to an underestimation of the problem. The example can be expressed similarly in SRL.

7.2 Tasks

The ability to run sequences in parallel as well as quasi parallel is common to tasks and CO-blocks. The most important characteristics of tasks are:

- They usually have local data storage.
- They may be started, halted, continued or terminated from outside (by a system command or another task) and they may halt and finish themselves or even restart cyclically as new **incarnations.**
- Various systems allow multiple calls of the same task; it may be executed as a task incarnation.
- A more or less comfortable **intertask communication** is available according to the operating system.

A task concept can be implemented in such a manner that tasks form programs of their own.

```
SYSTEM_SPECIFICATION
  ROBOT:          r56 = ROBOT(1);
  EFFECTOR:       roundgrip = GRIPPER(0) OF r56;
  SENSOR:         detector = CHANNEL(5);
                  STRUCTURE detector = RECORD
                                            switch: BOOLEAN;
                                            length: REAL;
                                         END;
  INTERRUPT:      empty FROM CHANNEL(7);
END_SYSTEM_SPECIFICATION;

PROGRAM  monitoring(INFILE, OUTFILE);
  (*    .
        .
        .       *)
  SECTION distribution;
    VAR
      magazin_list:              ARRAY[1..9] OF FRAME;
      get_pos:                   FRAME;
      current_magazin_no:        INTEGER;
      set_exit:                  BOOLEAN;
      ch:                        CHAR;

      SECTION break(magazin_number: INTEGER);
        BEGIN_SECTION "break"
          WRITELN (' Switch on after magazine ',magazin_number);
          set_exit := TRUE;
        END_SECTION "break";

  BEGIN_SECTION "distribution"
    WRITELN (' Distribution started! ');
    set_exit := FALSE;
    (* Initialisation of the values for get_pos *)
    (* and magazin_list                         *)
    ALWAYS WHEN detector.switch = TRUE
      DURING
      FOR current_magazin_no := 1 TO 9 DO
        PTPMOVE r56 TO get_pos;
        CLOSE roundgrip;
        SMOVE r56 TO magazin_list[current_magazin_no]
              WITH DEP = 10;
        OPEN roundgrip;
        IF set_exit THEN
          BEGIN
          WRITE ('Switch is set! Continue? (Y/N): ');
          READLN (ch);
          IF ch = 'N' THEN
```

```
          EXIT_SECTION
        ELSE
          SUSPEND 500 MS
        END_IF
        END
      END_IF;
    END_FOR;
    END_DURING
      DO
        START break(current_magazin_no);
  WRITELN('All magazines filled !');
  END_SECTION "distribution";

BEGIN_PROGRAM
  WRITELN (' Program for monitoring started !');
  (*    .
        .
        .       *)
  START distribution;
  (*    .
        .
        .       *)
END_PROGRAM.
```

Program section 7.2: SRL program with a monitor section activated by a program

SRL contains a task concept which controls processes in various ways: in parallel, cyclically, offset in time or initiated by interrupts. The tasks are identified as **sections** and declared the same as procedures. Like variables, they have a span, hence they can only be called from within the section in which they have been declared, or from the main program, which can be regarded as a higher-level section. If the higher-level section terminates, all other sections (which have been declared inside it) are also ended. Thus the situation is clear for the programmer, and no tasks which might be started by side effects somewhere in the program will run during the program execution (or even beyond). Sections can be declared or called with parameters, just as with procedures. In practice, they represent procedures which can be executed in parallel, possibly-priority controlled.

Program section 7.2 shows the declarations and nesting of sections. Initially, robots, grippers, sensors and interrupts are defined in the specification part of the main program `monitoring`. After the usual declarations of constants, types and variables, the section `distribution` is declared without parameters. It contains the local variables `magazin_list`, `get_pos`, `current_magazin_no` and `set_exit`. These may also be referred to during the section `break`, which is declared inside `distribution`.

The messsage "Program for monitoring is started!" is printed after the start of the program and some actions executed. After the start of the section `distribution` an-

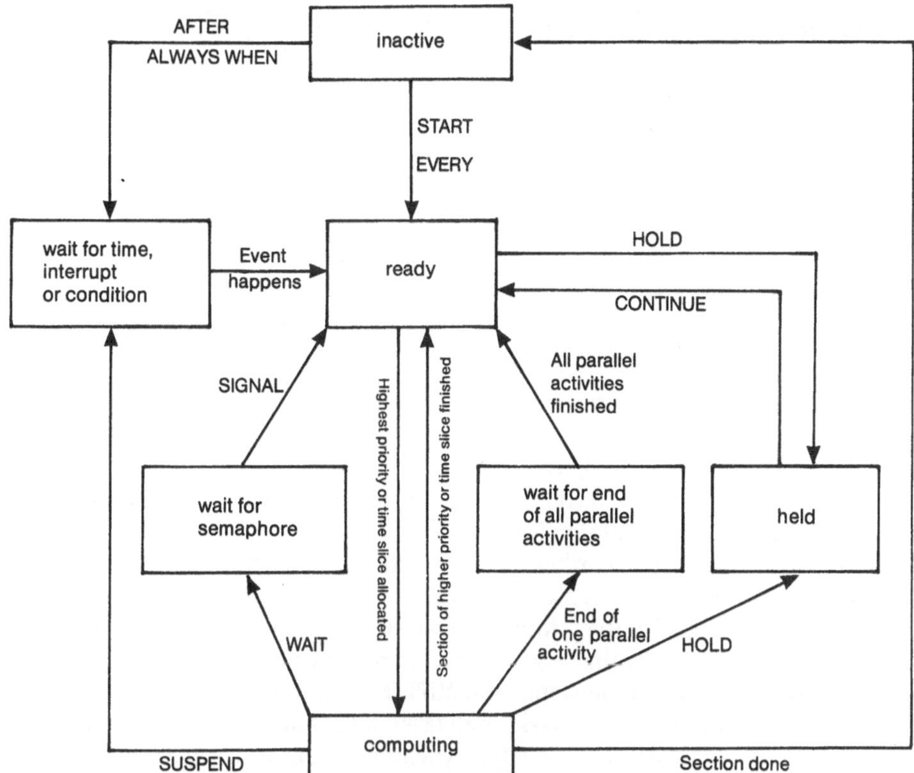

Fig. 7.2: Task status diagram in SRL

other starting message is given and set_exit set to FALSE. The initialization of variables with their respective values has been omitted for the sake of clarity. The following instruction ALWAYS WHEN causes the monitoring of the setting of the sensor switch detector during the execution of the instructions between DURING and END_DURING. If this condition becomes TRUE at any time, the section break is started with the current magazine number as its parameter. The DURING-part consists of a FOR-loop, in which the robot r56 picks up an object and moves it to the relevant magazine. It is then tested whether the switch has been activated in the mean time. Should this be the case, the Boolean variable set_exit has been set in the section break and the test is true. The user is then asked whether he would like to continue. If he would not, the section distribution is terminated via EXIT_SECTION. Thus the monitoring for the sensor detector is also terminated and the interrupt is disabled again.

As already mentioned in the example, one section can override another, which then enters the state "ready" whereas before it was "computing". Figure 7.2 illustrates the possible states in SRL and their transitions.

In AML, interrupt routines can be defined by the MONITOR system subroutine. A

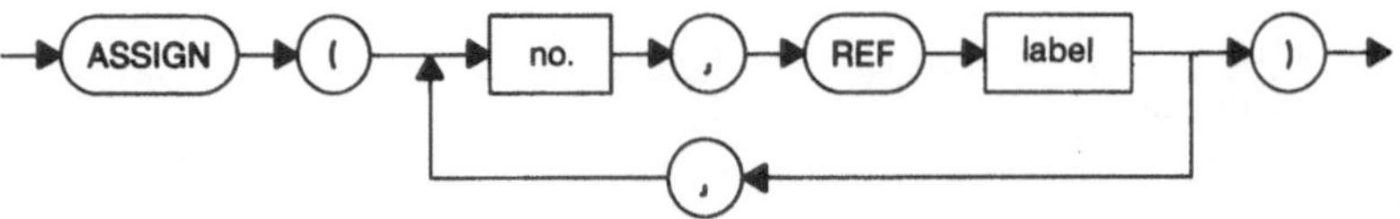

Fig. 7.3: Task definition of HELP

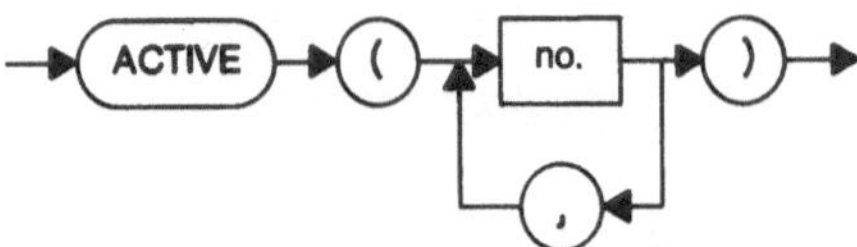

Fig. 7.4: Activating a task in HELP

time interval for the monitoring can be specified as well as upper and lower bounds for up to 16 sensor inputs. The MONITOR can be bound to MOVE instructions in order to interrupt them if the condition is fulfilled.

The HELP instruction according to Fig. 7.3 allows the declaration of the subroutine determined by <label> as a kind of task, to which the number <no> is assigned. This number also signifies the priority of the task, where 1 indicates the highest **priority**, which is reserved for the "end process" terminating a program run.

Tasks are activated with the command ACTIVE shown in Fig. 7.4. The command ERASE (Fig. 7.5) stops a specified task, or all tasks, if the number <no> is omitted. The SIGNAL and WAIT instructions serve the synchronization of tasks, as discussed in Sect. 4.7.3. An ACTIVE instruction starts a task, which has been halted by an ERASE, from the beginning. HELP only contains global variables; thus subroutines which have been declared as tasks have no local data area, and they probably cannot form incarnations. Nevertheless, they are mentioned here, as they come closest to the task concept because of their calling mechanism. Unfortunately, the inexact description does not allow the presentation of a status diagram here. If the subroutine representing the task is programmed as a loop and contains the operation sequence of the robot, then manipulation programs are obtained, called "cycle programs" in HELP.

SIGLA supports the synchronization of parallel processes through the semaphore instructions described in Sect. 4.7.3. Processes may be started in the form of subroutines from within a SIGLA program.

7.3 Coroutines

Contrary to CO-blocks and tasks which may be executed in parallel, **coroutines** are organized such that they - or certain parts of them - run sequentially, and the data areas belonging to them are preserved even during the inactive phase. The first call of a coroutine establishes the data area. Thereafter the instructions of the body are executed until it is stopped by a halt instruction (e.g., DETACH or RESUME in the lan-

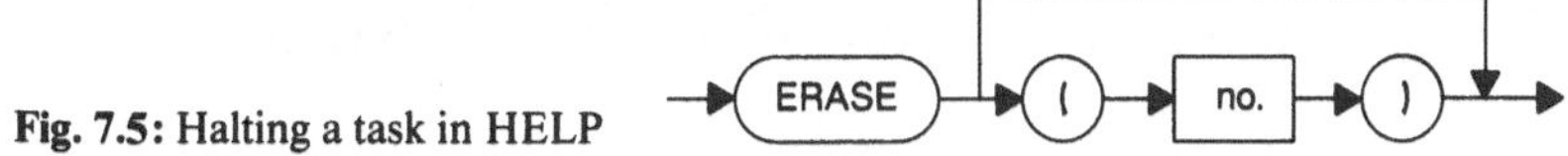

Fig. 7.5: Halting a task in HELP

guage SIMULA 67). DETACH returns the control to the main program, while previously started coroutines are reactivated by RESUME. The main program itself can restart coroutines which have been previously suspended.

The fact that coroutines always continue from the place at which they were interrupted is peculiar to coroutines. The working cycle of a robot may be described by a coroutine which consists of a loop, in which interruptions by other coroutines may deal with error situations, or, especially, execute tasks concerning several arms.

No coroutine concept is contained within the languages discussed here, nor is any robot language known at present which offers coroutines. Nevertheless, this concept has been introduced here since it is suitable for dealing with parallel processes and also offers higher security than the CO-blocks of AL because of local data storage and clearer synchronization. However, it does not permit parallel execution of independent tasks, which is a major disadvantage.

SIMULA 67 may be mentioned as a language which has implemented coroutines. It belongs, like AL, to the family of ALGOL and PASCAL, i.e., it is also block-orientated, supports structured programming and allows recursive procedures and functions.

8 Programming and Run-Time Systems

A system of software components is necessary for the generation, maintainance and housekeeping of programs in all areas of computation. This includes an **editor, filing system, compiler, interpreter, debugger**, etc. Some of these aids - such as filing systems and text editors - belong more in the category of general resources, which are not provided for a specific programming language but are intended for more general usage. The other components are more or less robot-specific. The entirety of such collated software components is called the **programming system** (BLUME [8.1]; SANDEWALL [8.2]). The consistency of this system, the necessary components and their capabilities are determined by the programming method, the language, the underlying hardware and the required operating comfort. Such user friendliness should not be regarded as comfort for the user or a more or less superfluous luxury, but rather as an aid for the support and improvements of *program security, maintainability* and *modifications.* These are important characteristics which influence **software costs** considerably in the long term, and the significance of early investments in this area is very often underestimated.

The command interpreter for the language and the **interactive component**, if it is present, belong to the run-time system of the robot language. In addition, the run-time system utilizes a number of functions from the operating system, particularly when organizing parallel processes. If the operating system does not contain such functions (single user systems), they have to be realized in the run-time system.

Figure 8.1 shows an exemplary general view of the interaction between various software components. The **simulator**, which has not been mentioned yet, but is also shown, simulates the robot action as a testing facility (Sect. 8.6) and can thus replace a large part of the testing phase on the robot. The compiler may contain facilities for the generation and administration of a more or less complicated world model according to its capabilities and the system concept.

This chapter introduces the components for programming and run-time systems, which are significant in programming robots.

8.1 Editor

Editors are auxiliary programs intended to generate or change large amounts of data, for example, texts or frames including their values (cf. Fig. 8.2).

Text editors belong to the standard auxiliary programs and thus will be mentioned here only briefly. Simple editors allow the manipulation of text line by line, which can be quite cumbersome especially for corrections. Thus most editors today

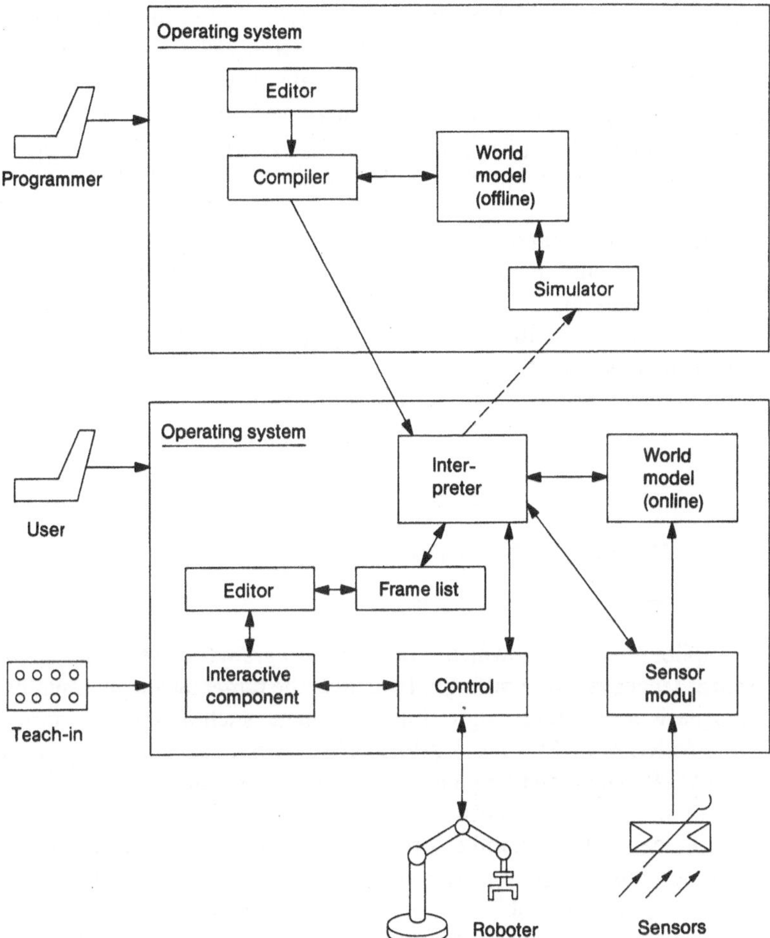

Fig. 8.1: Structure of a programming and run-time system for a high-level language

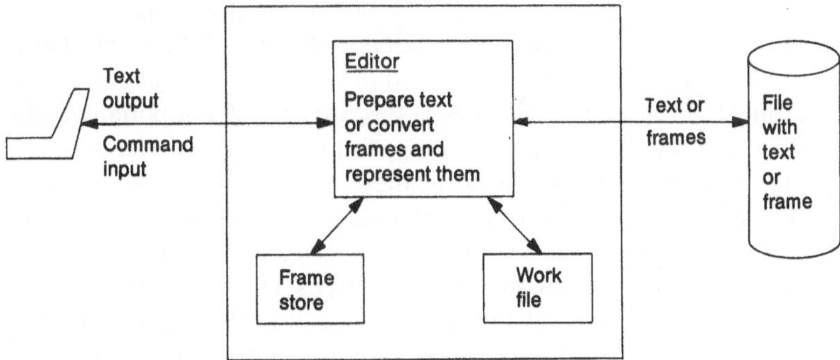

Fig. 8.2: Structure of a text or frame editor

contain commands which permit a character-oriented manipulation on a screen, such that the cursor is positioned arbitrarily at any place in the text and characters are inserted, replaced or erased. Additionally, they contain search functions and permit the merging of text files or dividing one file into several.

Frame editors are typically robot-specific components of the programming system. They allow the manipulation of frame lists and enable the user to

- list frames in a form comprehensible to humans,
- input new frames textually or via the teach-in procedure of the interactive component (Chap. 5),
- correct values of individual frames by specifying, for example, a new position and orientation, and
- output edited frame lists to external storage devices.

8.2 Compiler and Processor

A compiler is a program which translates a program (or a section thereof) written in a **source language** into a different language, the **target language**. Usually, the source language is a high-level, algorithmic language, such as the ones introduced here SRL, PASRO or PASCAL, AL AML, HELP and RAIL, and, with some limitations, also VAL-II. The target language may be **machine code**, which can be executed directly by the computer hardware, or it may be *intermediate* or *pseudo code*, which has to be executed by another program, the interpreter. Translation programs for NC-programming, called processors, will be discussed later. (The processor in an NC-language should not be confused with the hardware component of a computer, which is generally called the central processing unit, or in short the processor.)

Most tasks of a compiler do not vary from those usually present in normal computing, so that they may be dealt with briefly here and the reader may be referred to the relevant literature [8.3 - 8.6].

To begin with, the entered program, which can be seen as a finite sequence of characters, has to be analyzed *lexically* and *syntactically*. The lexical analysis identifies the *basic symbols*, such as *word symbols* (BEGIN, IF etc), numbers, *identifiers* and special characters (or groups of characters) such as "=", ":=" etc. During the syntactic analysis phase, these are checked to see whether they form a valid sentence according to the grammatical rules of the language. An important quality criterion is the way in which the compiler deals with syntax errors. The simplest and worst way is simply to announce the error: then so many consequent errors result that it would have been better to stop the compiler run immediately. Good compilers restrict the error to one statement, generate a meaningful error message which points to the cause and then continue with the analysis. Under certain circumstances, subsequent errors are unavoidable, as in wrong variable declarations, for example. There are even some compilers which are able to correct simple syntax errors such as a missing comma or semicolon. The quality of a compiler is an important criterion in economic considerations because the development time depends on it. After the syntax analysis, semantic checks are run, such as the compatibility of operands with operators or the correspondence of actual parameters with their types in a procedure call. Now the compiler can organize the allocation of storage for variables in block-orientated languages (Sect. 2.2.3).

The subsequent phase of *code generation* evaluates target addresses which are generated by jump instructions or by structured commands for program flow control.

This may also include an optimization of the code in various ways. For example, code may be changed in such a way that values which are used frequently are loaded into registers at the beginning of the program or procedure, or that relative branch instructions with less expense in time and code may be used for short distances. The whole process can be organized so that the compiler stores all of the program intermediately, in an internal form, and scans it several times from beginning to end. According to the number of times a compiler passes through the input text, it is referred to as a one-pass or multi-pass compiler. One-pass compilers are usually more compact and faster, as long as the program is syntactically correct. If this is not the case, a one-pass compiler has to generate unnecessary code, which may be suppressed in a multi-pass compiler, thus making it faster.

The *processor* in NC-languages fulfils tasks similar to those performed by a compiler, with the difference that it usually generates intermediate code (pseudo code). It then has to be processed by a machine-dependent **postprocessor** in order to make it suitable for the corresponding control of the NC-machine or robot, as in the ROBEX system (Fig. 8.3).

A further difference to compilers is the presence of the results of all calculations in the pseudo code as they appeared in the source program. Hence code generated by the processor consists mainly of instructions which may be translated into movements by the controller directly, without the possibility of recalculating new targets (such as those caused by sensory data) during run-time. Meanwhile, some NC-systems do exist which contain beginnings of information processing, for example the language MCL from the company McDonnell Douglas.

As far as is evident from the documentation and the concept of SIGLA (SALMON [12], [13]), Olivetti apparently uses an interpreter for the source language directly, thus departing from traditional NC-techniques. The interpreter communicates directly with the NC-controller of SIGMA units. SIGMA units are taken to be systems which "contrary to conventional NC-machines can make decisions in the event of abnormal situations", and Olivetti counts industrial robots as such machines.

Specialities of compilers for robot languages are firstly the inclusion of an interface to an interactive component, however fashioned (see Chap. 5), and secondly the generation of a world model (Sects. 2.3.7 and 2.3.8) in languages which have been designed with reference to implicit programming. The first is present in nearly all robot languages developed up to now. The latter will gain more and more importance as time goes on.

Most of the data together with the robot position has to be kept up-to-date during program execution. The compiler needs to extract the data relevant to the task from a static and general world model which is appropriate for the individual task. Thus

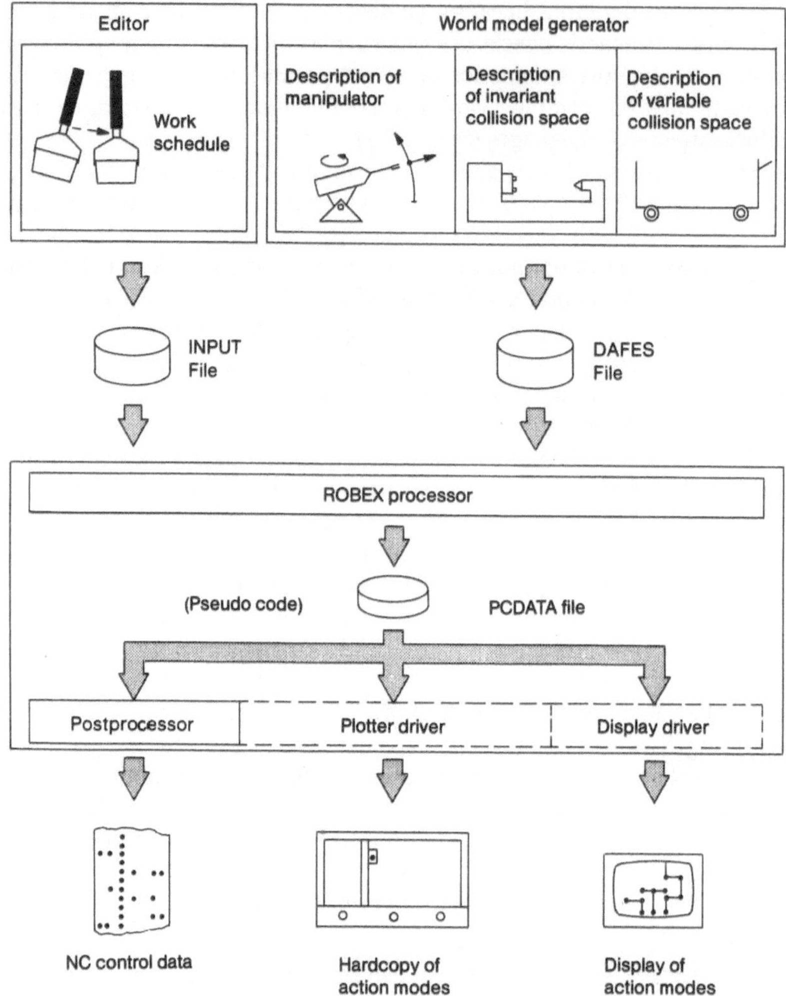

Fig. 8.3: Structure of the ROBEX programming system (WECK, et al. [15])

it can build up a problem-orientated world model and use this information for the program generation.

This modern concept of robot programming is still in the development phase, and none of the languages introduced here contains even part of it. The world model generator of ROBEX included in Fig. 8.3 should be seen as an aim rather than a reality. Development systems pointing in that direction can be found in the AUTOPASS system of IBM (LIEBERMAN [2.10], [8.7]) or LAMA of the Artificial Intelligence Laboratory of Massachusetts Institute of Technology (LOZANO-PEREZ [8.8]).

A further special feature worth mentioning is the *world modelling facility* in the American AL compiler. It administers all frames and their corresponding affixment structures, i.e., the way in which the frames are attached to one another, and attempts to calculate probable motion instructions as early as compile time. These calculations have to be repeated and corrected at run time, because only a limited knowledge exists at compile time about individual positions. Indeed, the more such features in high-level languages such as AL are exploited, the less it is possible to know such data before program execution. One reason for this method is the relatively demanding task of path calculations to be located in the background computer, so that the controller is not burdened with such tasks. However, since this is only possible to a certain degree, this argument is gradually diminishing in importance, and as the power of microprocessors increases and their prices decrease, this idea is generally being abandoned.

8.3 Interactive Component

The interactive component serves to execute the teach-in procedure, which has been described in Chap. 5. This necessitates the cyclic interrogation of the teach box, to determine whether a function key has been pressed - for example, motion in the x-direction of the base coordinate system or revolving the second robot axis. In such a case, the new robot position and orientation are evaluated with respect to the current robot position and required speed and are then approached (Fig. 8.4). Apart from the cyclic call to the teach-in component, the terminal has to be monitored for any input. As soon as the programmer has specified a frame identifier, the present robot position and orientation is read and corresponding coordinate values are entered into the frame list under a frame name (Fig. 8.5).

The interactive component may belong to the **run-time system**, if - as in the case of ROBEX - the frames are defined via a teach-in procedure during the first program execution (see also Chap. 5). Should it be used to generate a frame list, then it does not belong to the run-time system, although it is only executable with the robot connected, that is, online.

8.4 Run-Time System

The **run-time system** comprises all those software components which are necessary for the normal run of the user system. This restriction in normal operation is necessary, because some systems also need the interactive component for the compiler or editor (VAL), or during the first program run, thus making the programming and run-time systems nearly identical. Figure 8.6 show these connections, where those components which are only needed in exceptional circumstances are indicated by broken lines. The **world model** presents a special case, because its realization is only rudimentary during the run-time of the user program (see also Sect. 4.1).

An example for the close intermeshing of individual components is shown in Fig. 8.7. The editor and interpreter of the VAL programming system run in parallel, and any program code entered by the editor is executed immediately by the interpreter.

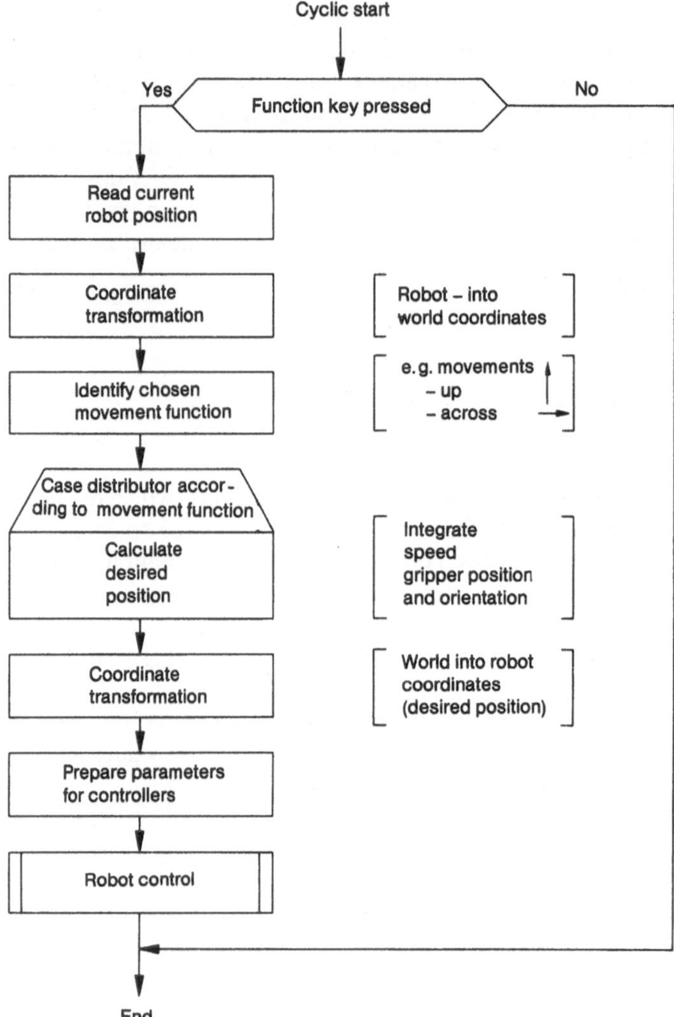

Fig. 8.4: Flow structure of the teach-in component

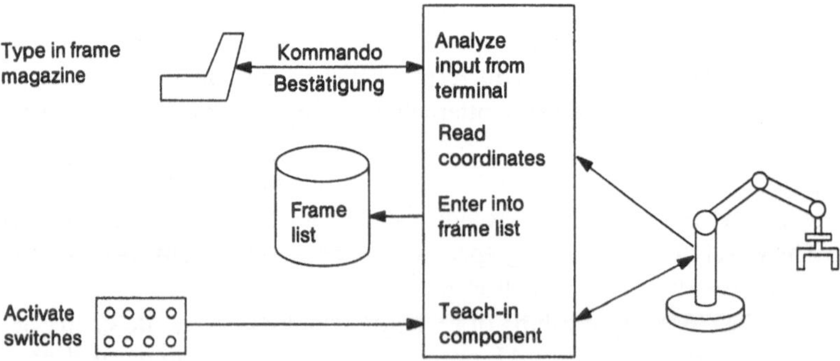

Fig. 8.5: Interactive component for frame list generation

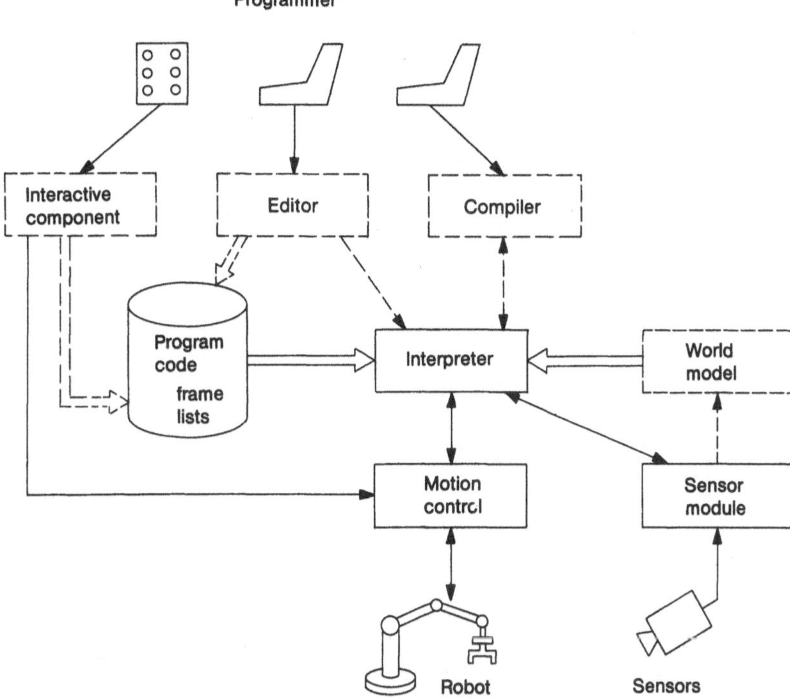

Fig. 8.6: Components of the run-time system for industrial robots

The editor also works in conjunction with the interactive component, so that frames can also be corrected during the program run.

The following sections describe both components, the interpreter and the motion control. A **sensor module**, which is a component for automatic management of various sensors, is currently in the development stage.

8.4.1 Interpreter

Since there is no processor which realizes the special instructions for robots in machine code, an **interpreter** is needed for the execution of user programs. This is a program which reads the code generated by the compiler, decodes individual instructions and calls appropriate routines (Fig. 8.8). Management of variables, that is, the reservation and release of memory, is carried out by the memory management according to the scheme described in Sect. 2.2.1. The arithmetic module evaluates formulae using a stack. The sensor interface, which is also shown, refers to the (normal) case in which the interpreter contains special drivers for the corresponding sensors in the programming system, but no sensor module for arbitrary sensors is present. Interpreters for robot languages are usually *stand-alone programs*, hence they run on their own on the control processor without the support of an operating system. They contain modules which realize the functions of an operating system,

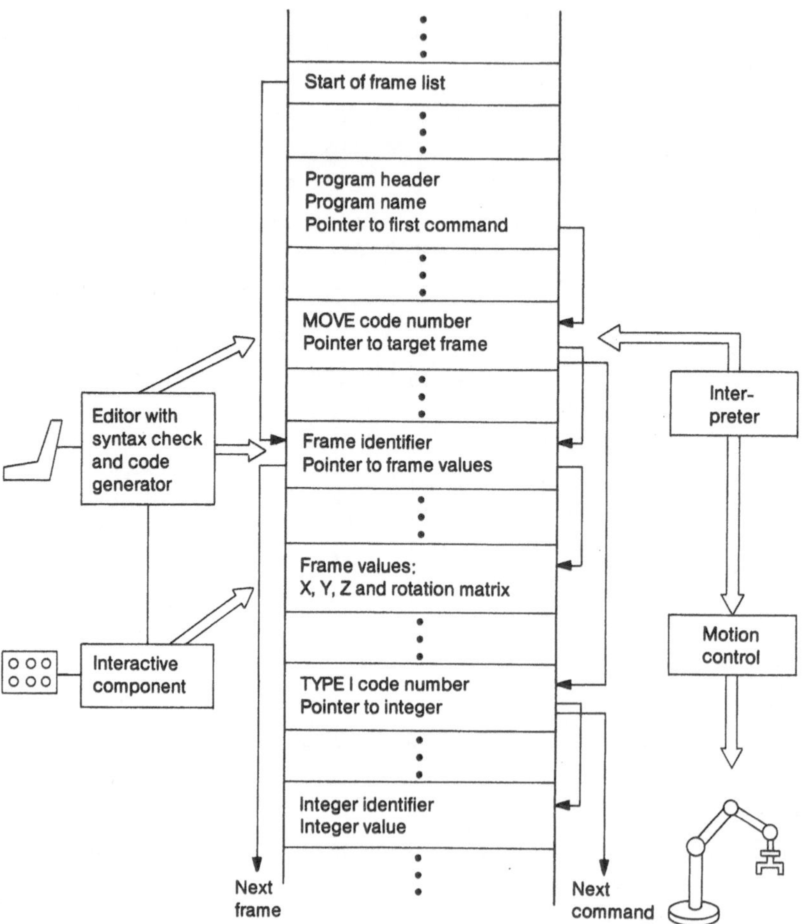

Fig. 8.7: Interaction of pogramming and run-time system in VAL

such as the file management of floppy disks, upkeep of the system clock, announcements of interrupts or an interface for user commands.

8.4.2 Motion Control

The purpose of the **motion control** is to drive the robot to target frames according to the motion instructions. Thus any starting, intermediate and target frames are passed to the module "pathcalculator", which evaluates the frames that have to be approached within certain time intervals - 28 ms in the case of VAL. Then each frame has to be transformed into the values of individual axes of rotation or translation by the module **coordinate transformation** according to the algorithm of Sect. 2.3.2 (see Fig. 8.9).

The actual **robot control** transforms the angles or translations of the axes into motor increments or motor voltages or currents. These values are then output at prede-

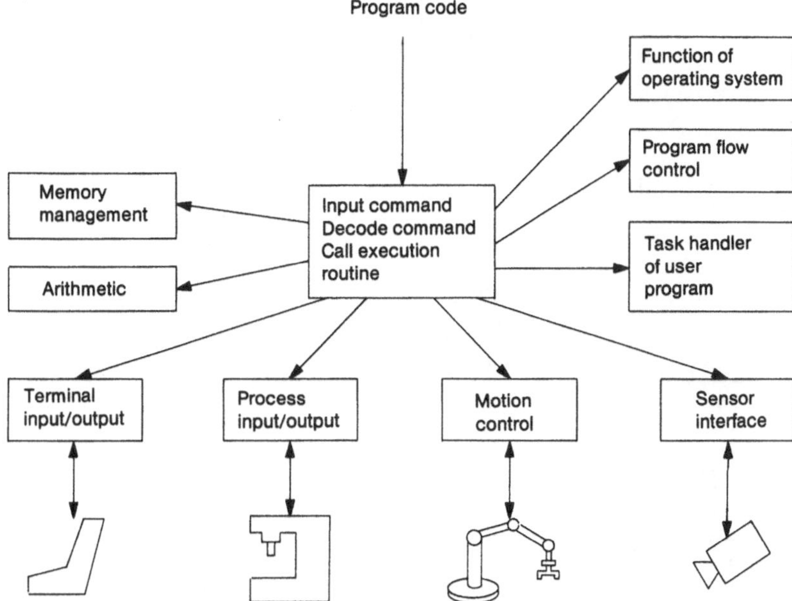

Fig. 8.8: Components of interpreters

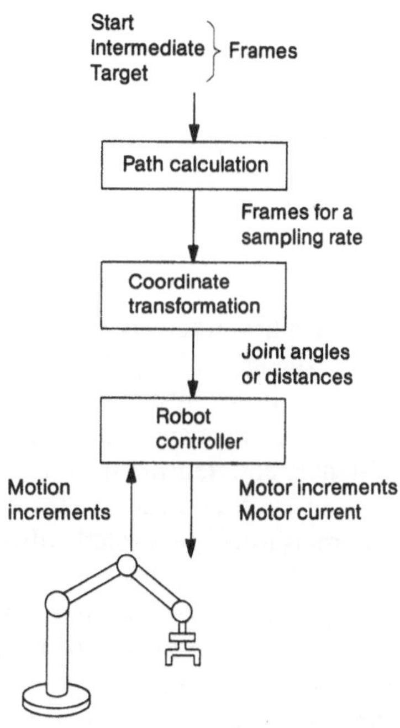

Fig. 8.9: Components of motion control

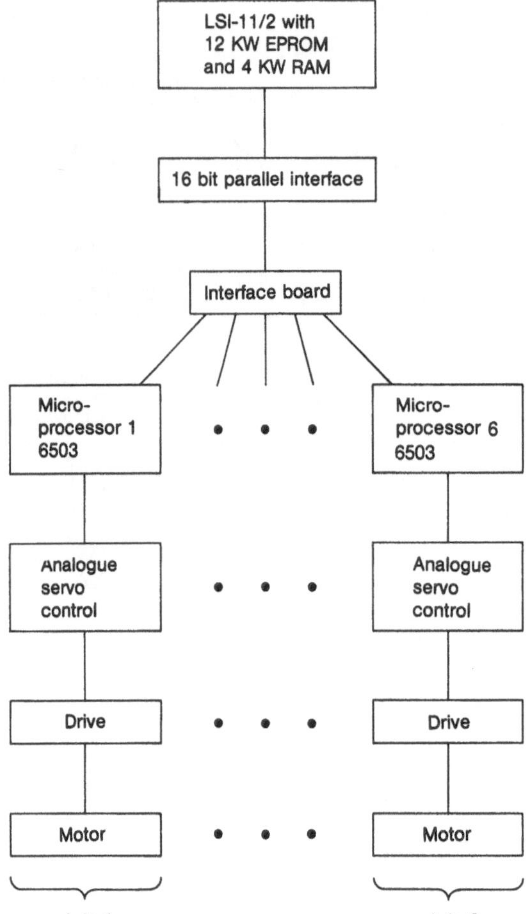

Fig. 8.10: Control hardware for the PUMA 600 of Unimation

termined time intervals, for which the processor needs a *system clock* or *timing interrupts.* The approach to the position of the axis according to the control demand is then monitored by a comparison between actual and demanded values. A **position measurement system** is mounted on each axis for this; it can determine the covered distances and transmit the information back to the controller. The *controls* then adjust the drives according to the differences in the actual and demanded values. This control is usually executed individually, leaving the servo-control to a microprocessor. Each axis thus has a *joint processor* because of time demands, and is supplied with parameters by the control processor. For example, Fig. 8.10 shows the control hardware for the PUMA 600 from Unimation. Current developments tend to execute path evaluations and coordinate transformations on one or more *separate arithmetic processors* in order to reduce the load on the control processor running the interpreter.

8.5 Software Interface IRDATA

Programming systems consist of various modules because of the increasing complexity in programming and control. Such modules are usually organized in a hierarchical manner, for example, the compiler can be regarded by the programmer as being on a higher level than an interpreter. The definition of modules requires a definition of how these modules work together. If modules are specified according to the sequence in which they are defined, a hierarchy of modules is obtained. *Interfaces* between them are structured accordingly.

8.5.1 Different Levels of Programming and Control

An interface between programming modules can also be regarded as a language (or representation) since it is used to pass information from one module to another. It contains a set of defined elements which are placed into a well-defined structure.

Figure 8.11 shows the different levels for robot programming: implicit programming, explicit programming, and robot control, and their respective elements. Explicit programming requires the user to write down all move statements and destinations explicitly to fulfil a robot task. In implicit programming this will be carried out by an intelligent planning module which automatically generates an explicit programming part. Software interfaces for a system based on SRL and IRDATA are placed between these levels:

1. Robot task description
2. SRL and
3. IRDATA

An additional interface will be introduced inside the robot control; this will be mentioned later, in Sect. 8.5.3.

The interfaces SRL and IRDATA are influenced by the upper level as well as the the lower ones. If the interface does not include features for multitasking, the planner can generate tasks which should be executed in parallel. However, there is no possibility of passing this information to the explicit programming level. Therefore, SRL and IRDATA should include elements which are powerful enough to satisfy any requirements from the upper level as well as from the lower level.

The elements of interfaces can be distinguished as **descriptive** and **action elements,** as will be explained in the next sections.

8.5.1.1 Descriptive Elements in Software Interfaces

Some information passed from one programming level to another has a descriptive character. This means the code element of the software interface does not result in an action such as the changing of program data or the moving of a robot. The receiving software facility (for example, the IRDATA interpreter) decodes these descriptive elements as information for the parameterization of its facilities.

Descriptive elements are:

1. Information of hardware addresses and program data
2. Selection of drivers, effectors, robots, sensors coordinate transformation modules, etc.
3. General values to define speed and accuracy
4. Definition of data structures
5. Description of the robot world (obstacles, trajectories etc.),
6. Data management (memory allocation, calculation of data space and access, etc.)

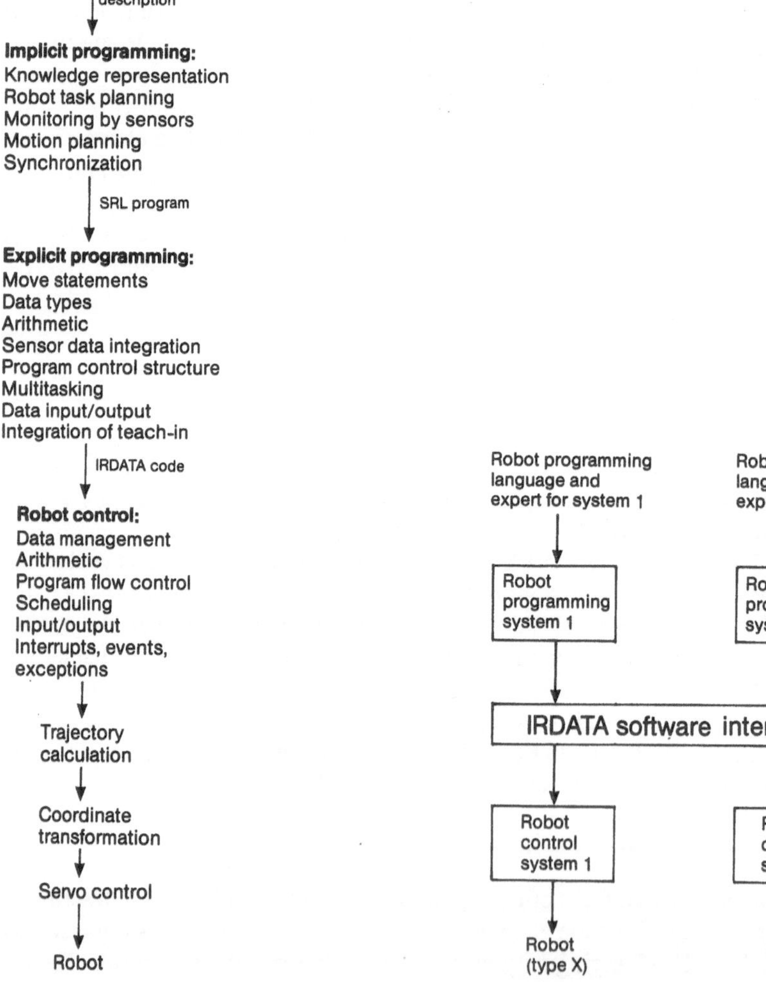

Fig. 8.11: Software levels and interfaces for robot programming

Fig. 8.12: IRDATA software interface between robot programming system and robot control system

7. Test information (break points, variable names, etc.)
8. Hardware description of robots, sensors, effectors, etc.

However, the reception of a descriptive parameter can cause action by the interpreter. For example, the interpreter will calculate the required size of the memory for program variables and reserve it.

In most cases, descriptive information will determine values for parameters which are used in subsequent action elements. Some descriptive elements are used in both directions of the software interface. When in test mode, the interpreter may send back current robot positions to the compiler.

8.5.1.2 Action Elements in Software Interfaces

Most of the information which is passed from a programming facility to the interpreter will initiate an action on data or robot-moves as soon as it is decoded. Action elements may include parameters used for performing this action. A result of an action element is a new state of the robot, the peripheral devices, program data or program flow control. Action elements are used for:

1. Arithmetic calculations
2. Assignments
3. Program flow control
4. Robot move or effector actions
5. Data input/output
6. Parallel, cyclic or time delayed program execution
7. Synchronization

The execution of an action element can result in a call of different interpreter modules, such as trajectory calculations, coordinate transformations and servocontrol for a robot move. These modules can be implemented on different processors or computers and may run in parallel to the execution of other action elements.

8.5.2 Intention and Structure of IRDATA

There are many different robot manufacturers today, all of them using their own programming systems with special robot-programming languages. These languages differ in various aspects concerning syntax, program structure and features. This fact results in the following disadvantages to the user of different robots:

1. Every robot system used in a facility needs a specially trained programmer.
2. The process of transferring an existing robot program to another system is equivalent to implementing the robot task a second time, even if the robot has a similar kinematic structure.
3. There is no possibility to transfer taught locations from one robot controller to another.

A solution to this problem could be obtained by standardizing interfaces between the robot-programming system and the robot control system, as is shown in Fig. 8.12. This would result in the following advantages:

1. Programs written on a special programming system could be used to control robots of different types.
2. A robot might receive programs or data from any programming system.
3. An expert for the programming of a special robot would become a general expert for robot programming.

Since such an interface covers the problems of robot programming, different experts in the areas of mechanics and control have joined the working group of the Association of German Engineers (VDI) with the aim of finding a common national standard. It is called IRDATA (Industrial Robot DATA) and is based on a former CLDATA standard for NC machines. IRDATA will be proposed for ISO.

An IRDATA program is a sequence of records which contain a certain number of arguments. The records are classified according to the special requirements of robot programming. For example, there are classes for:

1. The description of
   a) Robots
   b) Tools
   c) Sensors
   d) Working space
   e) Frame lists
2. Motion specification and execution
3. Arithmetic operations
4. Program flow control
5. Input/output operations

To achieve flexibility, the arguments may be either constants or variables. The following data types are provided:

1. BOOLEAN
2. INTEGER
3. REAL
4. VECTOR (three real numbers for x-, y- and z-coordinates)
5. ORIENTATION (three real numbers representing angles)
6. WORLD (a position vector and an orientation)
7. JOINT (a set of real numbers representing robot coordinates)
8. CHARACTER
9. STRING (a one-dimensional array of characters)
10. POINTER
11. ARRAY

A type conversion between some of the different types has been provided and a large set of standard operations has been defined to handle Boolean, arithmetic and geometric calculations. This allows a universal description of the robot position and movement for different types of robots. IRDATA is powerful enough to be a code interface for all introduced languages (see also Appendix K).

IRDATA is a numerical code, but the digits of the code numbers are represented as ASCII characters. Such a sequence of characters can be easily transferred between different computers. The hardware interface (such as RS232) is not part of the IRDATA definition. IRDATA can also be represented by keywords for documentation reasons.

At the University of Karlsruhe a so-called PR-IRDATA (PRogrammable IRDATA) has been developed. It provides formula notation, labels and variable declarations. A PR-IRDATA transformer checks the syntax and semantics, e.g., type equality or subroutine declarations, and generates IRDATA code. PR-IRDATA is useful for short programs which need to be translated quickly, or for corrections within (longer) IRDATA code.

8.5.3 IRDATA Interpreter and Move Control Interface

Tasks performed by the IRDATA interpreter can be divided into robot-independent tasks such as arithmetic operations, assignments, program flow control or multitasking, and into robot dependent tasks such as move and gripper control. Therefore, the interpreter can also be divided into a part for performing robot-independent tasks and a number of interfaces to different underlying robot controllers (see Fig. 8.13). This structure will allow for an easy and quick transfer of the IRDATA interpreter from one control computer to another. This is independent of robot manufacturers, because only the interfaces to the robot controllers may change. The interpreter can be written in a high-level programming language, which can be

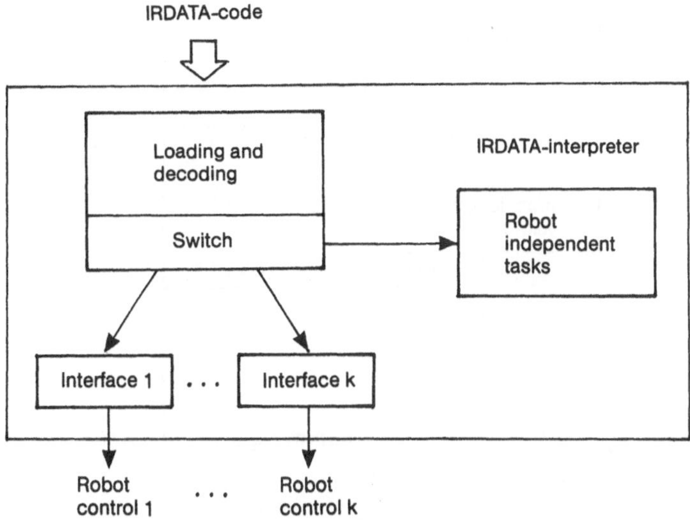

Fig. 8.13: Structure of IRDATA interpreter for different robot controls

found on many microcomputers, for example PASCAL or C. It is then possible to operate a robot from a manufacturer who does not offer IRDATA. An IRDATA interpreter can be used which was developed by a software house or institute and can be executed on a microcomputer like the LSI. It has to be extended by the interface to the robot controller only. Facilities needed in this interface concern only the trajectory planning, coordinate transformation and servo control.

Let us now take a closer look at these control interfaces. Depending on the level of control, they may be added to the interpreter interfaces or, if they are included in the robot control, they will be parameterized by the interface. The modules performing the higher levels of control and possible interfaces are shown in Fig. 8.14. If the robot control from the producer X includes features such as trajectory calculations and interpolation, coordinate transformation and cyclic output to the servocontrol, the interface only needs to pass the parameters like target frame or speed to the robot controller. If the robot control does not perform trajectory calculations, the interface has to include this module and the output to the controller are the frames of the robot trajectory.

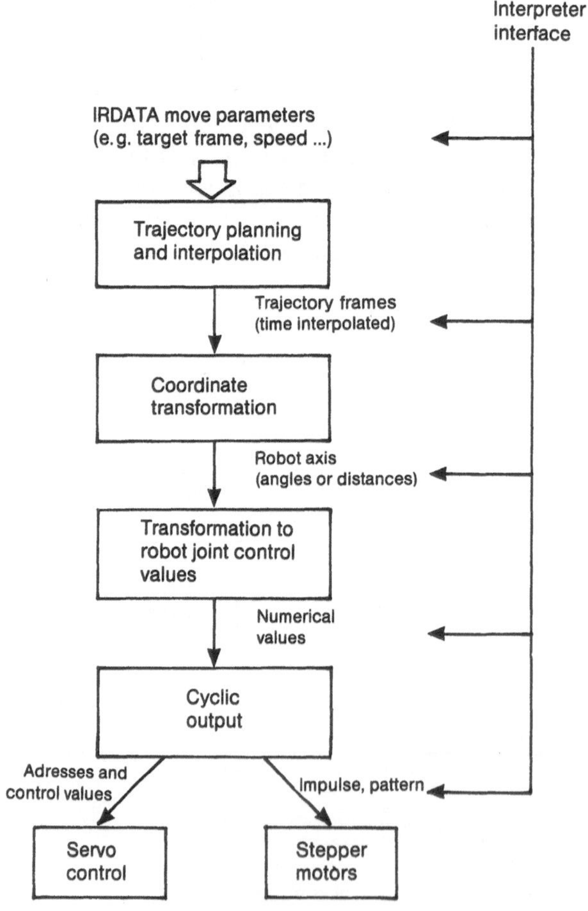

Fig. 8.14: Possible interfaces to underlying robot controllers

The next interface level is the level of values for the robot axes, if the robot controller does not perform coordinate transformation form world to joint coordinates. If the robot controller only executes servocontrol or accepts only stepper motor pulses, the interface consists only of numerical motor values or pulse patterns.

8.6 Simulators and Program Test

Before a robot program can be released for practical applications, it has to be tested thoroughly so that the expensive robot or any other installations are not damaged because of programming errors. The problem is highlighted if humans have to work in close proximity even for only part of the time and additional safety measures are necessary. Despite its importance, this subject of accident prevention in robot installations cannot be discussed here, as it lies outside the scope of this book.

Unfortunately, there is no practical algorithm to date which can prove the correctness of a program, although this has been investigated for some time now (ELSPAS [8.9]). Thus only the program test remains, in which the program is tried with a set of randomly or, better, systematically chosen data. However, according to Dijkstra, only the presence, and not the absence of errors can be proven. The best-known aids for testing are listed below:

- Tracing of the program execution
- Dumping of the main memory contents
- Reading and possibly changing of variable values which can be referenced by their symbolic name during program execution
- Interrupting programs at break points which can be set or reset arbitrarily
- Restarting programs with or without new initial data
- Retrying program steps or changing program flow control

Dump and trace are commonly used means of analyzing the causes after a possible program break. However, the remaining four points describe fundamental properties of a **symbolic debugger**, which is used as an auxiliary test during program execution. It is obvious that a symbolic debugger enables faster and more thorough testing.

Nevertheless, even the best debugger still includes the danger of damaging the robot or objects in its working envelope during the test procedure. Testing may even generate its own dangerous moments.

Let the robot be at position A, as shown in Fig. 8.15, and further let the program execution contain a collision avoidance movement via the positions B, C, D and E. Suppose an error occurs after position E has been reached; the programmer suspects the cause before the movement to B and sets program execution back. The next instruction is now "approach position B". Thus the robot would move directly from position E to position B and collide with the object. Hence robot programs should always be restarted during testing, and never set back. On the other hand, this means that a restart also executes a program part which has already been tested,

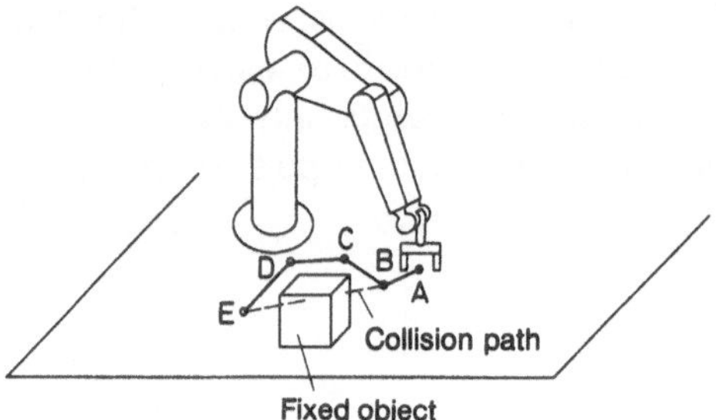

Fig. 8.15: Problems of testing a collision avoidance move of a robot

until the point of interest is reached. This can be rather time-consuming, considering the time needed for movements and other operations.

These two reasons - danger of physical damage and time expended - make the use of simulation programs valuable. The simulator allows the testing of all robot motions at varying velocities, and, even more important, at speeds faster than the robot could move in reality. The processes in the simulator should also be visualized graphically, so that any testing personnel can observe the relevant objects and the robot. As shown in Fig. 8.1, the simulator also accesses the world model. This data forms the basis for its own model with object-relevant information for graphical representation. The decisions to which objects are relevant can be taken by either the user or the compiler. It includes in any case

(a) any robots, and
(b) all objects involved in movements.

These may be determined by the compiler on its own. The user has to provide information about further objects - e.g., obstacles, magazines, etc. - especially if he wants to include them. Such a model, which has been built up during the initialization phase, is changed with each robot reaction and with each instruction via the logical relations of objects (for example, the affixments in SRL or AL). This also enables the checking for collisions or the keeping of safety distances. Additionally, graphical representation can be suppressed in order to speed up the simulation.

So far, the use of simulations has been mainly applied to the design of new robot types (BROOKS [8.10]). The programming systems AUTOPASS (LIEBERMAN [8.7]), SRL in connection with the IRDATA environment at the University of Karlsruhe (DILLMANN and HUCK [8.11]), ALFA (ZAMBUTO [8.12]) and TL (KUNO [8.13]) contain simulation modules.

AML/E provides for the IBM SCARA robots an advanced simulation tool which runs on the IBM PC and emulates the robot control. Thus, logic errors can be detected and the time needed for each motion as well as for the whole production cycle can be measured. Another feature is the simulation of peripheral signals, which allows logic tests but, of course, no checks of race conditions under real-time considerations. The motions are visualized by a 2D graphic animation, so that a rough collision check can be performed by the user.

Simulation is also feasible from the economical point of view, because a robot is about twice as expensive as any necessary graphic stations, which are available in sufficient quality as monochrome systems. Firstly, the robot is available for other tasks during the use of the simulator, secondly, it is not in any danger, and thirdly, time in testing the program is reduced.

A certain alternative to simulation is the use of new micro-robots, which cost only a fraction of the industrial counterparts. However, they are usually much smaller and can lift much less, and thus the environment and objects will have to be reduced as well. It also needs adaptation software, which translates the commands for the interpreter on the micro-robot. The time problem cannot be solved with the micro-robots, either.

8.7 Implementation

The effort of implementation and the structure of a programming language for industrial robots depend largely on the underlying concepts and the complexity of the language itself. Individual components of the programming system can be designed as part of the system right from the start, which of course would be an ideal case. In practice, a simple system is formed which will be extended later as needed. The specific problem for high-level robot languages arises in integrating the industrial robot control into the overall system. The interfaces described in Sect. 8.5 are particularly suitable for this purpose.

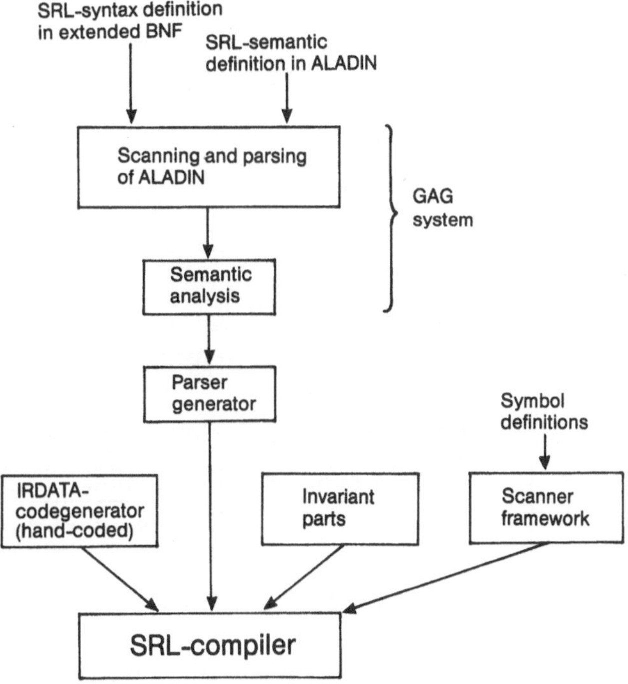

Fig. 8.16: Generation of the SRL-Compiler

The actual robot controller is usually implemented on specially devised hardware. A development system containing software tools such as editors, assemblers or compilers may also be employed. The robot control with its robot motion, sensor input/ouptut, etc., is usually implemented in assembler, FORTRAN or C as a stand-alone system (that is, without an underlying operating system on the controlling hardware). After it has been compiled or assembled and pretested on the development system, it is down-loaded into the controller or burnt into EPROM.

Simple languages such as VAL are implemented in the manner described above. Consequently, VAL can only drive a single robot, and the interpreter is divided into a part for the robot independent functions and the interfaces or modules for actual robot control (cf. Fig. 8.13). Because the robot independent functions are usually not time-critical in relation to the robot movements, they can be written in a high-level language like PASCAL. Apart from a faster, more secure implementation, a higher degree of portability is also achieved.

The implementation of the compiler depends primarily on the definition of the language. If the language has its own definition and is based on new syntax and semantics, as in SRL, a new compiler has to be generated. If the new robot language is not based on a specially defined syntax, but on an existing language such as PAS-

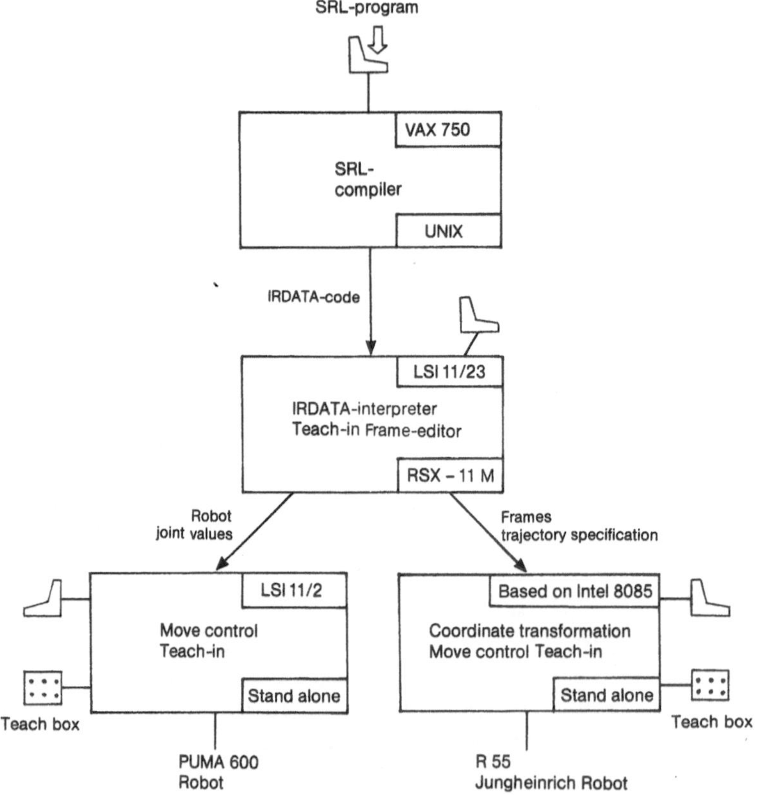

Fig. 8.17: Overall structure of the SRL programming system and interfaces

CAL, no new compiler is necessary. Two brief descriptions of implementations will highlight this.

The language SRL, which has been introduced in previous chapters, has been defined as a new language with complex and extensive syntax and semantics. Based on this definition, a compiler has been implemented under UNIX on a VAX 750. Normally, such a powerful language would have taken at least seven to nine man-years to complete. In order to reduce this to only one man-year, a *compiler generator* (GAG) has been used, see KASTEN et al. [8.14]. GAG (compiler Generator based on Attributed Grammars) generated the compiler for SRL, which has been defined by an attributed grammar. Figure 8.16 shows the process of generation in a simplified form and without the underlying theory. First, the syntax is formally defined in an expanded BNF (Backus Naur Form) and the semantics of SRL is described in the input language ALADIN (A Language for Attributed DefINitions). Then the GAG system generates the necessary definitions and instructions for semantic analysis in PASCAL. According to the definitions of the BNF notations, the parser is also generated in PASCAL by a Parser Generating System. It consists of the parsing routines and tables which control the syntactic analysis. Apart from the invariant part of the code, such as reading input files, a scanner frame is made available for the lexical analysis. Only the SRL-specific symbol definitions have to be added to the PASCAL program for the symbol recognition. The SRL compiler is then completed by the addition of IRDATA code generator, which has been written in PASCAL by hand. The total SRL compiler has been implemented in PASCAL on the Siemens 7765 under the operating system BS 2000 and on the VAX 750 under UNIX (see Fig. 8.17).

The IRDATA interpreter has been implemented containing all robot-independent functions such as the administration of variables, arithmetic, multitasking, program flow control and input/output modules, in PASCAL on the LSI 11/23 under the operating system RSX-11M. According to Sect. 8.5, different interfaces exist for the PUMA 600 from Unimation and the R55 from Jungheinrich. Program section 8.1 shows the listing of an SRL compiler output of a simple program and program section 8.2 shows the IRDATA code in mnemonic form. As the used implementation was made before the final version of IRDATA was defined and decided at the end of October 1985, there are some minor differences to the published and international proposed version of IRDATA.

The PASRO programming system is almost completely implemented in PASCAL, and the system procedures are written in standard PASCAL. Only parts like the I/O to the robot control are implemented in nonstandard PASCAL or assembler.

The minimum hardware requirements for a PASRO implementation are a CPU, 64k bytes of memory, a peripheral device like a floppy disk of 200k bytes, and an interface to the robot controller. Normally, the robot interface is an 8-bit parallel port and handled like a printer interface. This system is fairly small and thus mostly useful for cheap teaching and educational systems.

Each PASRO implementation requires the following software components:

- An operating system for the support computer, such as CP/M or DOS
- A PASCAL compiler, e.g., PASCAL/MT+
- A linker (e.g., LINKMT)

```
***************************************************************************
*                                                                         *
*                               S R L                                     *
*                                                                         *
* PROTOCOL of the SRL-Compiler                                            *
*                                                                         *
*                                                                         *
* DATE :  23 Oct 85                                                       *
*                                                                         *
* NUMBER of MESSAGES:                                                     *
*                                                                         *
*     Informations  :   0 (I)                                             *
*     Errors        :   0 (E)                                             *
*     System errors :   0 (S)                                             *
*     System Limits :   0 (L)                                             *
*                                                                         *
*                                                                         *
***************************************************************************
```

```
PROGRAM sample (INFILE,OUTFILE);
VAR
   magazin,waitposition: FRAME;
   gripper_down:         ROTATION;
   obstacle,save_pos:    VECTOR;
   cyclenumber,i:        INTEGER;

BEGIN_PROGRAM
   obstacle:= VECTORC(50,100,71);    (* Frames could be defined   *)
   save_pos:= VECTORC(50,0,100);     (* by teach-in also          *)
   gripper_down:= ROTC(VECTORC(0,1,0),180); (*Gripper points down*)
   waitposition:= FRAMEC(VECTORC(60,-20,10),gripper_down);
   magazin:= FRAMEC(VECTORC(100,100,20),gripper_down);
   WRITE (' Give cycle number : ');
   READ (cyclenumber);
   IF ((cyclenumber >= 1) AND (cyclenumber <= 10)) THEN
     FOR i:= 1 TO cyclenumber DO
       PTPMOVE TO waitposition;
       CLOSE GRIPPER;
       SMOVE TO magazine;
          WITH VIAFRAMES (obstacle,save_pos)
          WITH VELOCITY = 0.5
          WITH ACCELERATION = 0.33;
       OPEN GRIPPER;
       WRITE (' Part ', i:2);
     END_FOR
   ELSE
     WRITE (' Wrong part number !')
```

```
      END_IF
END_PROGRAM.

```

Program section 8.1: Compiler listing of a sample SRL program

- PASCAL system libraries (PASLIB, real arithmetic, transcendental functions and others)
- An assembler (optional for the robot interface)

The PASRO implementation consists of two libraries:

- PASROLIB (robot-independent part with geometric operations and input/ouptut)
- ‹robot short form›LIB with control procedures (e.g., MRLIB for the Microrobot and JRLIB for the Mitsubishi robot)

System procedures are written reentrant and have clear interfaces for special adaptations or extensions. The modular structure is necessary for an easy change of coordinate transformations and robot control modules for other types of robot.

The PASCAL programming system for a PASRO implementation should include the following facilities:

- Declaration of external procedures (indispensable)
- Possibility of compiling modules or single procedures (indispensable)
- Including of source files by the PASCAL compiler ("include" commands)
- Standard functions `sin`, `cos`, `arctan`, `sqrt`, `trunc` and `round` (indispensable)
- Optional nonstandard functions `length` and `concat`; e.g., for the output to the interface of the Mitsubishi robot
- User libraries
- Linker selection of referenced external procedures from the PASRO libraries
- Optional linking of routines written in assembler with parameter integration of procedure calls in the PASCAL program

The user writes his PASRO program, which is then translated by the PASCAL compiler. The compiler includes the necessary declarations and variables of PASRO according to the include directives. The program has to be structured in a given scheme. The translation generates a machine code program and the linker automatically inserts the necessary PASRO system procedures from the PASRO libraries (see Fig. 8.18).

The user does not need to write down the PASRO specific types, variables and procedures. They are included in the following files:

`PASROTYP.SYS`	PASRO type declaration
`PASROVAR.SYS`	PASRO system variables
`PASROEXP.SYS`	PASRO system procedures
`PROGBEG.SYS`	PASRO program initialization

Mnemotechnical IRDATA-code (preliminary version, February 1985) generated by the SRL compiler, Universität Karlsruhe, Institut für Informatik III, Roboter-Gruppe.

```
1,PBEG,2,45,4,'sample',2,INFILE,OUTFILE;
2,BLBEG,0,1,I,1,FRA,1,I,1,I,1,VEC,1,VEC,1,ROT,1,FRA,1,FRA,1;
3,GEN,VEC,%ST,I,#,50,I,#,100,I,#,71;
4,MOVDAT,%BR,0,4,VEC,%ST;
5,GEN,VEC,%ST,I,#,50,I,#,0,I,#,100;
6,MOVDAT,%BR,0,5,VEC;%ST;
7,GEN,VEC,%ST,I,#,0,I,#,1,I,#,0;
8,GEN,ROT,%ST,VEC,%ST,I,#,180;
9,MOVDAT,%BR,0,6,ROT,%ST;
10,MUL,I,%ST,I,#,-1,I,#,20;
11,TYPCON,R,%ST,I,%ST;
12,GEN,VEC,%ST,I,#,60,R,%ST,I,#,10;
13,GEN,FRA,%ST,VEC,%ST,ROT,%BR,0,6;
14,MOVDAT,%BR,0,8,FRA,%ST;
15,GEN,VEC,%ST,I,#,100,I,#,100,I,#,20;
16,GEN,FRA,%ST,VEC,%ST,ROT,%BR,0,6;
17,MOVDAT,%BR,0,7,FRA,%ST;
18,DATOUT,2,#,1,STR,' Give cycle number : ';
19,DATIN,2,#,2,I,%BR,0,2;
20,GE,B,%ST,I,#,1,I,%BR,0,2;
21,LE,B,%ST,I,#,10,I,%BR,0,2;
22,AND,B,%ST,B,%ST,B,%ST;
23,IF,%ST,42;
24,MOVDAT,%BR,0,3,I,#,1;
25,FOR,0,3,#,1,%BR,0,2,41;
26,MOVE,V,A,FRA,%BR,0,8;
27,CLOSE,#,0;
28,REMARK,'  CLOSE is not an order of IRDATA-level-1';
29,FEDRAT,1,R,#,5.00000000000000E-01;
30,ACCEL,1,R,#,3.30000000000000E-01;
31,GEN,FRA,%ST,VEC,%BR,0,4,FRAORI,%BR,0,7;
32,GEN,FRA,%ST,VEC,%BR,0,5,FRAORI,%BR,0,7;
33,MOVE,L,V,A,FRA,%ST;
34,MOVE,L,V,FRA,%ST;
35,MOVE,V,FRA,%BR,0,7;
36,OPEN,#,0;
37,REMARK,'  OPEN is not an order of IRDATA-level-1';
38,DATOUT,2,#,1,STR,' Part ';
39,DATOUT,2,#,1,I,%BR,0,3,#,2;
40,FOREND,25;
41,GOTO,#,43,0;
42,DATOUT,2,#,1,STR,' Wrong part number !';
43,BLEND;
```

```
44,PSTOP,1;
45,PEND;
```

Mnemotechnical representation of IRDATA:

- \# means constant
- %ST means operand on stack
- %BR indicates a block relative address, which is given as 2 integers for block nesting count and index (instead of one code number)

Non-standard IRDATA elements:

OPEN	open gripper (in IRDATA an output to a port)
CLOSE	close gripper, see OPEN
MOVEDAT	type corresponding to the target address is left out
GEN	a vector can be generated by integer values

There are one integer and one frame variable predefined in the program block for internal reasons.

Program section 8.2: IRDATA code generated by the SRL compiler

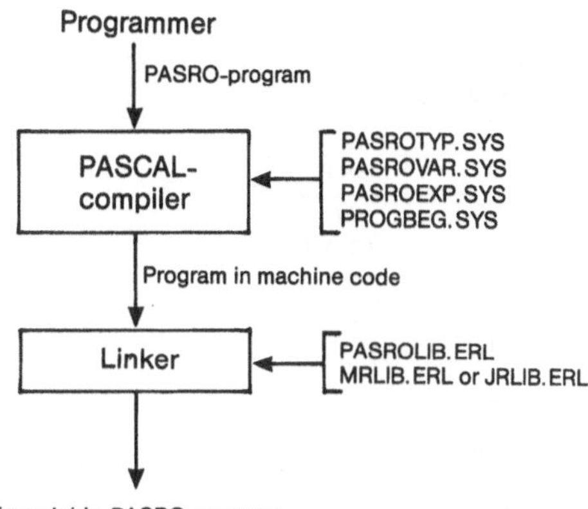

Fig. 8.18: Structure of compiling and linking in PASRO

If a PASCAL/MT+ compiler is used, the include command `$I <source file>` may be utilized. It reads the given source files from disk and includes them in the compilation. The following sequence has to be included by the programmer:

```
PROGRAM <program_name> ( <parameter> );

< user constants >

(*I PASROTYP.SYS*) (* Read PASRO system types *)
< user type declarations >
```

```
(*I PASROVAR.SYS*)  (* Read PASRO system variables *)
< user variable declarations >
(*I PASROEXP.SYS*)  (* Read PASRO system procedure
                       declarations *)
< user procedures >
(*I PASROBEG.SYS*)  (* Begin of a PASRO program *)
      .
      .
      .
< PASRO and PASCAL statements >
      .
      .
      .
END. (* program end *)
```

PASRO has been implemented for different robots and different computers. Fig. 8.19 shows the implementation on an IBM PC for the Mitsubishi robot RM-501 under the operating system CP/M and DOS. The interface between the IBM PC and the robot controller is a Centronics-like parallel printer port.

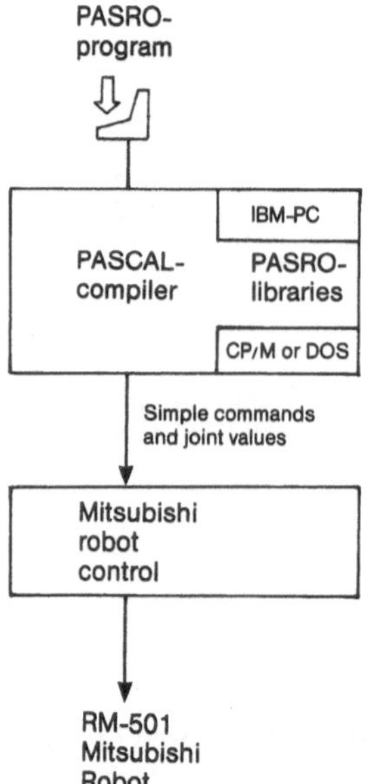

Fig. 8.19: Structure of a PASRO implementation

References

I. General References

[1] Jensen, K., Wirth, N.: *PASCAL, User Manual and Report.* Springer, New York Berlin Heidelberg Tokyo (1975)

[2] Mujtaba, S., Goldman, R.: *AL User's Manual.* Stanford University, Palo Alto, California (1979)

[3] Blume, C.: *AL - ein textuelles Programmiersystem für Industrieroboter* (AL - A Textual Programming System for Industrial Robots). PDV-Berichte, KfK-PDV 187, Karlsruhe (1980)

[4] Blume, C.: *Vorstellung der Sprache AL und die Besonderheiten der Implementierung an der Universität Karlsruhe* (Presentation of the Language AL and the Specialities of Its Implementation at the University of Karlsruhe). In: Verbesserte Programmierung von Robotern in der Montage (In: Improved Programming of Robots in Assembly). PFT-Entwicklungsnotiz, KfK-PFT-E2, Karlsruhe (1981), pp. 17-53

[5] Jakob, W.: *Der AL-Compiler aus der Sicht des Benutzers* (The AL-Compiler from the User's Viewpoint). In: Verbesserte Programmierung von Robotern in der Montage. (In: Improved Programming of Robots in Assembly). PFT-Entwicklungsnotiz, KfK-PFT-E2, Karlsruhe (1981), pp. 54-76

[6] Jakob, W.: *Entwurf und Implementierung eines Compilers für eine höhere Manipulatorsprache zur Programmierung eines Roboters* (Design and Implementation of a Compiler for a high level Programming Language for Industrial Robots). Thesis, Universität Karlsruhe (1980)

[7] Blume, C.: *A Structured Way of Implementing the High Level Programming Language AL on a Mini- and Microcomputer Configuration.* Proceedings of 11th ISIR, Tokyo (1981)

[8] Blume, C.: *VAL - ein Roboterkontrollsystem von Unimation* (VAL - A Robot Control System from Unimation). Bearbeitete Übersetzung der VAL-Beschreibung (Abridged Translation of the VAL-Description). Universität Karlsruhe (1980)

[9] *User's Guide to VAL. A Robot Programming and Control System.* Version 12, Unimation Robotics Inc., Danbury, Connetticut (1980)

[10] *Guide to HELP-Language.* Digital Electronic Automation (DEA), Turin

[11] *PRAGMA A 3000 - Operating and Programming Manual.* Digital Electronic Automation (DEA), Turin

[12] Salmon, R.: *SIGLA - The Olivetti Sigma Robot Programming Language.* 8th ISIR, Stuttgart 1978

[13] *SIGMA/MTG - Handbuch und Programmierung* (SIGMA/MTG - Programming and Reference Manual). Release A 10, Olivetti NC-Systeme GmbH

[14] Weck, M., Zühlke, D.: *Short Reference Manual of the ROBEX-Language.* Technische Hochschule Aachen, (1980). In: Verbesserte Programmierung von Robotern in der Montage. (In: Improved Programming of Robots in Assembly). PFT-Entwicklungsnotiz, KfK-PFT-E2, Karlsruhe (1981), pp. 95-107

[15] Weck, M., Eversheim, W., Zühlke, D.: *ROBEX - Ein Programmiersystem für numerisch gesteuerte Handhabungsgeräte* (ROBEX - A Programming System for Numerical Controlled Robots). In: Verbesserte Programmierung von Robotern in der Montage (In: Improved Programming of Robots in Assembly). PFT-Entwicklungsnotiz, KfK-PFT-E2, Karlsruhe (1981), pp. 77-94

[16] *RAIL Reference Manual. AUTOVISION.* Automatix Inc., Burlington, Massachusetts (1981)

[17] *RAIL Software Reference Manual. ROBOVISION and CYBERVISION.* Automatix Inc., Burlington, Massachusetts (1982)

[18] Blume, C., Jakob, W.: *Design of the Structured Robot Language (SRL).* Proceedings of Advanced Software for Robotics, Liège (1983)

[19] Blume, C., Jakob, W.: *PASRO - Pascal for Robots.* Springer, Berlin Heidelberg New York Tokyo (1985)

[20] *A Manufacturing Language - Concepts and Users's Guide.* IBM Corporation, Boca Raton, Florida (1982)

[21] *A Manufacturing Language - Reference.* IBM Corporation, Boca Raton, Florida (1983)

[22] *AML/Entry, Version 3, Users's Guide.* IBM Corporation, Boca Raton, Florida (1984)

[23] *AML/Entry, Version 4, Users's Guide.* IBM Corporation, Boca Raton, Florida (1984)

[24] *User's Guide to VAL II - A Robot Programming and Control System.* Unimation Inc., Danbury, Connetticut (1983)

[25] *IRDATA, VDI-Richtlinie 2863, Blatt 1.* Beuth Verlag, Berlin, Köln, VDI-Verlag, Düsseldorf 1986 (in German, english translation (ISO proposal) in preparation)

II. Chapter Specific References

[2.1] Coole, A.: *Macro Processors.* Cambridge, University Press, Cambridge (1976)

[2.2] Rohlfing, H.: *SIMULA.* Bibliographisches Institut, Mannheim Wien Zürich (1973)

[2.3] Wettstein, P.: *Aufbau und Struktur von Betriebssystemen* (Design and Structure of Operating Systems). Hansa Verlag, München (1978)

[2.4] Denavit, J., Hartenberg, R.: *A Kinematik Notation for Lower Pair Mechanisms Based on Matrices.* Journal of Applied Mechanics, vol. 22, Trans. ASME Vol. 77 (1955)

[2.5] Blume, C.: *Sprachen und Programmiersysteme für Industrieroboter* (Languages and Programming Systems for Industrial Robots). Lecture notes, Universität Karlsruhe, (1982)

[2.6] Paul, R.: *Robot Manipulators.* MIT Press, Cambridge, Massachusetts (1981)

[2.7] Paul, R.: *Modelling Trajectory Calculation and Servoing of a Computer-Controlled Arm.* Stanford Artifical Intelligence Laboratory, AI Memo 177, Stanford University, Palo Alto, California (1972)

[2.8] Binder, D.: *Interpolation in numerischen Bahnsteuerungen* (Interpolation in numerical path control). Springer, Berlin Heidelberg New York Tokyo (1979)

[2.9] Weck, M.: *Werkzeugmaschinen* (Machine Tool). Band 3. VDI-Verlag, Düsseldorf (1979)

[2.10] Liebermann, L., Wesley, M.: *AUTOPASS: An Automatic Programming System for Computer-Controlled Mechanical Assembly.* IBM J. Res. Dev. 21.4 (1977)

[2.11] Bernorio, M., et al.: *Programming an Industrial Robot in Italian.* 7th ISIR, Tokyo (1977)

[2.12] Maroy, J., Berthod, M.: *Natural Language Understanding by a Robot: A Pattern Recognition Problem.* Pattern Recognition, vol. 10 (1978), pp. 63-71

[2.13] Taylor, R.: *The Synthesis of Manipulator Control Programs from Task-Level Specifications.* Dissertation, Stanford University, Palo Alto, California (1976)

[2.14] Blume, C.: *Implicit Programming based on a High Level Explicit System.* In: Rathmill, K. (ed.) Robotics Assembly. Springer, New York Berlin Heidelberg Tokyo (1985)

[3.1] Stoer, J.: *Einführung in die numerische Mathematik* (Introduction to Numerical Mathematics). Springer, Berlin Heidelberg New York Tokyo (1972)

[3.2] Wilkinson, J.H.: *Rundungsfehler* (Rounding Errors). Springer, Berlin Heidelberg New York Tokyo (1969)

[4.1] Paul, R., Luh, J., et al.: *Advanced Industrial Robot Control Systems.* 4th Report, Purdue University, Lafayette, Indiana (1980)

[4.2] Ambler, A., Beattie, R., Corner, D.: *RAPT1 - User's Manual.* University of Edinburgh (1982)

[4.3] Lundstroem, G., Glemme, B., Rooks, B.: *Industrial Robots-Gripper Review.* International Fluidics Services Ltd., Oxford (1977)

[4.4] Auer, H.: *Beitrag zur Steigerung der Flexibilität von Handhabungseinrichtungen im Bereich der Einzel- und Kleinserienfertigung* (A Contribution to the Improvement of the Flexibility of Robots in the Area of Customized and Low-Volume Serial Production). Dissertation, Technische Universität Berlin, West-Berlin (1977)

[4.5] Liebermann, L., et al.: *Three Dimensional Modelling for Automated Mechanical Assembly.* Research Report of the IBM Watson Research Center, New York (1979)

[4.6] Dijkstra, E. W.: *Structured Programming.* In: Software Engineering Techniques. Conference proceedings, Rome (October 1969). NATO-Science Committee, Brussels (1969)

[4.7] Dahl, O.J., Dijkstra, E.W., Hoare, C.A.R.: *Structured Programming.* Academic Press, New York (1972)

[4.8] Miller, E.F., Lindamood, G.E.: *Structured Programming: Top Down Approach.* In: Datamation 19 (1973) p. 55

[4.9] Schnupp, P.: *Systemprogrammierung* (System Programming). De Gruyter, Berlin (1975)

[4.10] Schnupp, P., Floyd, C.: *Software, Programmentwicklung und Projektorganisation* (Software, Program Development and Project Organisation). De Gruyter, Berlin (1976)

[4.11] Denning, P.J.: *Is it not time to define "Structured Programming"?.* ACM Operating System Review 8/1, (1974) p. 6

[4.12] Zelkowitz, M.V.: *It is not time to define "Structured Programming".* ACM Operating System Review 8/2, (1974) p. 7

[4.13] Faulk, C.R.: *Yet another attempt to define "Structured Programming".* ACM Operating System Review 8/3, (1974) p. 7

[4.14] Denning, P.J.: *Is "Structured Programming" any longer the right term?.* ACM Operating System Review 8/4, (1974) p. 7

[4.15] Dijkstra, E.W.: *GOTO-Satement Considered Harmful.* In Communications of the ACM 11 (1968) p. 147

[4.16] Nassi, I., Shneidermann, B.: *Flow Chart Techniques for Structured Programming.* In: SIGPLAN Notices 8 (1973) p. 12

[8.1] Blume, C., Dillmann, R.: *Frei programmierbare Manipulatoren - Aufbau und Programmierung von Industrierobotern* (Free Programmable Manipulators - Construction and Programming of Industrial Robots). Vogel Verlag, Würzburg (1981)

[8.2] Sandewall, E.: *Programming in an Interactive Environment: The "LISP" Experience.* Computing Surveys, Vol. 10, No. 1 (1978)

[8.3] Aho, A., Ullmann, J.: *Principles of Compiler Design.* Addison-Wesley, Reading, Massachusetts (1977)

[8.4] Aho, A., Ullmann, J.: *The Theory of Parsing, Translation and Compiling.* Prentice Hall, Englewood Cliffs, New Jersy, (1972-Vol. 1; 1973-Vol. 2)

[8.5] Gries, D.: *Compiler Construction for Digital Computers.* Wiley & Sons, New York Chichester (1971)

[8.6] Bachmann, P.: *Grundlagen der Compilertechnik* (Fundamentals of Compiler Design). München (1975)

[8.7] Liebermann, L.: *Model-Driven Vision for Industrial Automation.* Research Report, IBM Watson Research Center, New York (1978)

[8.8] Lozano-Perez, T., Winston, P.: *LAMA: Language for Automatic Mechanical Assembly.* Proceedings 5th International Joint Conference on Artifical Intelligence, Boston (1977)

[8.9] Elspas, B., Levitt, D., Waldinger, R., Waksman, A.: *An Assessment of Techniques for Proving Program Correctness.* In: ACM Computing Surveys 4 (1972), p. 87

[8.10] Brooks, R., Greiner, R., Binford, T.: *The ACRONYM Model-Based Vision System.* Proceedings of the 6th International Joint Conference on Artifical Intelligence, Vol. 1, Tokyo (1979) pp. 105-113

[8.11] Dillmann, R., Huck, M.: *Intelligent Simulation of Robot Applications.* Symposium on Robot Control 85, Barçelona (1985)

[8.12] Zambuto, D., Chaney, J.: *An Industrial Robot with Minicomputer Control.* 6th ISIR, Nottingham (1976)

[8.13] Kuno, T., et al.: *Robot Performance Simulator.* 9th ISIR, Washington (1979)

[8.14] Kastens, U., Hutt, B., Zimmermann, E.: *GAG: A Practical Compiler Generator.* Lecture Notes in Computer Science, Springer, Berlin Heidelberg New York Tokyo (1982)

Appendix A
SRL

I. Main Elements of SRL

System specification	SYSTEM_SPECIFICATION ROBOT EFFECTOR OPERATION SENSOR ADDRESS REGISTER DIGITAL and ANALOG PORT INTERRUPT SEMAPHOR and SYSFLAG TIME_OUT DATABASE ERROR_NOTICE
Program and module structure	PROGRAM MODULE INCLUDE WITH_MODULE WITH_SYSTEM_SPECIFICATION SECTION PROCEDURE FUNCTION BLOCK
Declarations	CONST TYPE LOCAL_TYPE VAR PERMANENT_VAR
Standard types	INTEGER REAL BOOLEAN CHAR
Geometrical data types	VECTOR ROTATION

	ORIENTATION FRAME
Geometrical data construction	VECTORC ROTC ORIC FRAMEC
Structured types	ARRAY RECORD FILE
Statements	BEGIN END assignement WITH SYSNOPAR PARALLEL
Move statements	DRIVE PTPMOVE SYNMOVE SMOVE LANEMOVE VIAMOVE CIRCLEMOVE MOVE DIRMOVE MOVECONT STOP CALIB FIRSTPOS
Move construction	MOVEDEF MOVEDO
Effector statements	OPEN CLOSE GRIPWIDTH OPERATE GRIP LET MOUNT UNMOUNT
Program flow control	IF IF_SIG FOR WHILE REPEAT CASE EXIT RETURN

Multitasking	START
	EVERY
	AFTER
	SUSPEND
	HOLD
	CANCEL
	CONTINUE
Monitor statement	ALWAYS WHEN
Synchronization	SIGNAL
	WAIT
	INTWAIT
	SEMINIT
Data management	AFFIX
	UNFIX
	ATTRIBUTE
Input/output	INPUT
	OUTPUT
	BUFFER
	OPEN_DATA
	CLOSE_DATA
Arithmetic	Base operations
	Higher mathematical functions
	Geometrical operations

II. Syntax Diagrams

The syntax is presented in diagram form. In some cases there are additional semantic information, e.g., sensor name instead of name.

Program component:

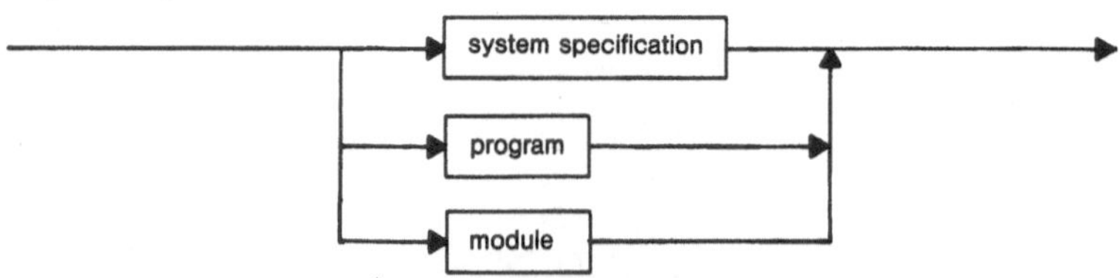

At the beginning of a SRL-program there can be a system specification and modules, which can be compiled separately (indicated by a point).

System specification:

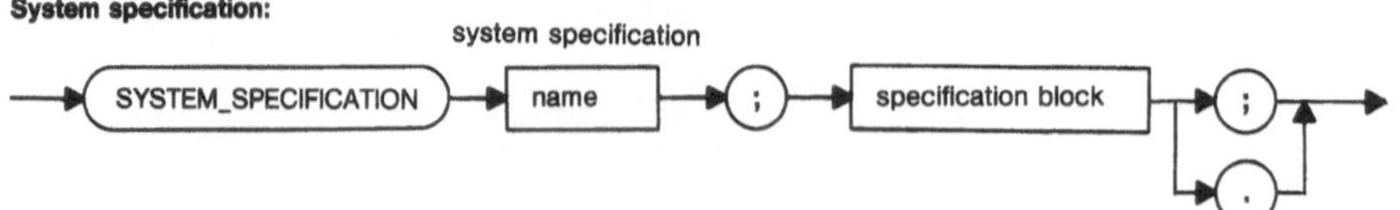

Spezification block:

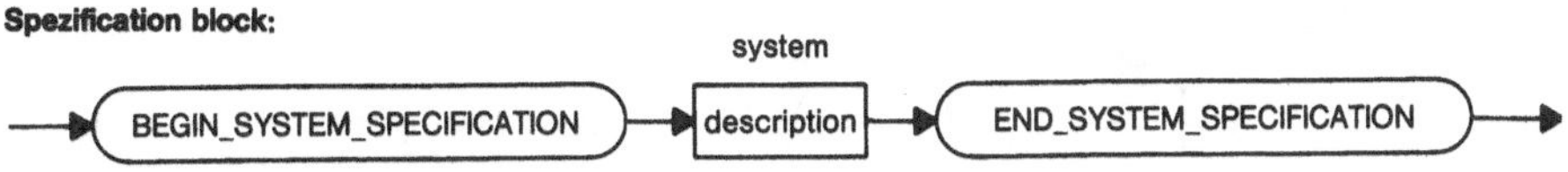

BEGIN_SYSTEM_SPECIFICATION → system description → END_SYSTEM_SPECIFICATION

Description:

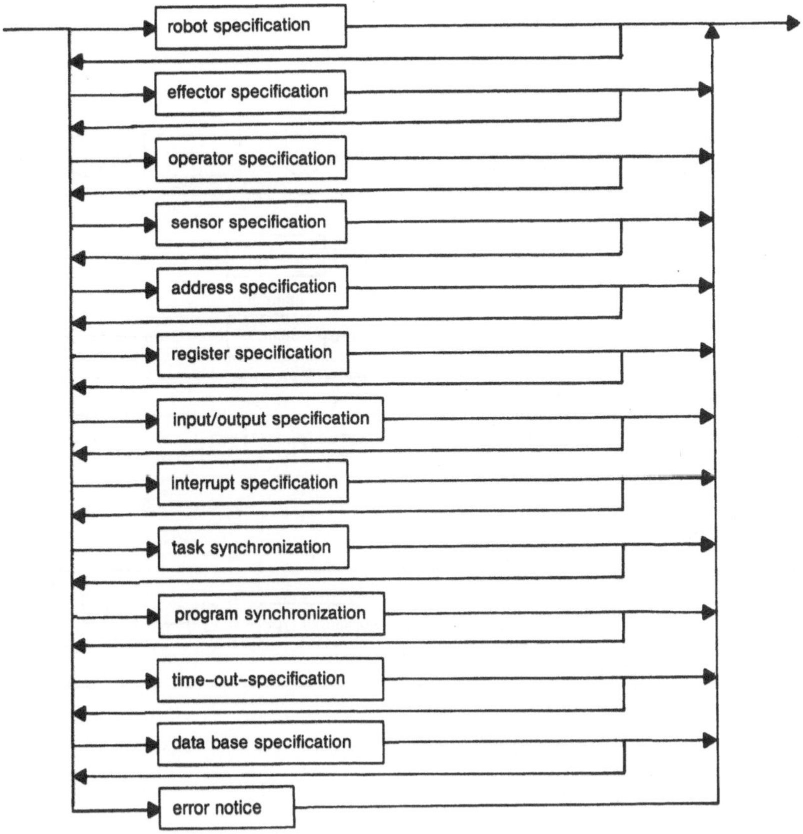

robot specification
effector specification
operator specification
sensor specification
address specification
register specification
input/output specification
interrupt specification
task synchronization
program synchronization
time-out-specification
data base specification
error notice

Robot specification:

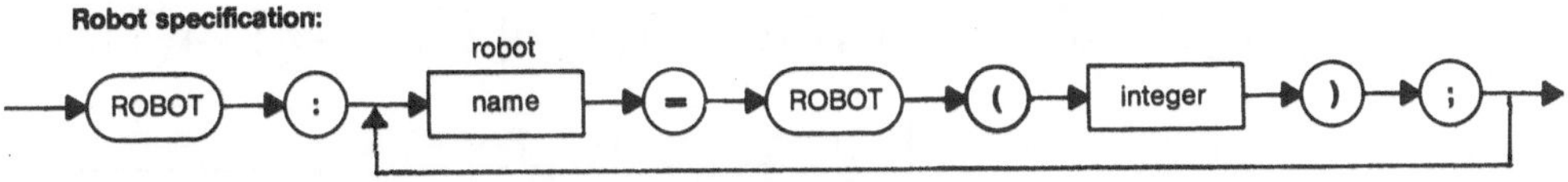

ROBOT → : → robot name → = → ROBOT → (→ integer →) → ;

The logical device numbers of robots, effectors and others are passed to the control unit and the implemented drivers.

Effector specification:

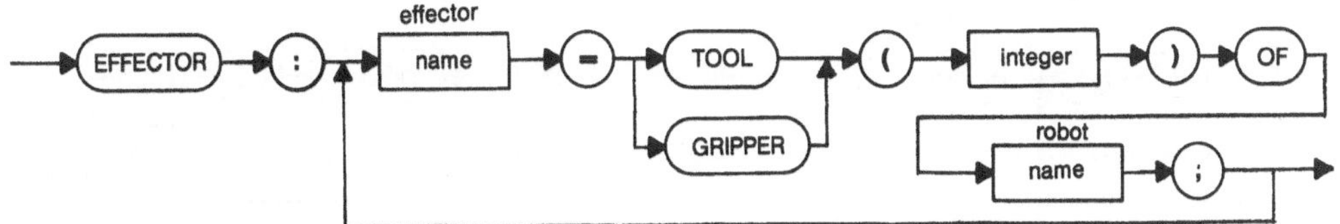

Operation specification:

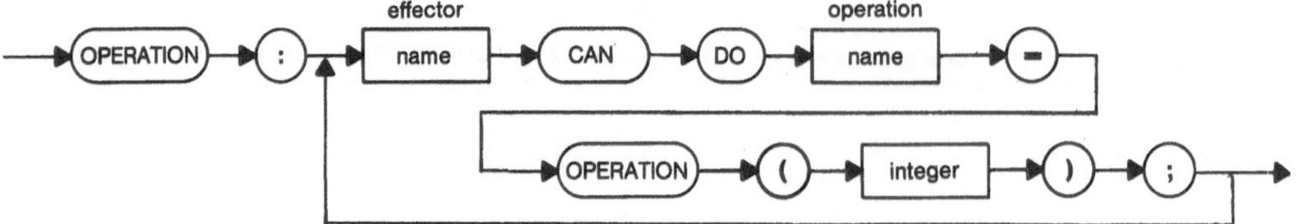

Sensor specification:

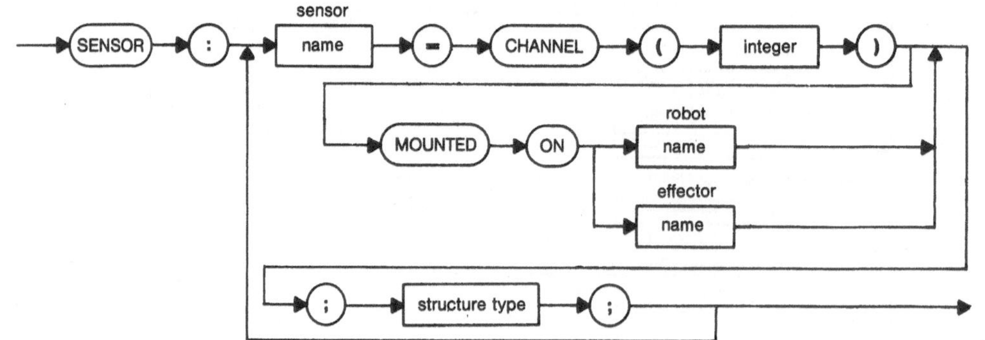

Structure type:

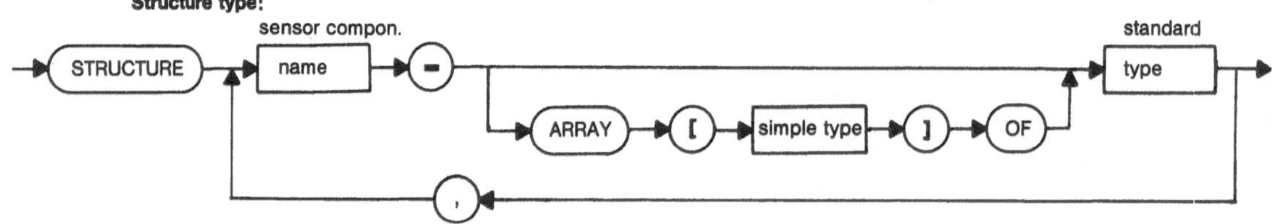

The structure of the sensor data is described which allows an access to the desired component like using a record.

Address specification:

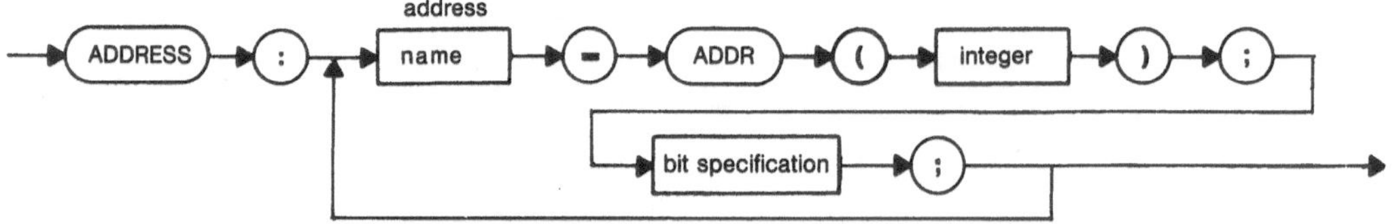

The specified bits of the absolute address can be read or written to a program variable and are then interpreted with respect to the data type of the variable.

Register specification:

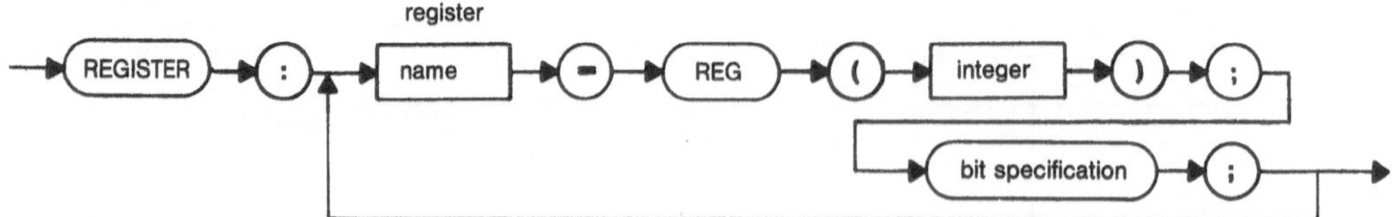

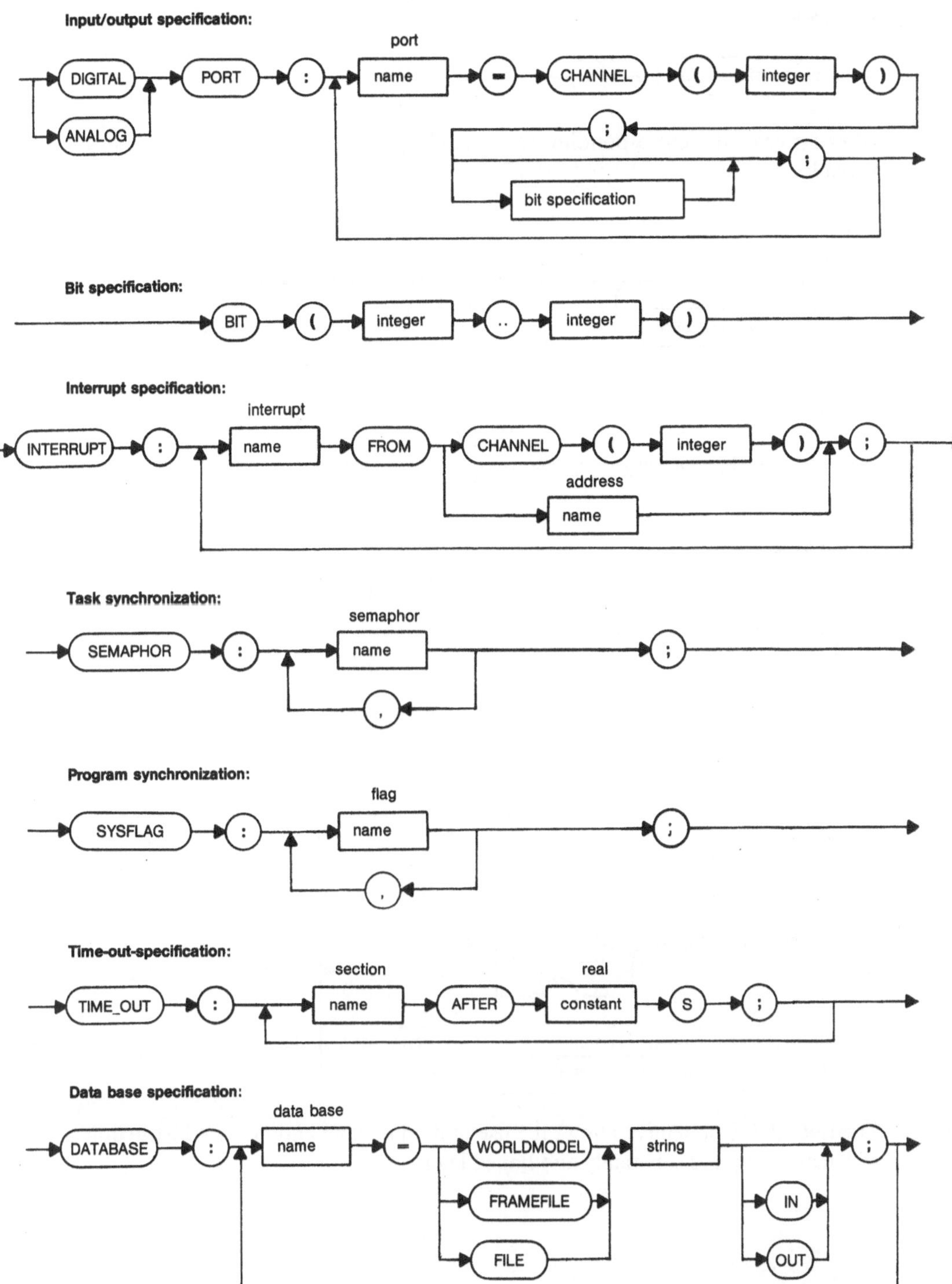

A file variable will be created and assigned to the external file name specified by the string. The type of the file elements is defined in the type declaration part.

Error notice:

After an error the corresponding error number is stored in the specified integer variable.

Program:

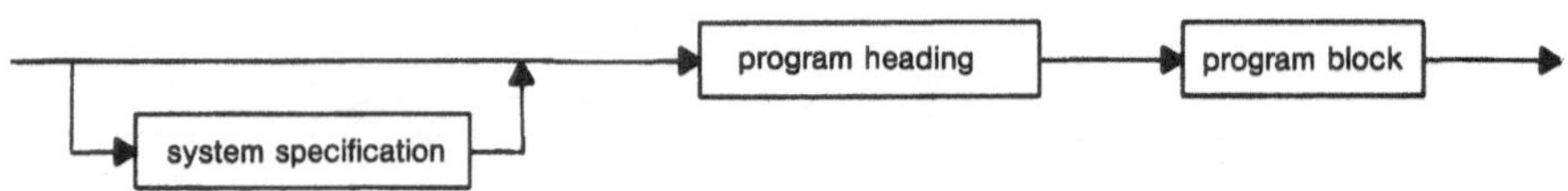

Program heading:

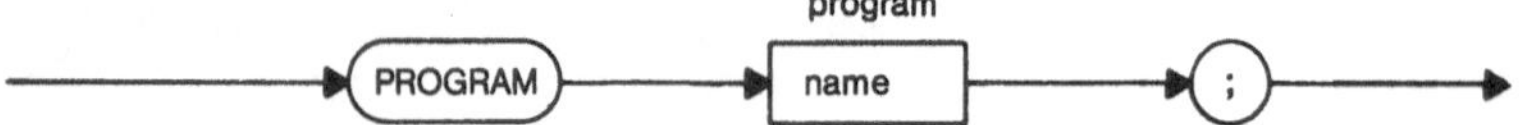

Program block:

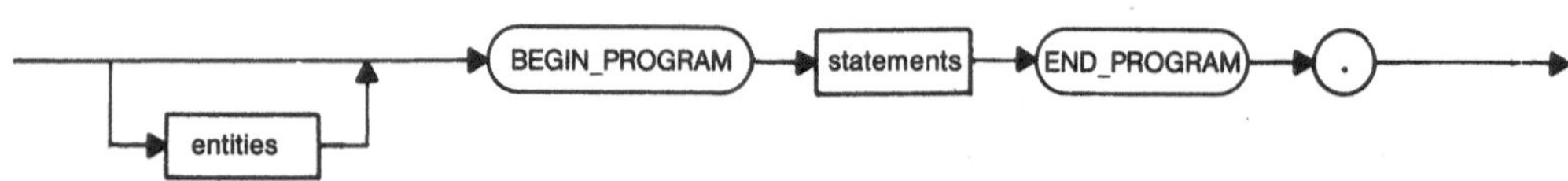

Entities:

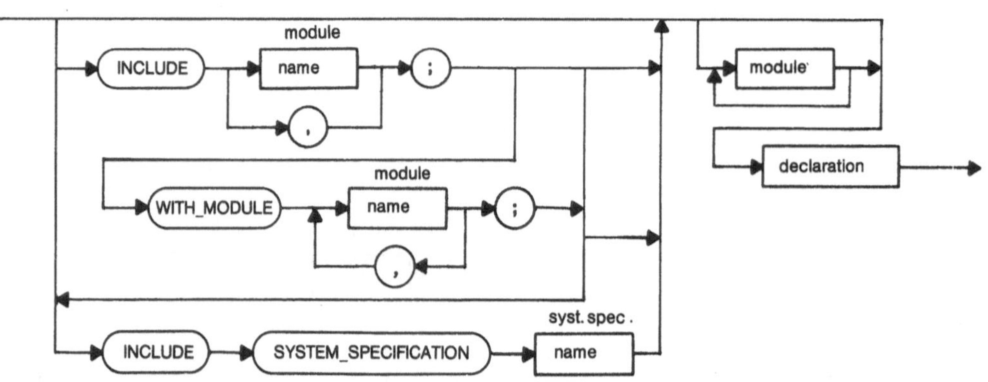

The INCLUDE-statement allows the use of module data and module instructions, which are compiled separately and linked later.

Declarations:

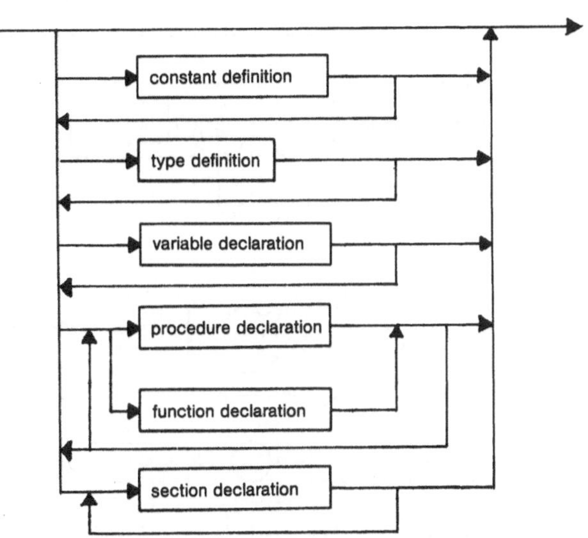

Constant definition:

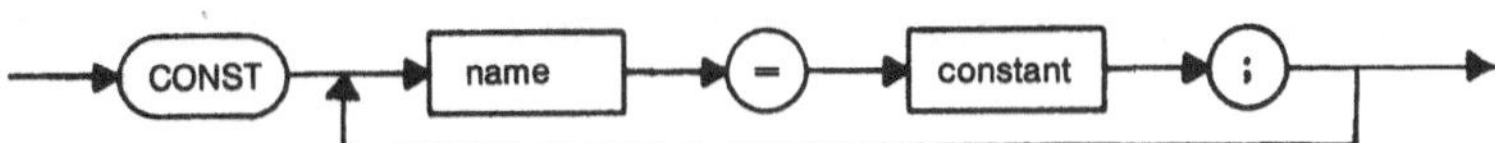

Constant:

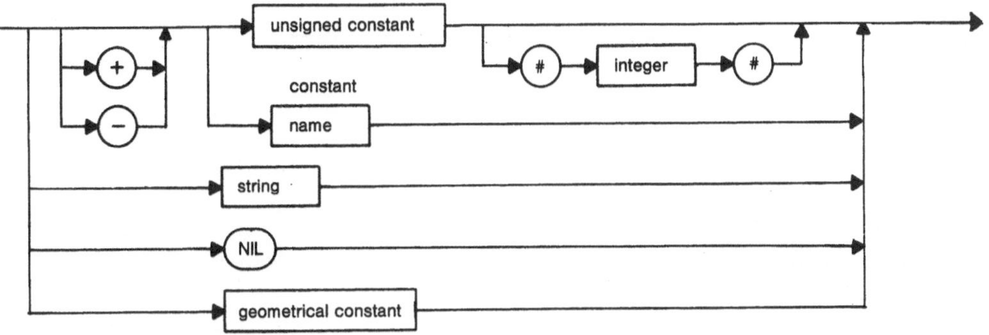

String:

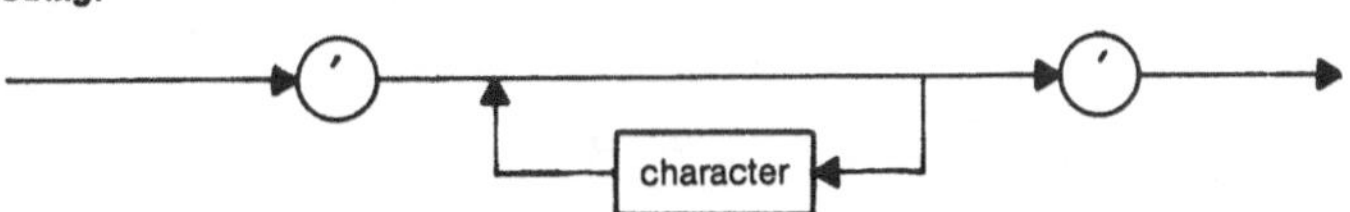

Unsigned constant:

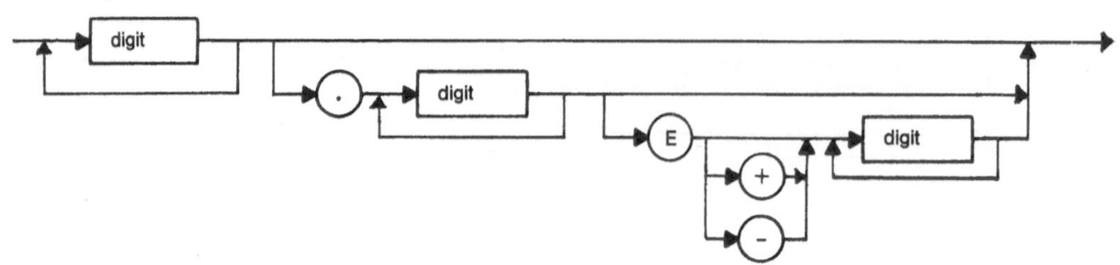

Geometrical constant:

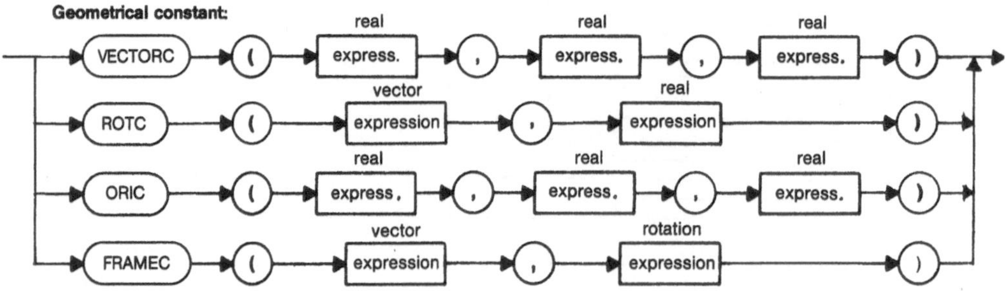

The internal representation of the structured geometrical data variables is calculated and stored in the components. Note that a rotation and a orientation are described by the same matrix.

Type definition:

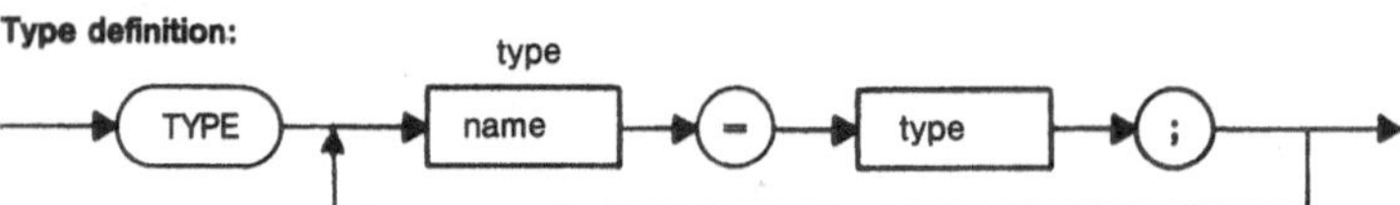

Type:

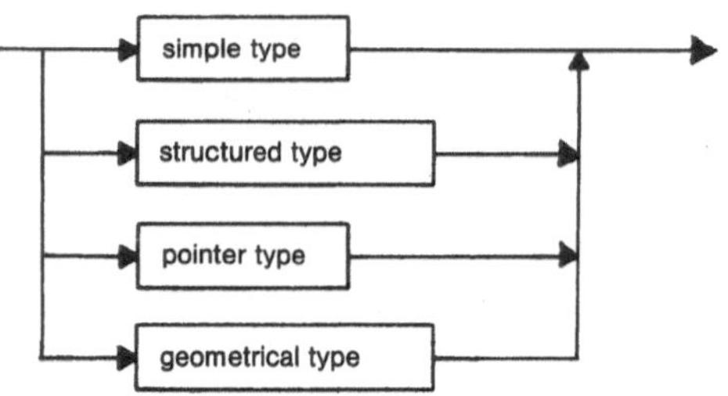

Simple type:

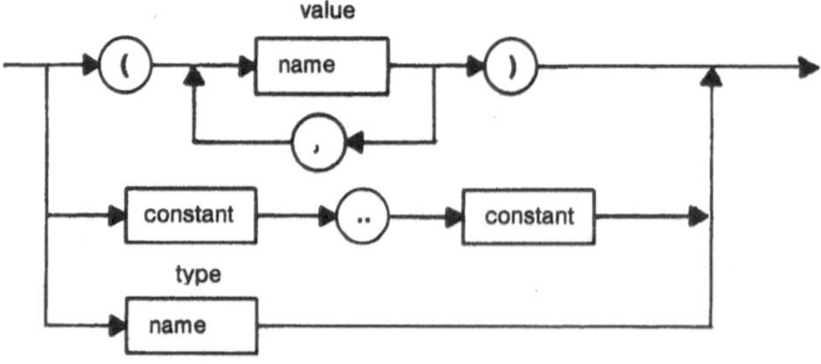

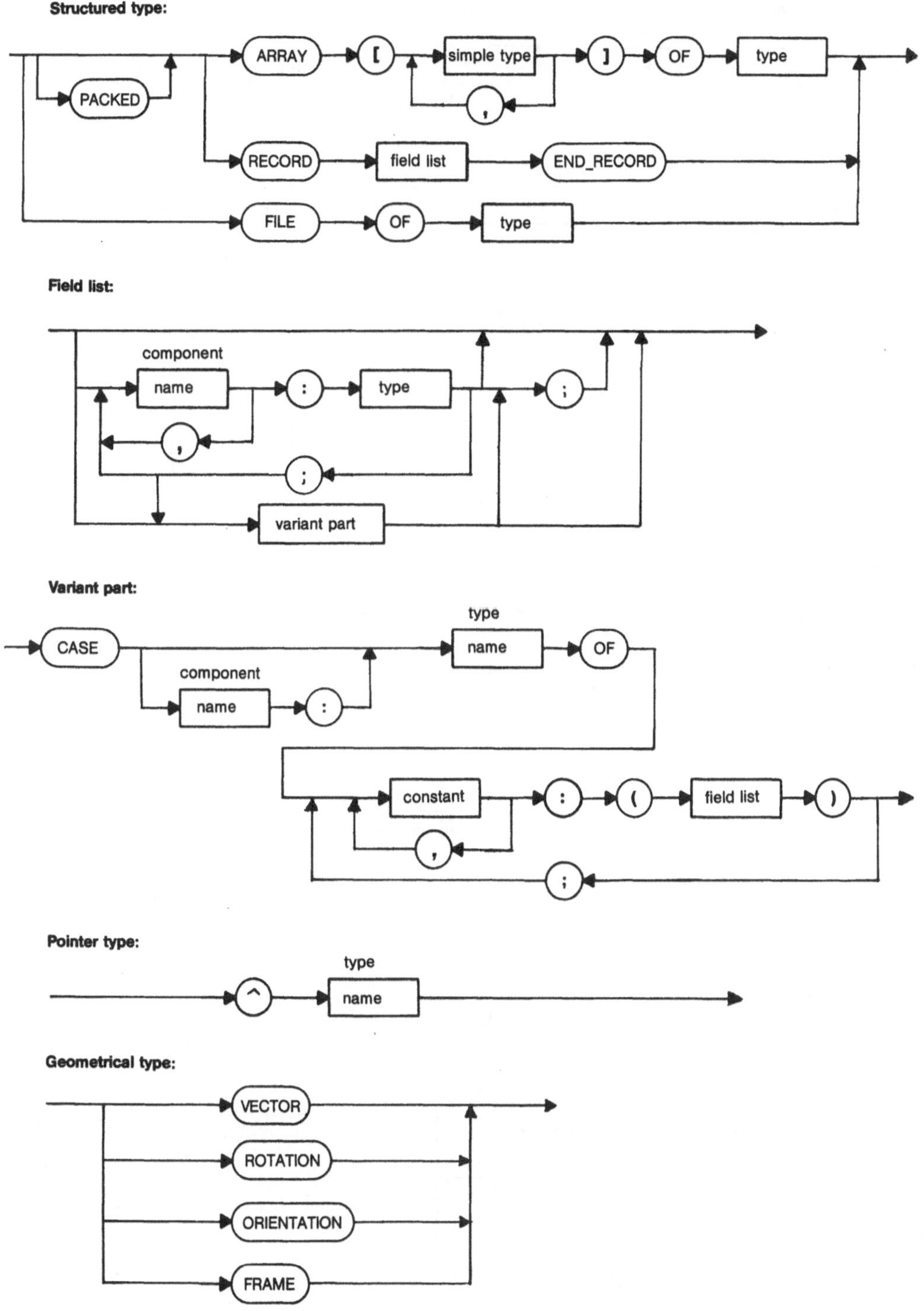

The geometrical data types are noted here for didactical reasons. They are, like INTEGER or CHAR, predefined standard data types.

Procedure declaration:

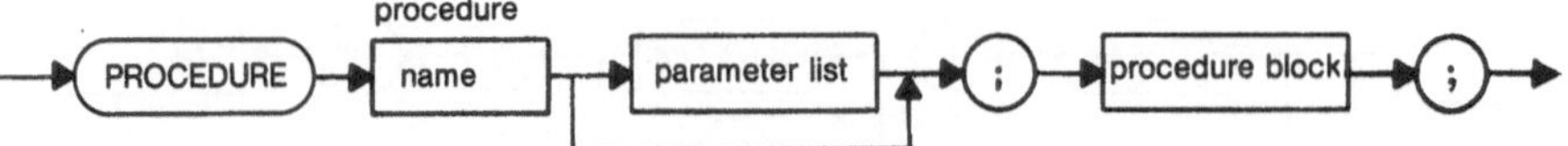

Parameter list:

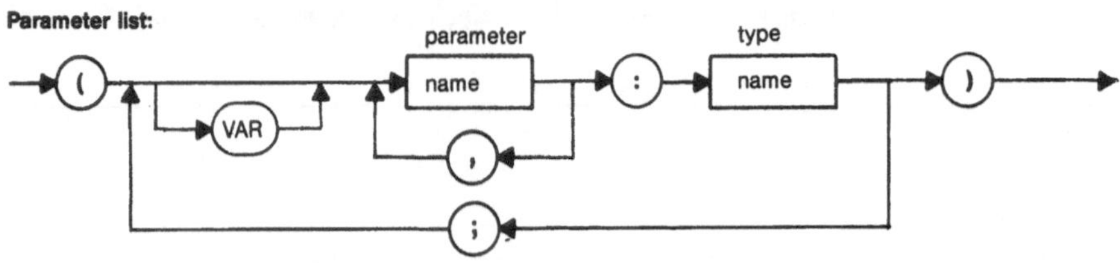

Function declaration:

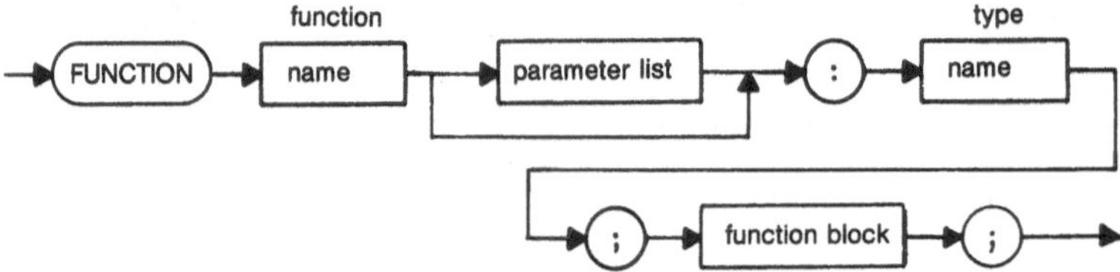

Section declaration:

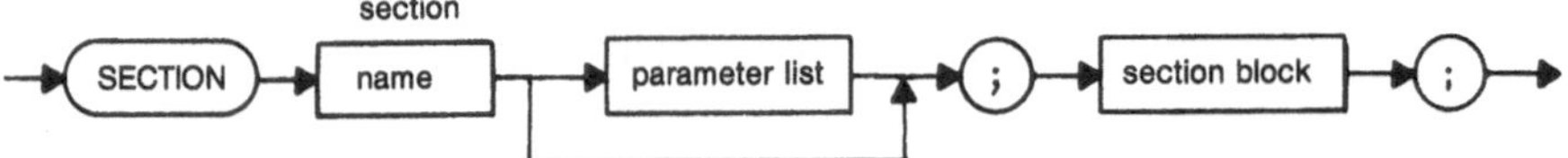

Block declaration:

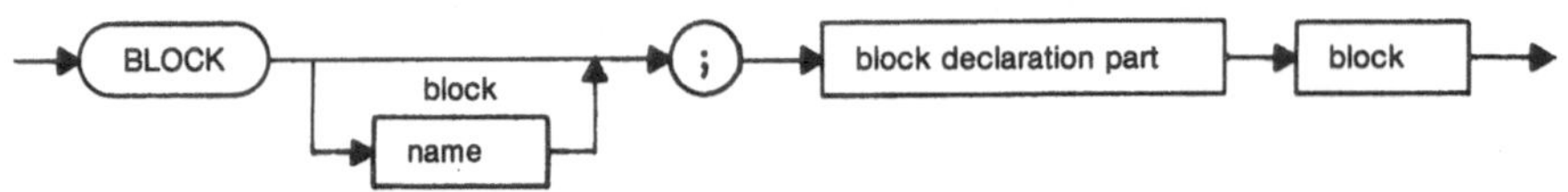

A block may include local data.

Block declaration part:

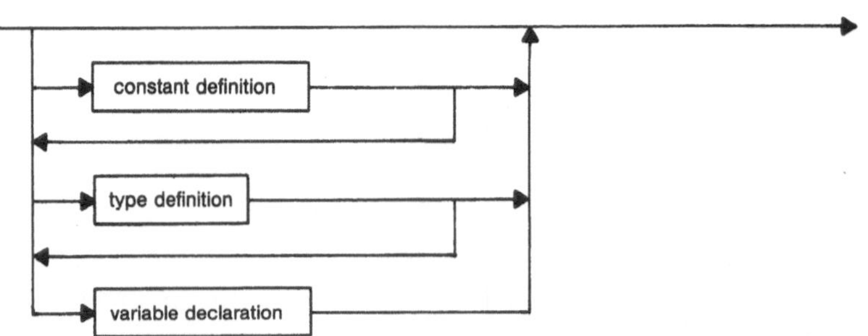

Procedure declaration part:

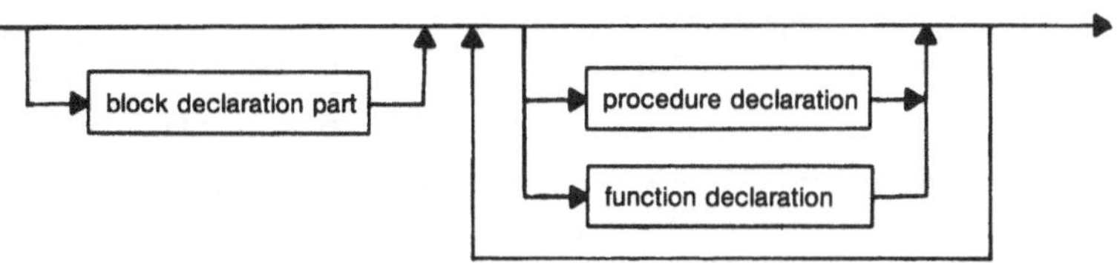

Section declaration part:

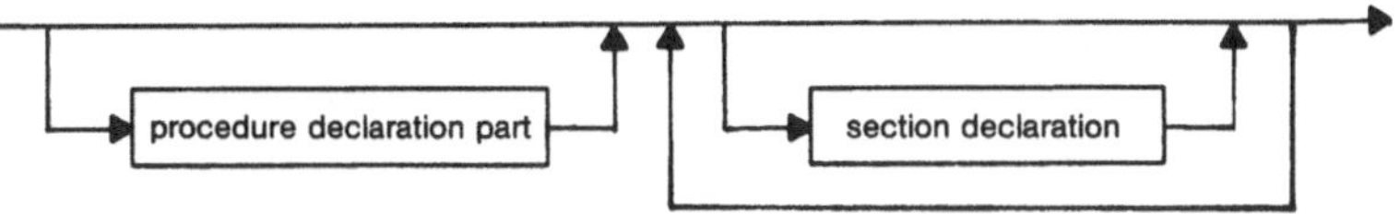

Block:

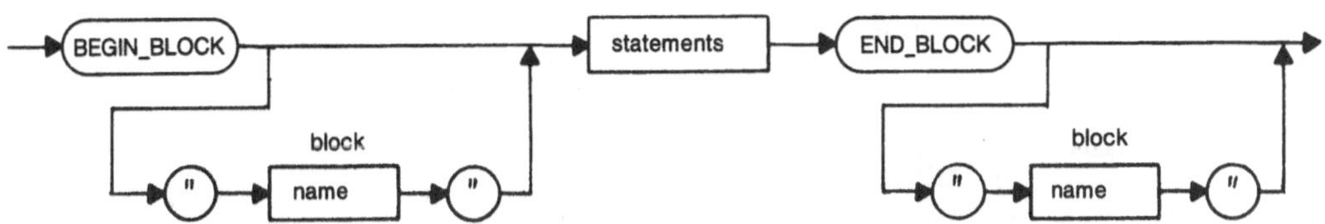

Procedure block:

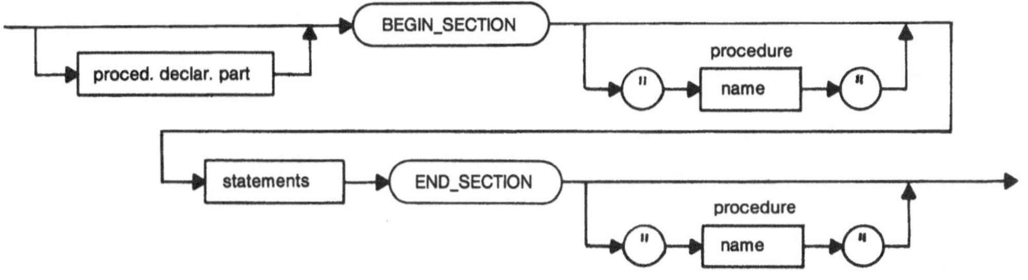

Section block:

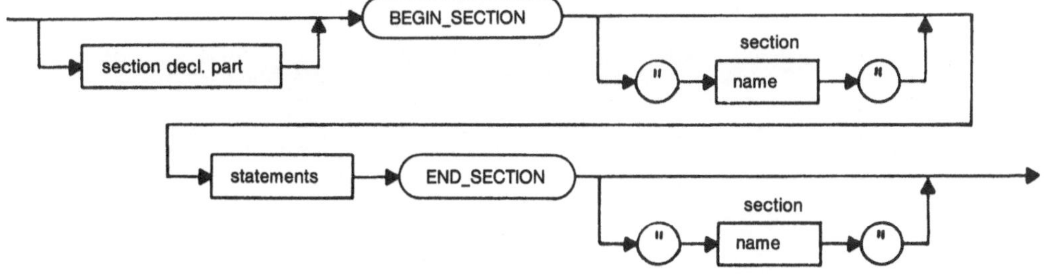

Function block:

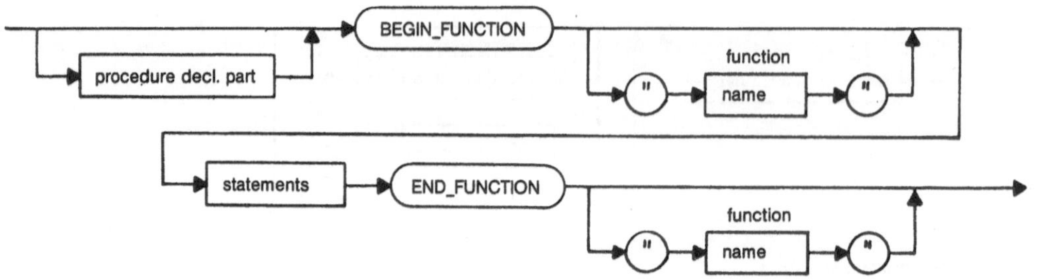

Variable declaration:

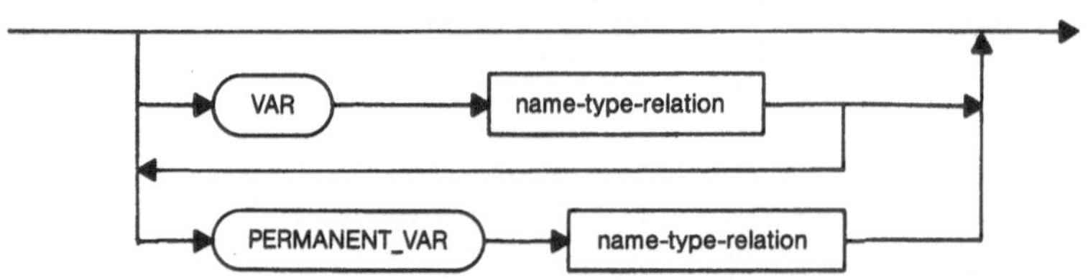

Name-type-relation:

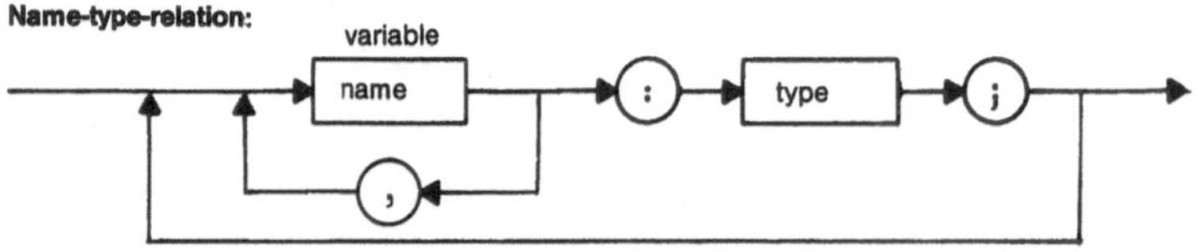

Statements:

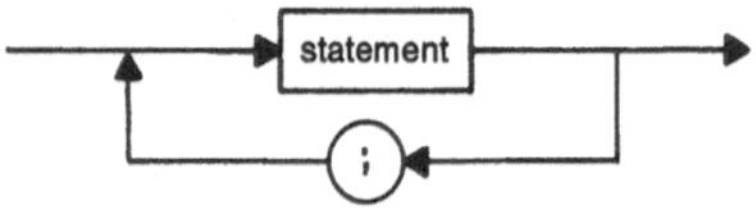

statement

;

Statement:

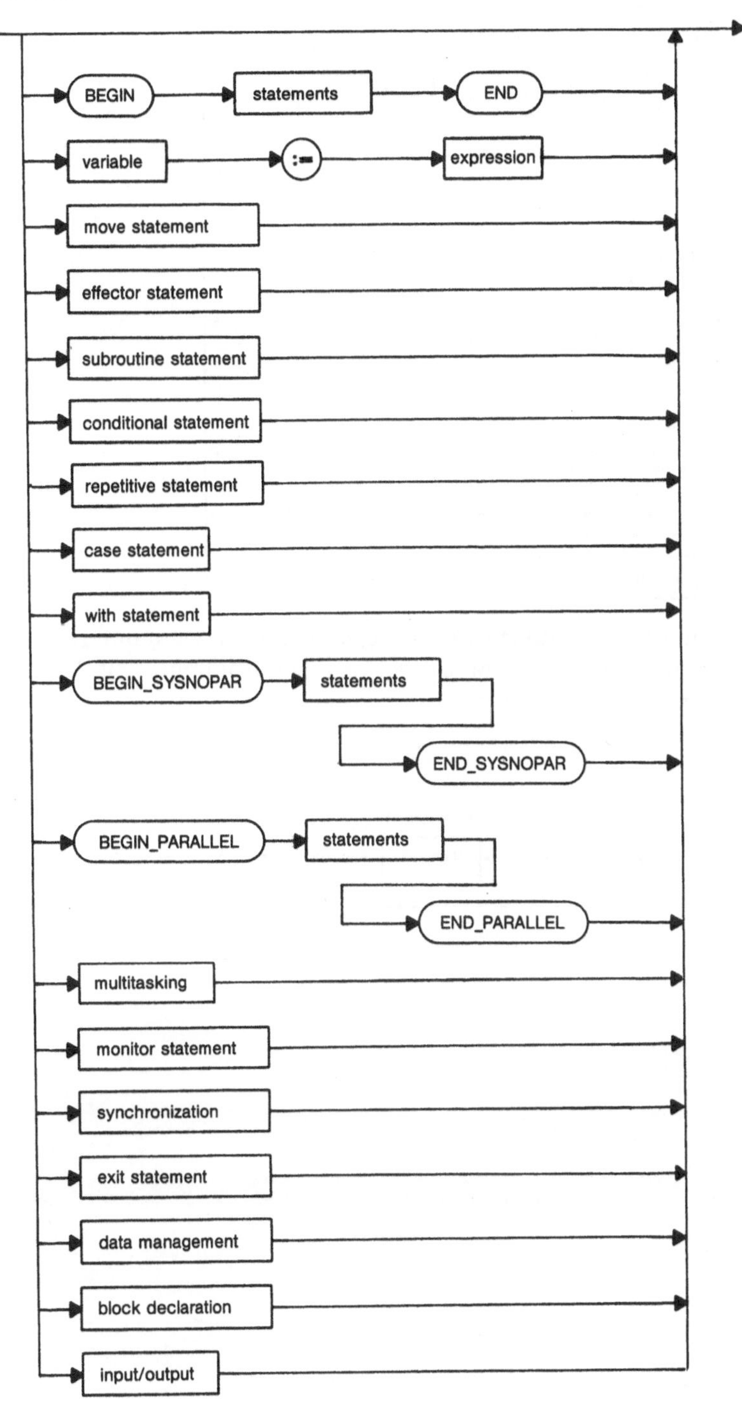
BEGIN
statements
END
variable
:=
expression
move statement
effector statement
subroutine statement
conditional statement
repetitive statement
case statement
with statement
BEGIN_SYSNOPAR
statements
END_SYSNOPAR
BEGIN_PARALLEL
statements
END_PARALLEL
multitasking
monitor statement
synchronization
exit statement
data management
block declaration
input/output

Variable:

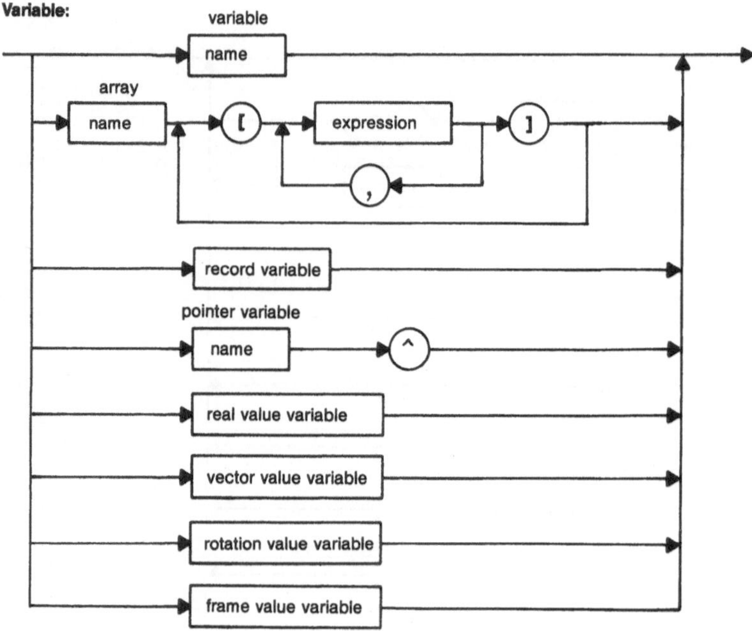

Real, vector, rotation or frame value variable means that the result of different variables and component descriptions is of real, vector, rotation or frame type. It is only mentioned here for didactical reasons.

Record variable:

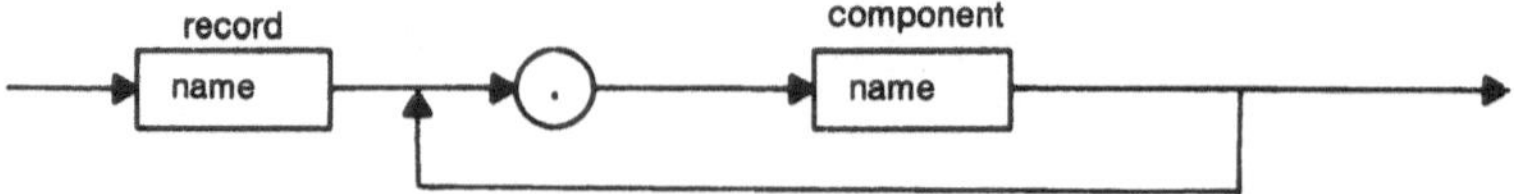

Real value variable:

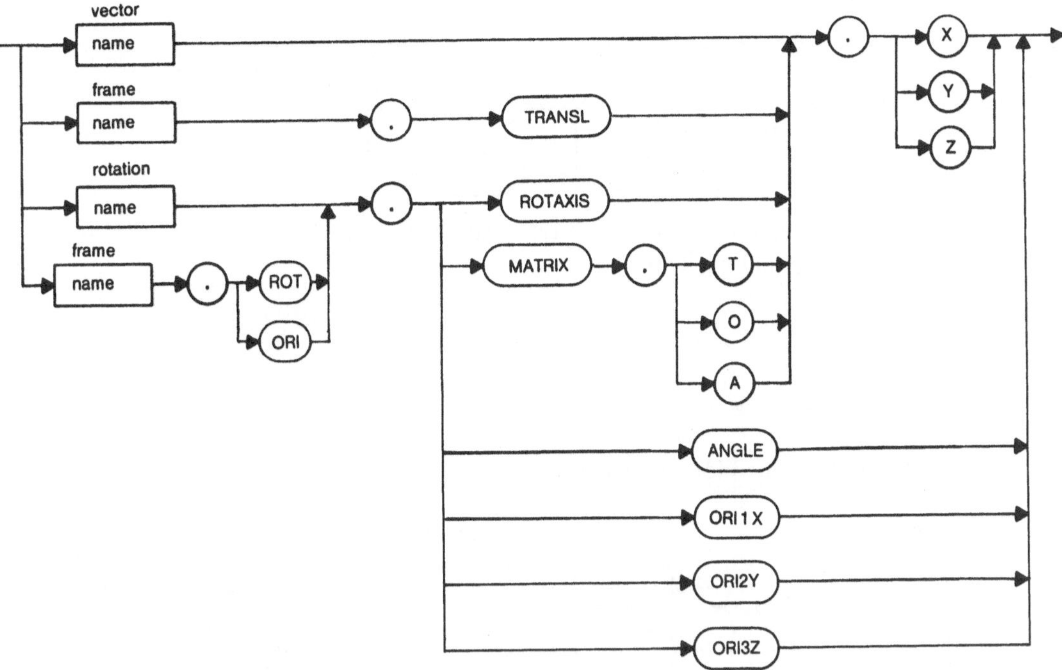

The result of an access to the different variables using the predefined component names results in a value of real type.

Vector value variable:

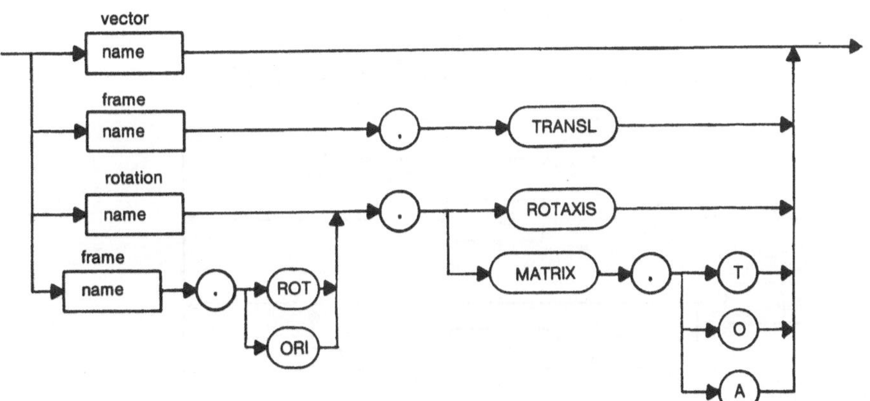

The result is of vector type.

Rotation value variable:

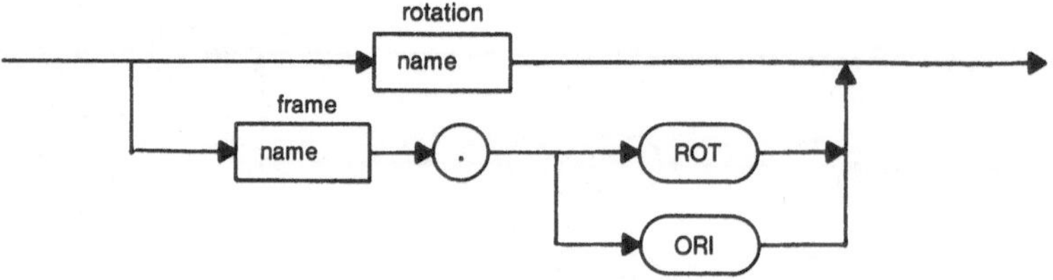

The result is of rotation or orientation type, which is mapped in any case to the internal representation by a matrix.

Frame value variable:

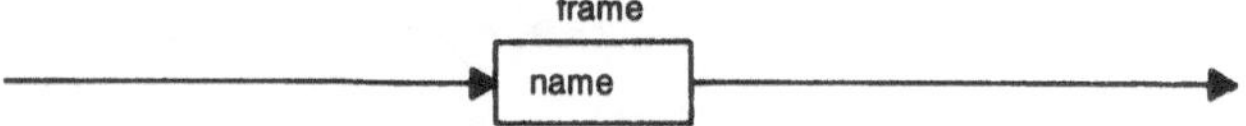

This is only mentioned for consistence.

Factor:

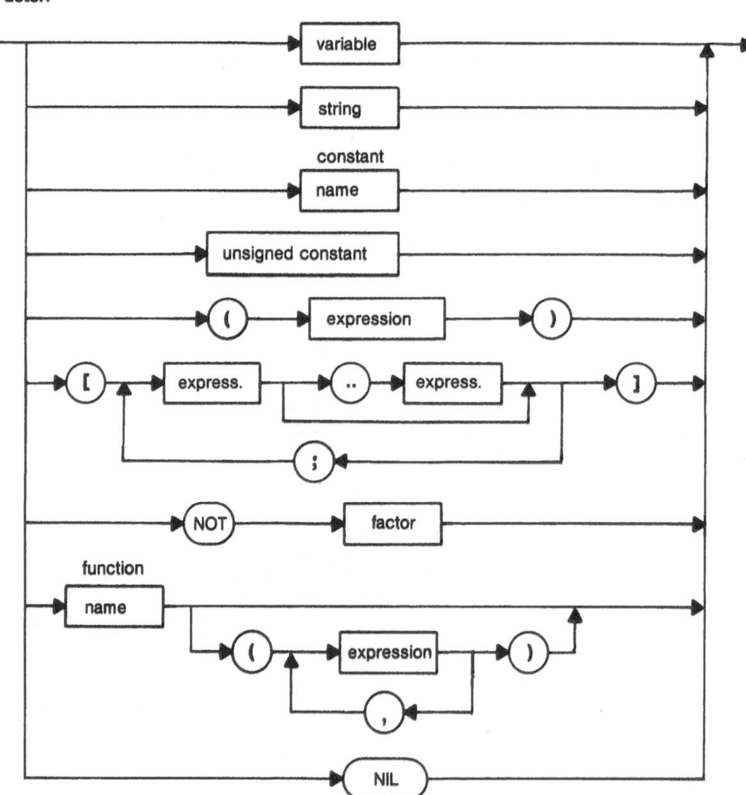

Term:

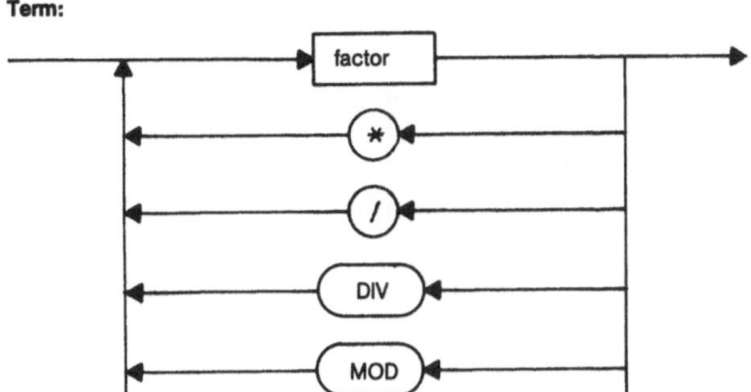

Single expression:

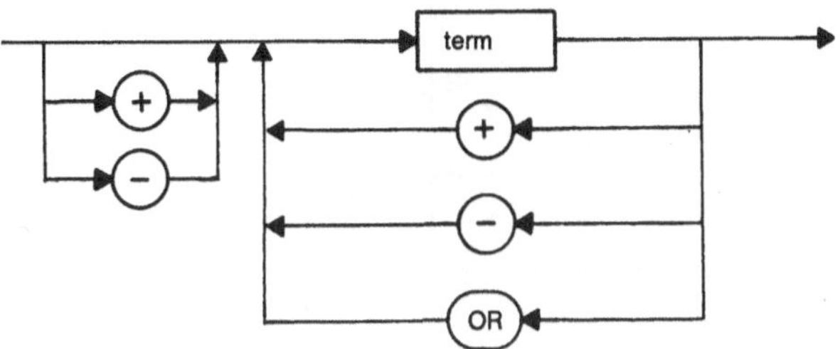

Expression:

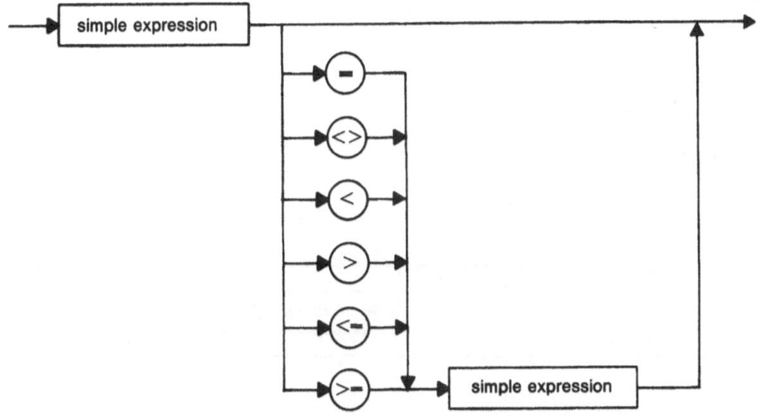

Move statement:

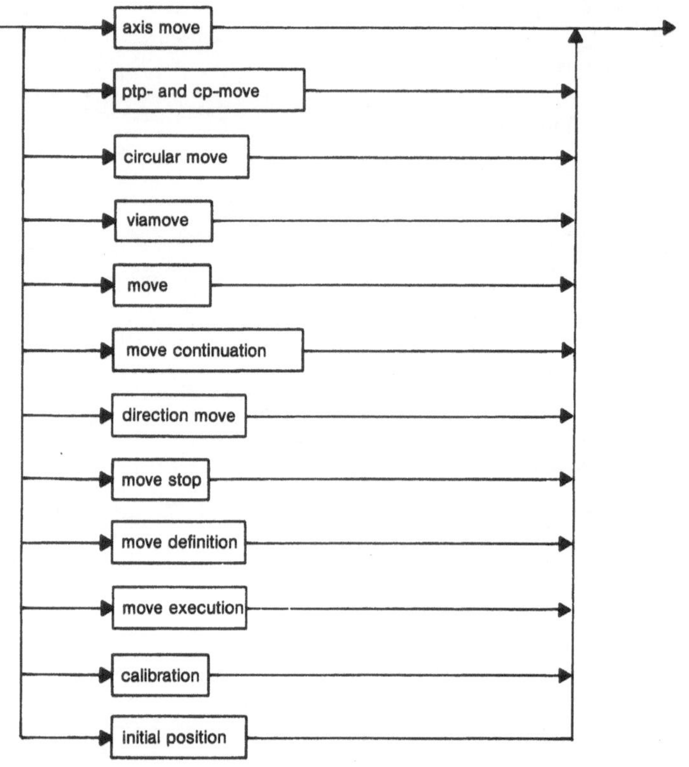

Axis move:

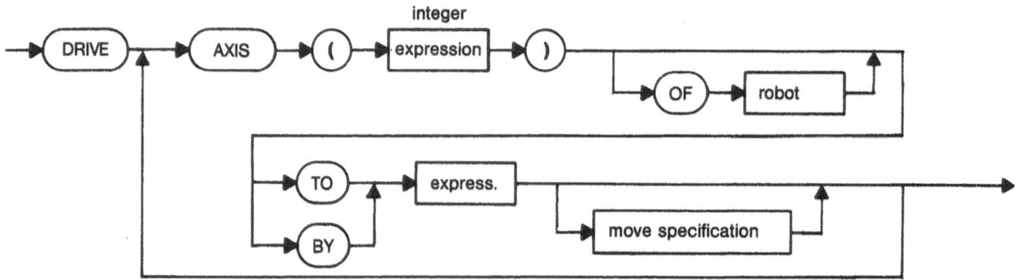

ptp- and cp-move:

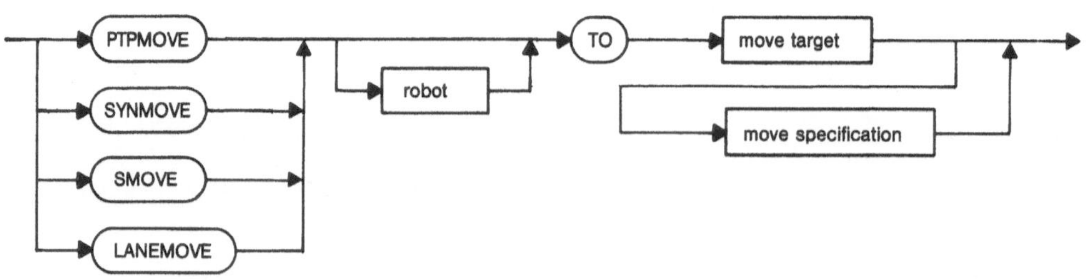

Robot:

robot
name
ROBOT
(
integer
express.
)

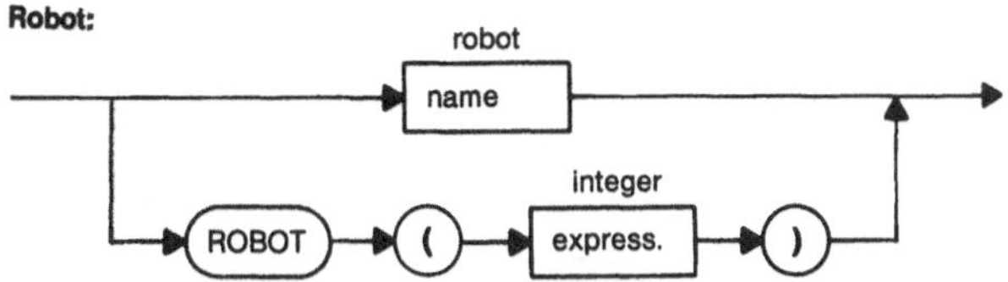

A robot name must be previously defined in the system specification.

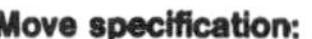

Move specification:

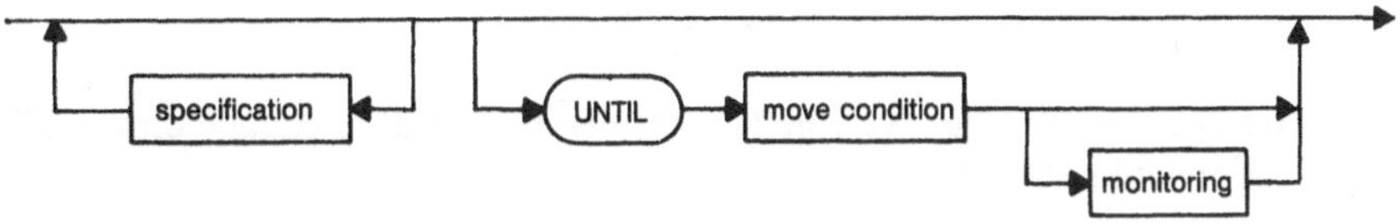

specification
UNTIL
move condition
monitoring

Specification:

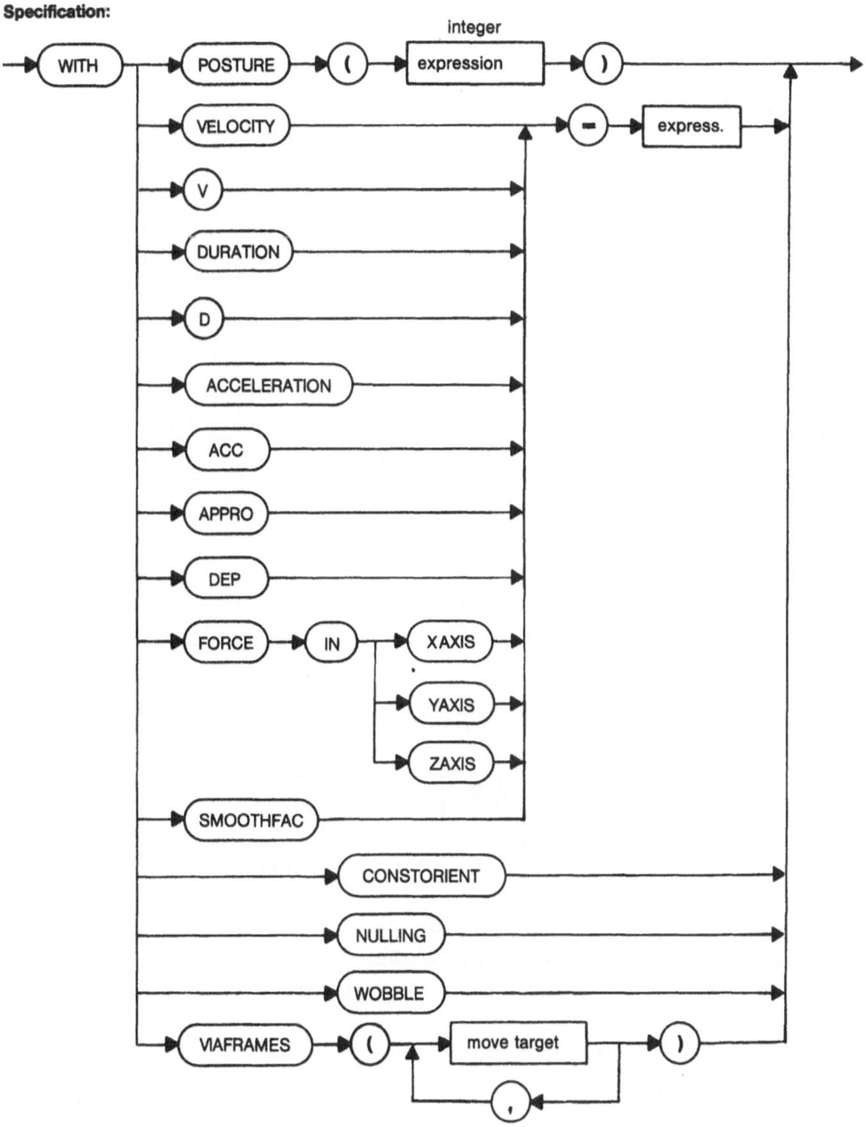

WITH
POSTURE
(
integer
expression
)
VELOCITY
=
express.
V
DURATION
D
ACCELERATION
ACC
APPRO
DEP
FORCE
IN
XAXIS
YAXIS
ZAXIS
SMOOTHFAC
CONSTORIENT
NULLING
WOBBLE
VIAFRAMES
(
move target
,
)

Move target:

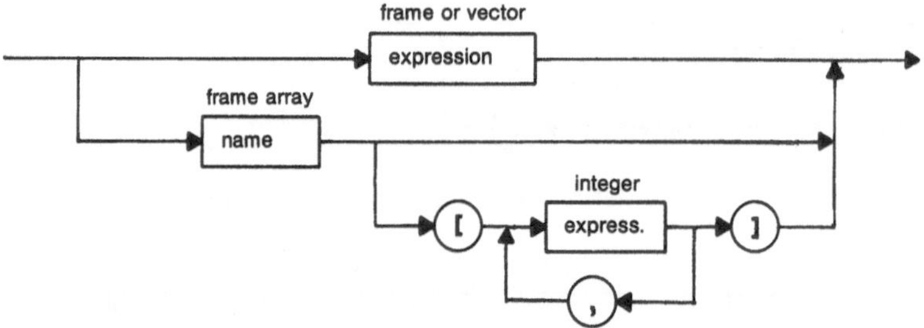

The move target can also be a vector. This results in a position change only, while the orientation will not be changed.

Viamove:

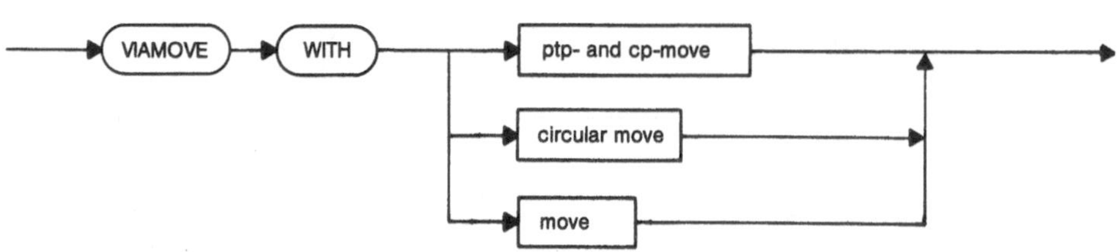

Circular move:

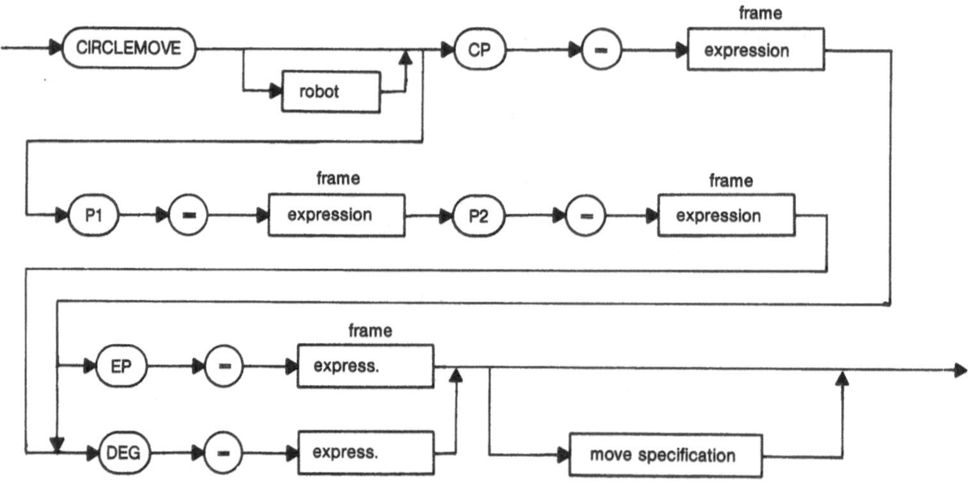

If the center point or frame is given (CP), the plane of the circle is the x-y plane of the frame. The move target is given by an endpoint (EP) or by specifying the degrees of a circle segment. The degrees are defined with respect to the z-axis of the center frame.
It is also possible to define the circle move by three points on the circle segment (P1, P2 and EP).

Move:

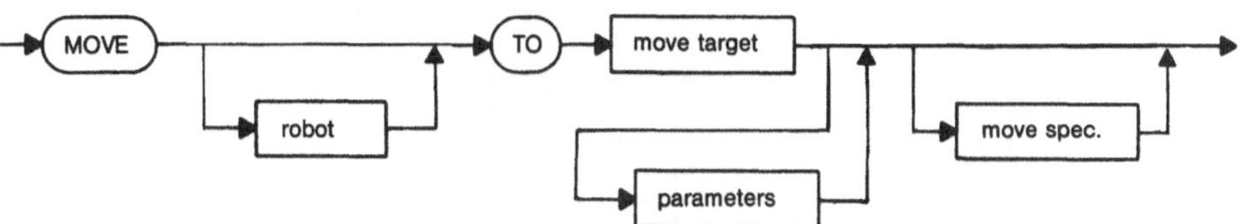

Parameters:

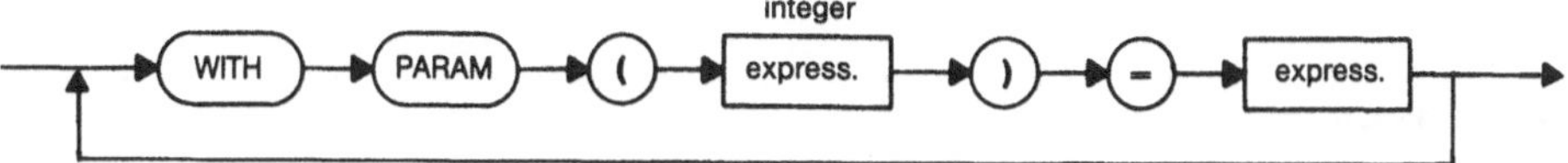

Move continuation:

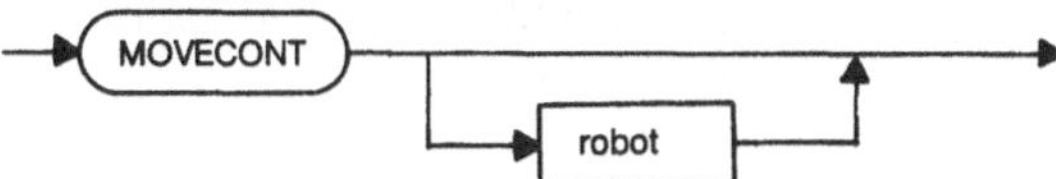

Direction move:

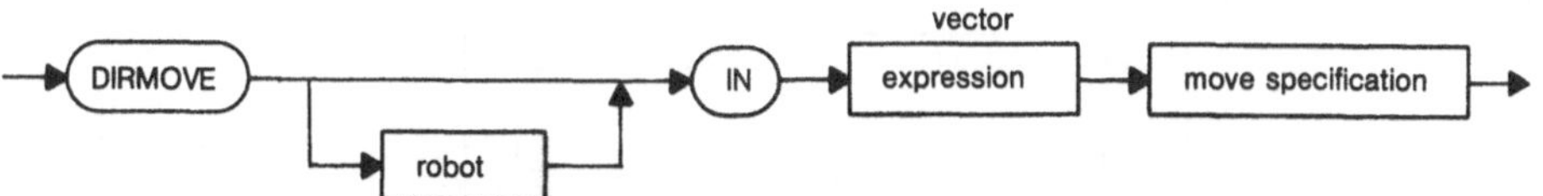

A robot move is started along the direction of the given vector until the specified move condition becomes true.

Move condition:

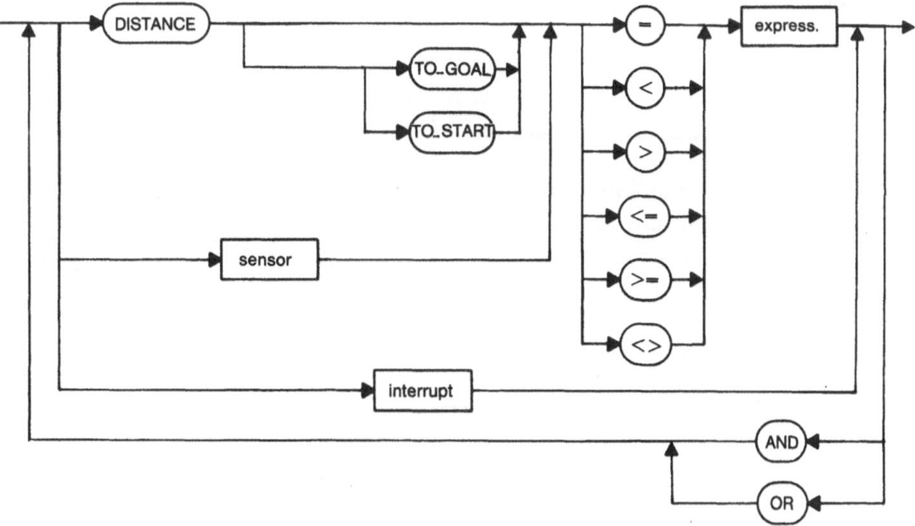

Move stop:

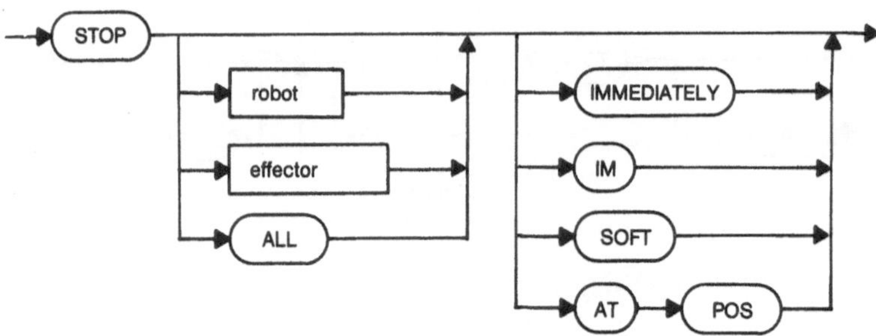

Move definition:

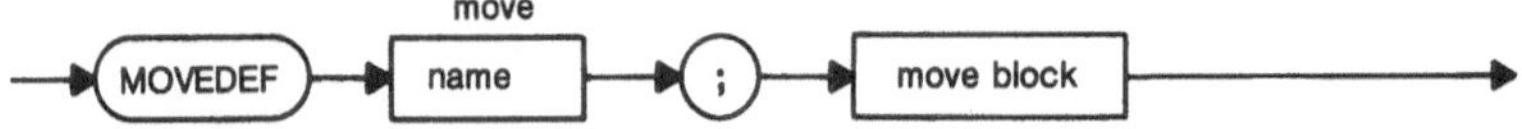

The complete move can be composed by different kinds of trajectory calculation.

Move block:

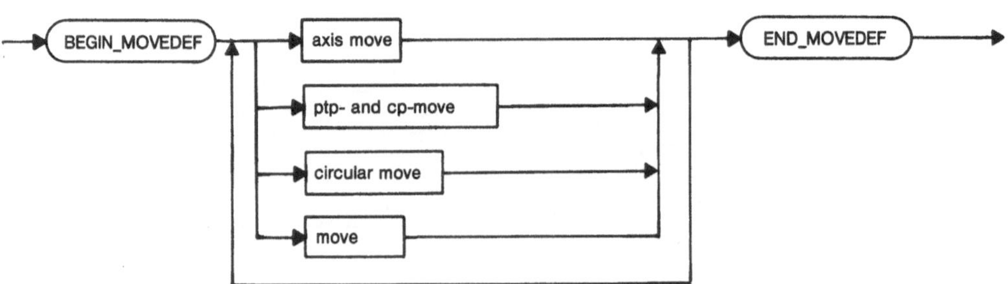

After the move definition the trajectory is calculated and the interpolated axis values are stored in an internal table.

Move execution:

Without any new evaluation the axis table is transferred to lower level of robot control and the move performed.

Calibration:

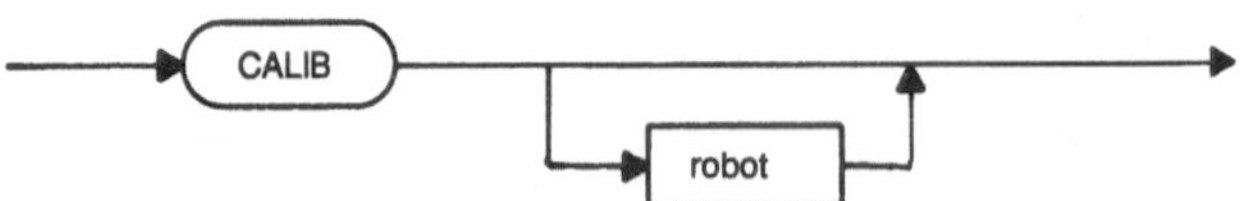

Initial position:

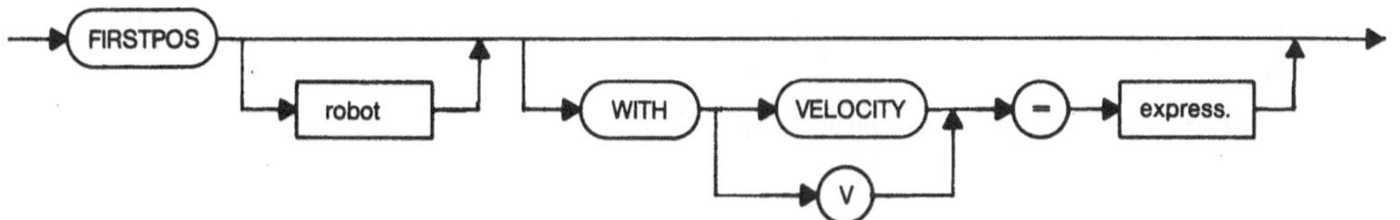

Effector statement:

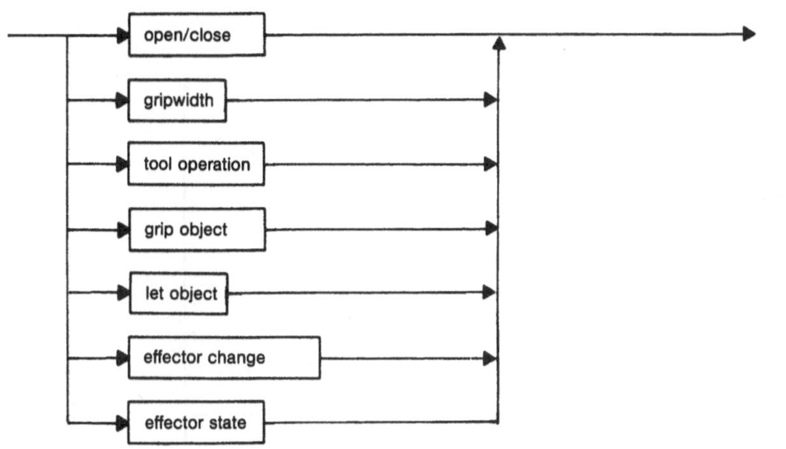

Open/close:

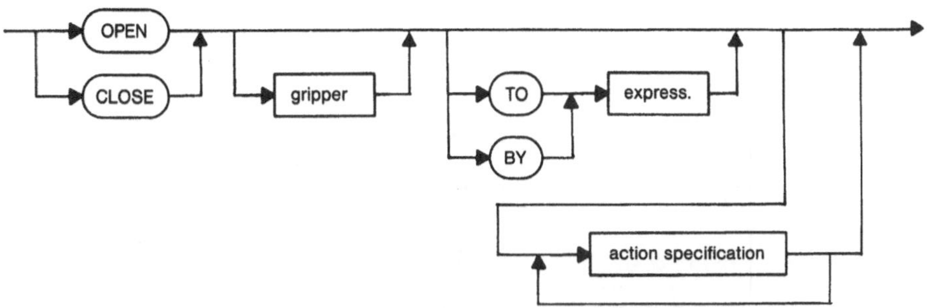

The move of the gripper brackets is checked as to whether it is really an open or a close.

Gripper:

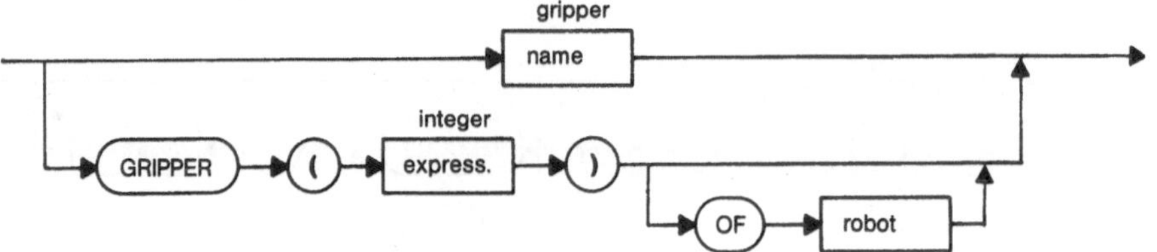

Gripper and robot name have to be defined in the system specification.

Action specification:

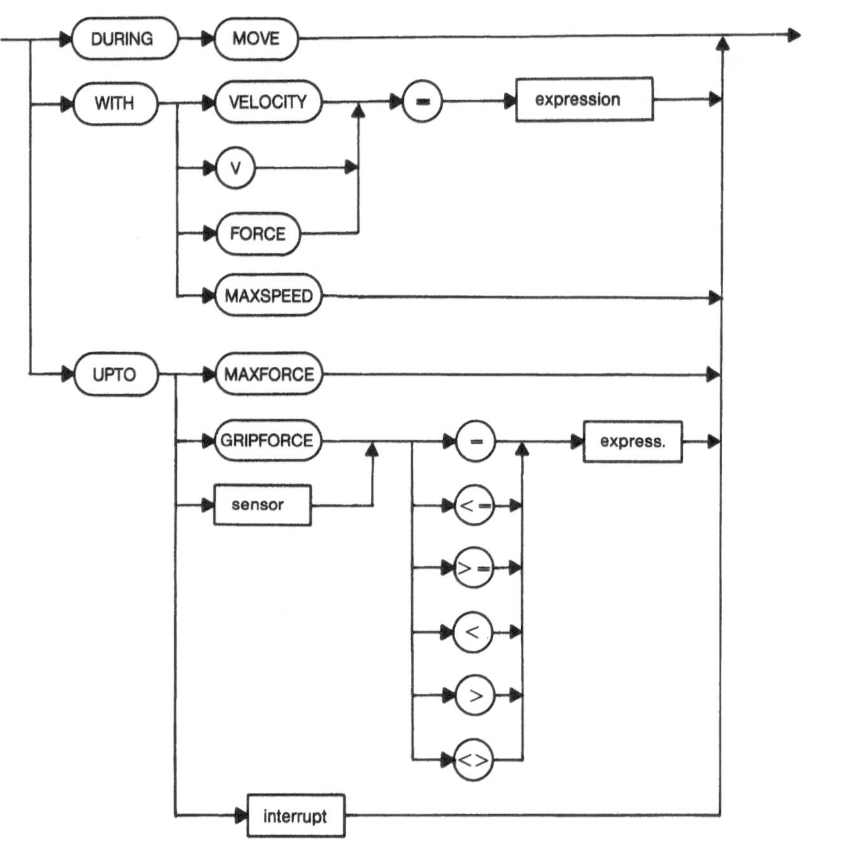

Gripwidth:

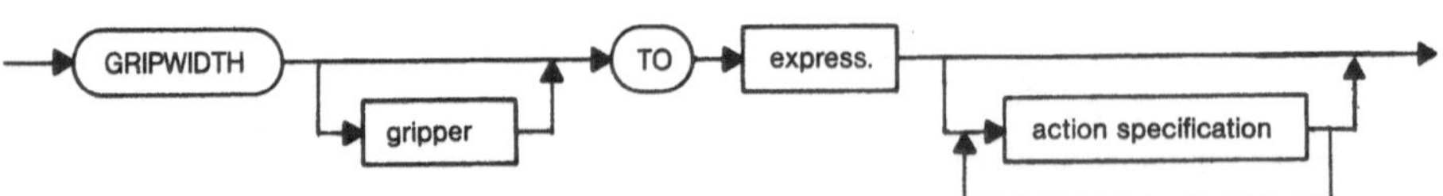

Tool operation:

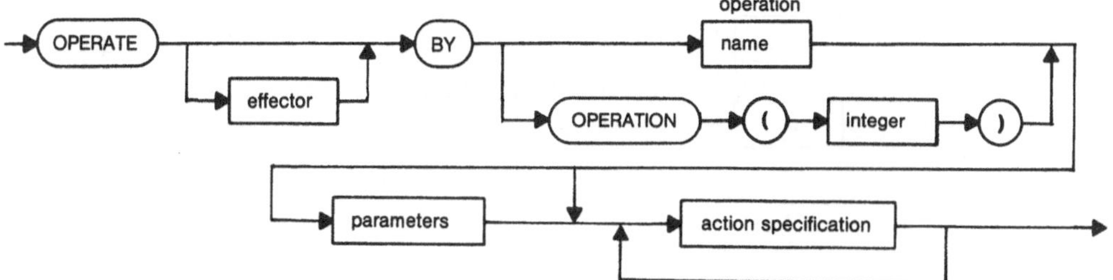

The operation name has to be defined in the system specification.

Grip object:

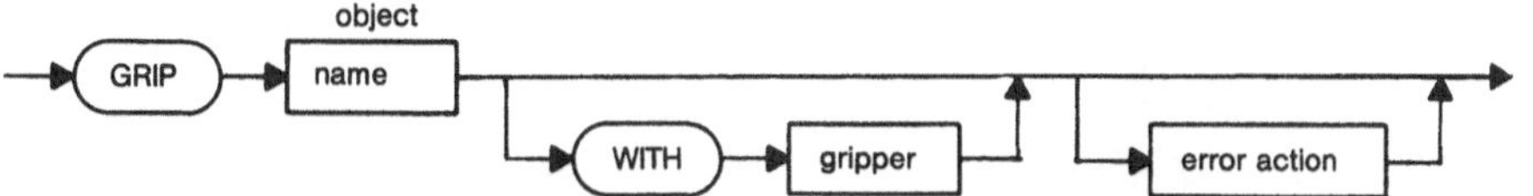

The object and all needed information are read from a world model defined in the system specification.

Error action:

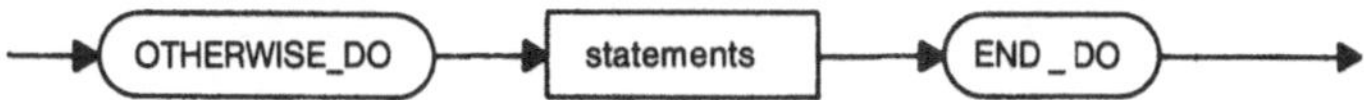

Let object:

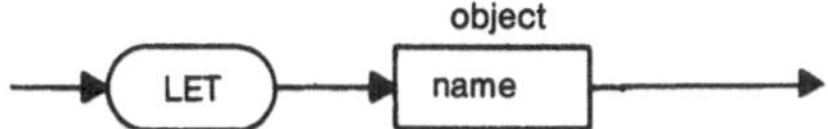

Effector change:

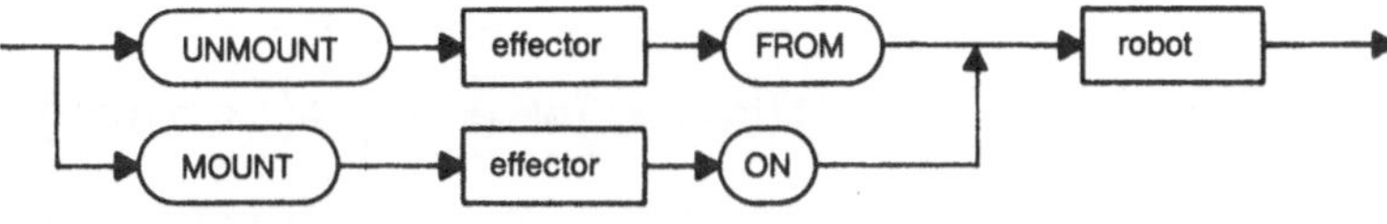

Effector:

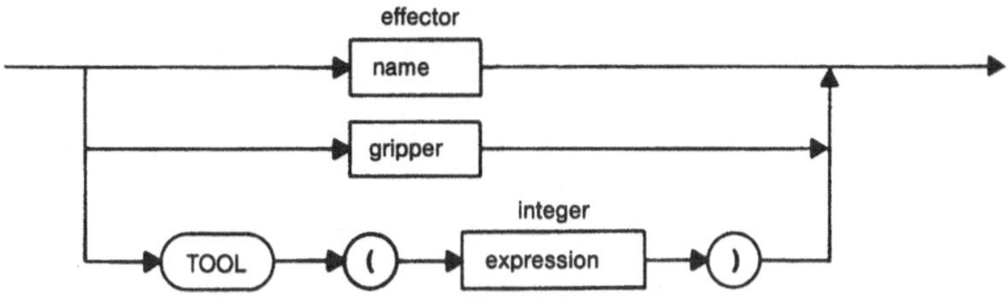

Subroutine statement:

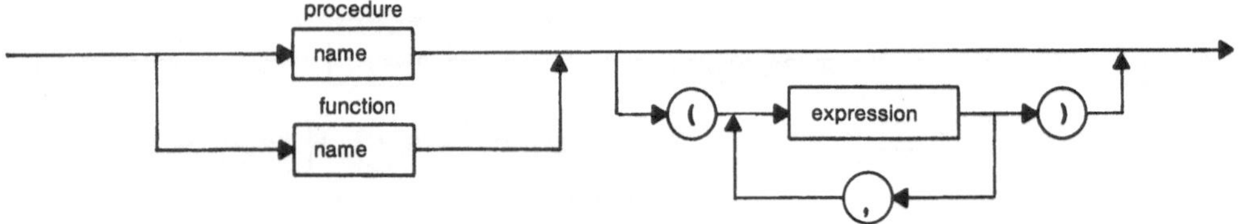

Conditional statement:

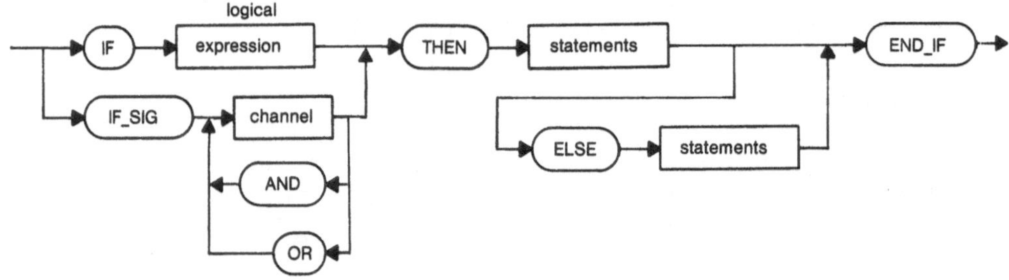

Channel:

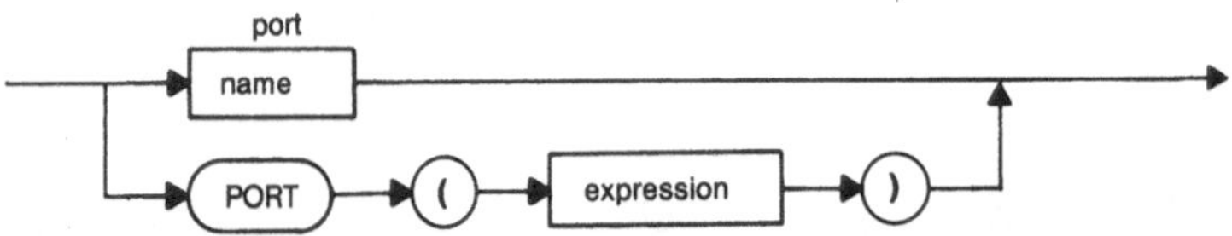

The port name must be defined in the system specification.

Repetitive statements:

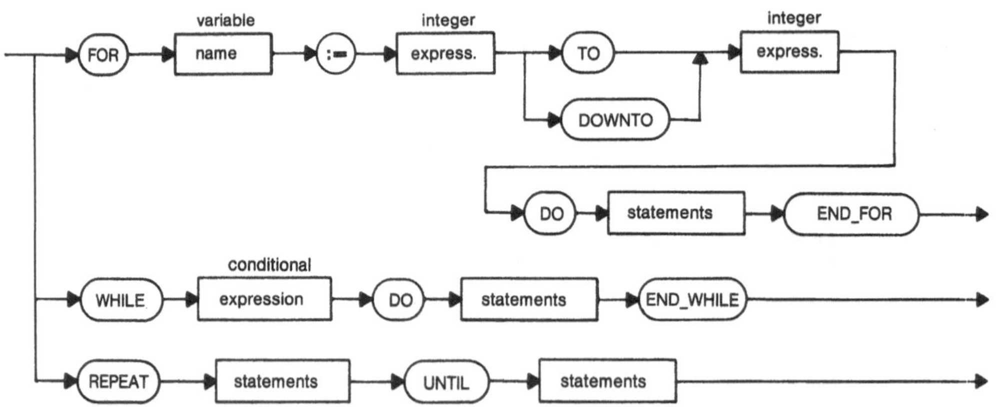

Case statement:

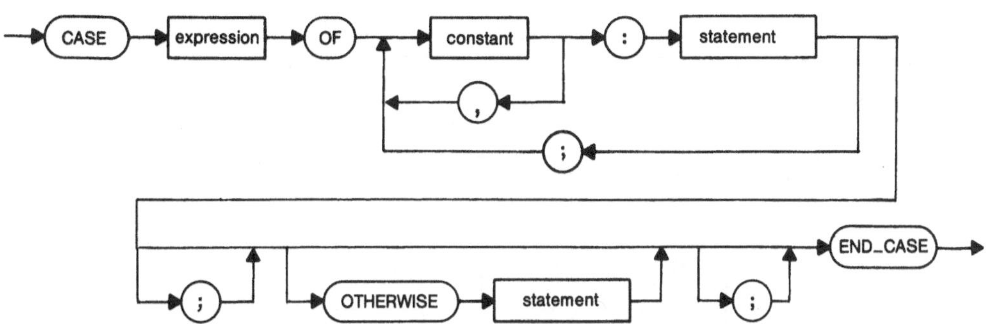

Multitasking:

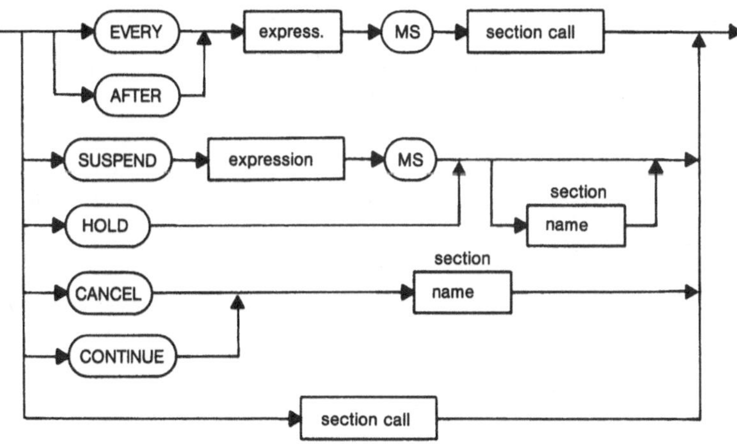

If no section name is given, the action will affect the section including the statement.

Section call:

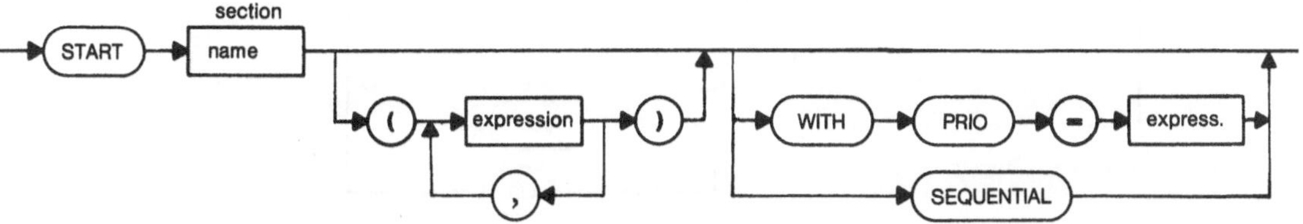

If a section is called with the specification SEQUENTIAL, it is executed like a procedure.

Monitor statement:

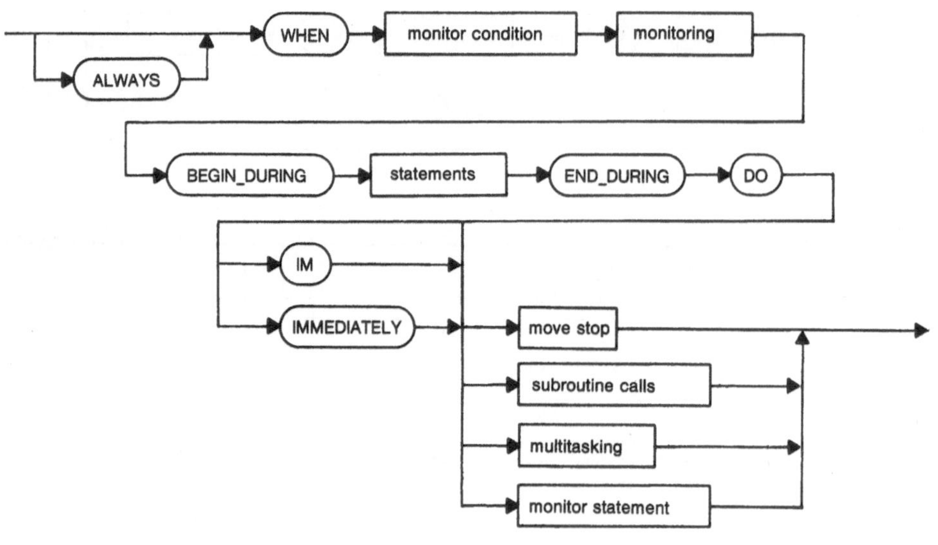

Monitor condition:

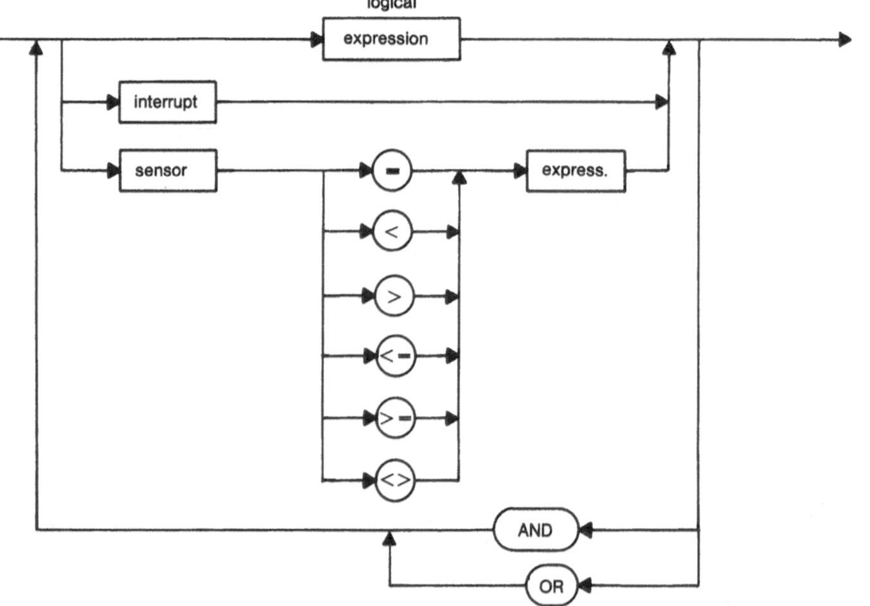

Interrupt:

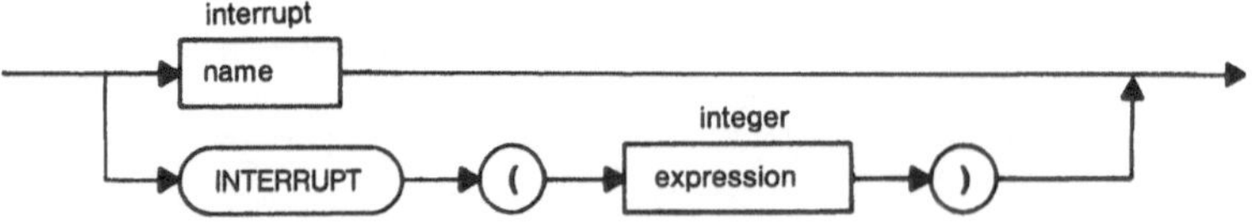

The interrupt name must be specified in the system specification.

Sensor:

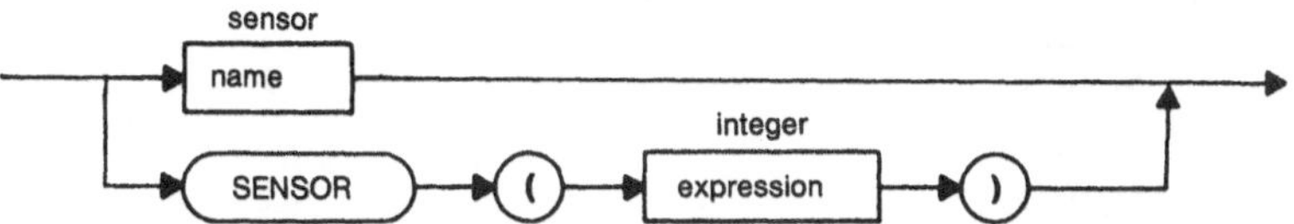

The sensor name must be specified in the system specification.

Monitoring:

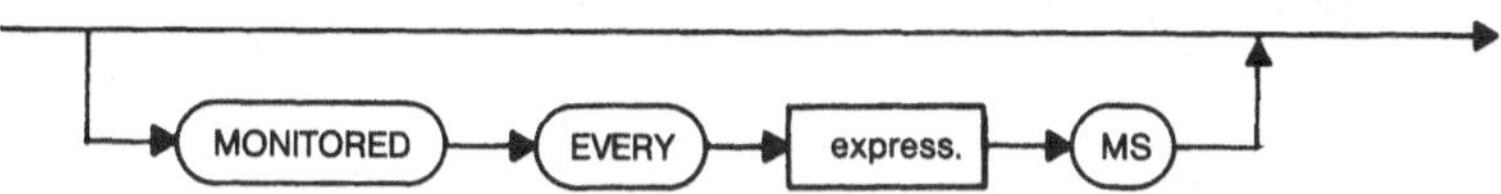

A condition is evaluated as cyclic by the given cycle time.

Synchronization:

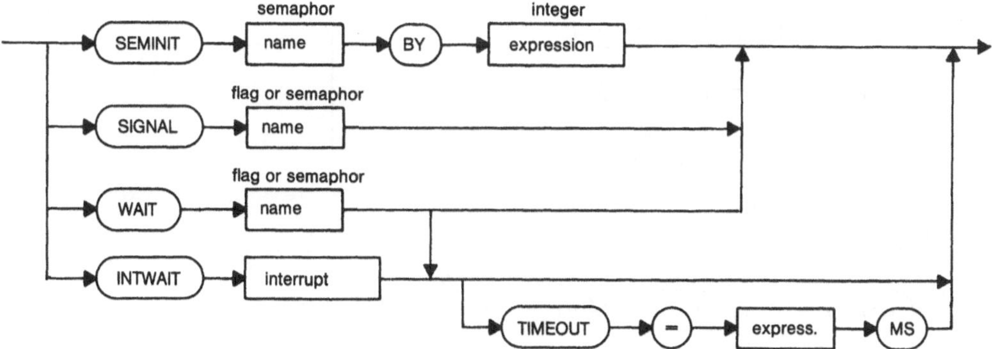

Exit statement:

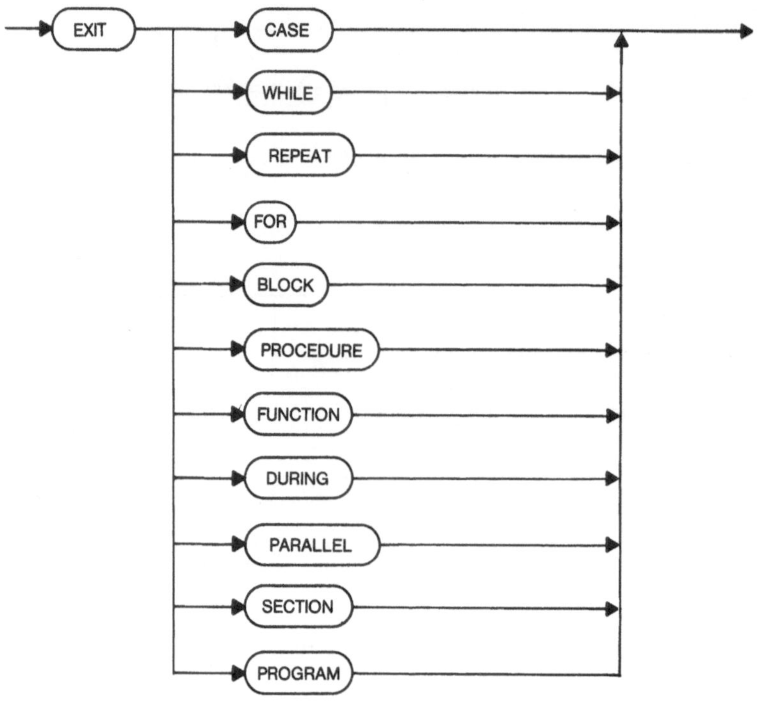

Data management:

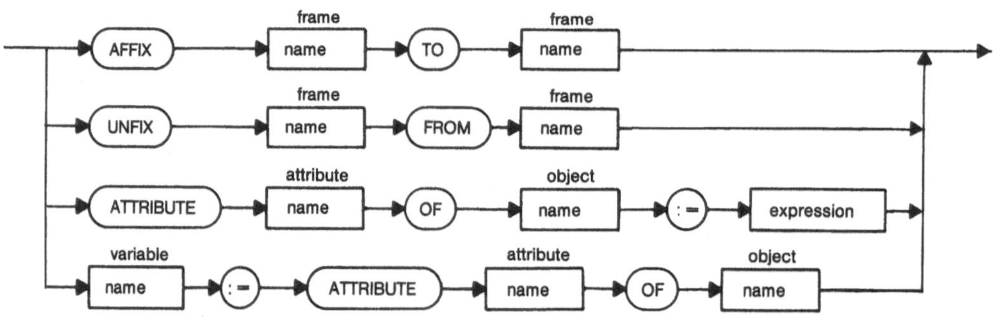

With statement:

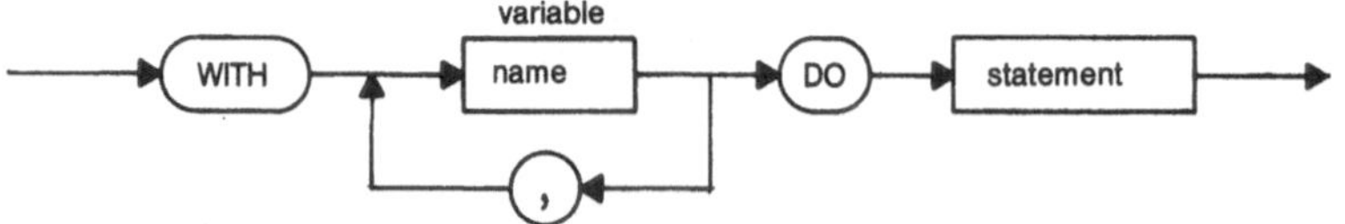

Input/output:

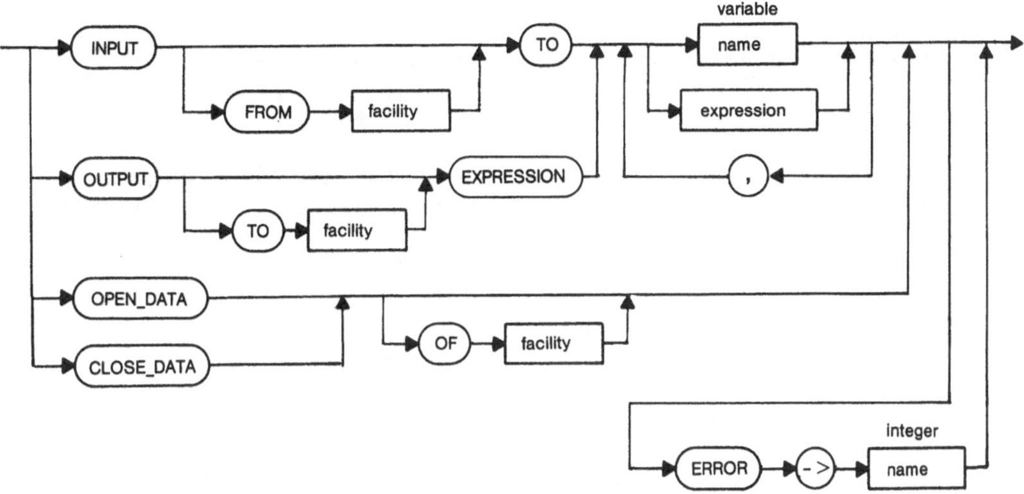

Facility:

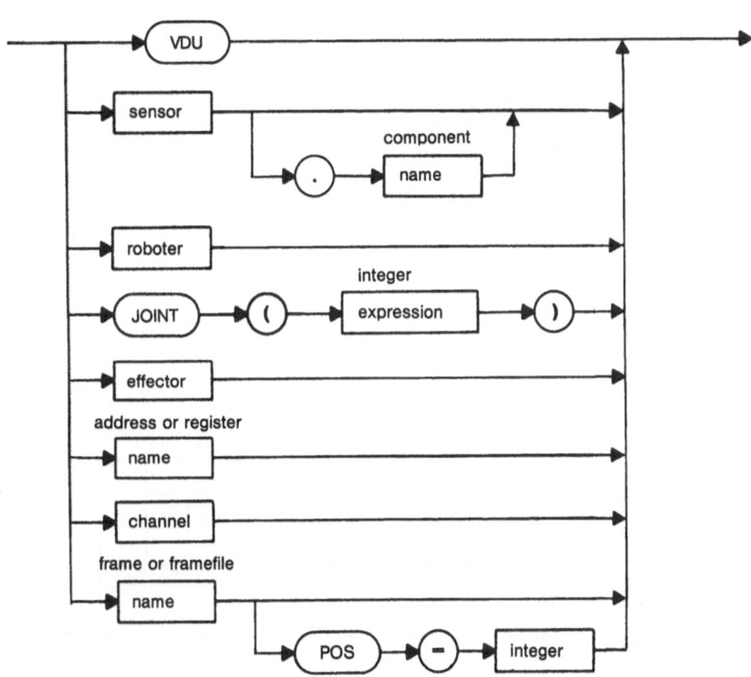

Module:

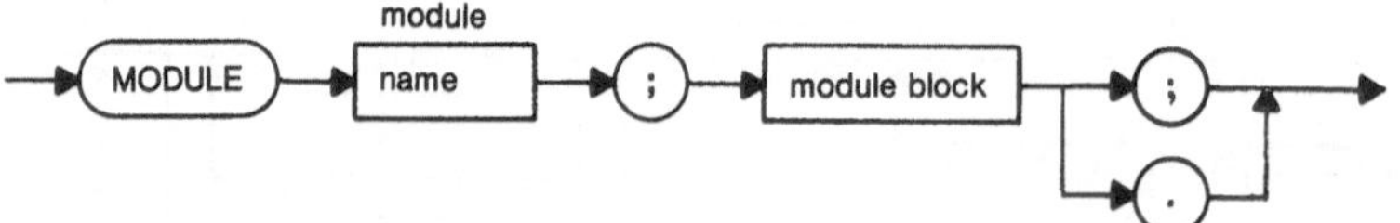

A module can be compiled separately. It provides data, procedures, functions or sections. It allows to build up libraries with often-used program elements or general interfaces.

Module block:

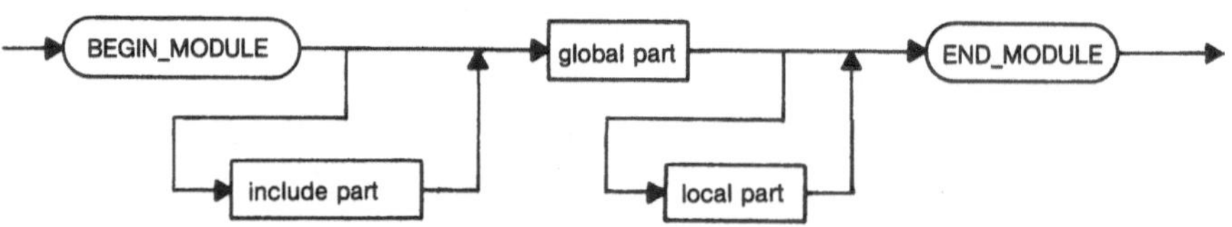

The local part can be used by procedures, functions or sections of the module only.

Include part:

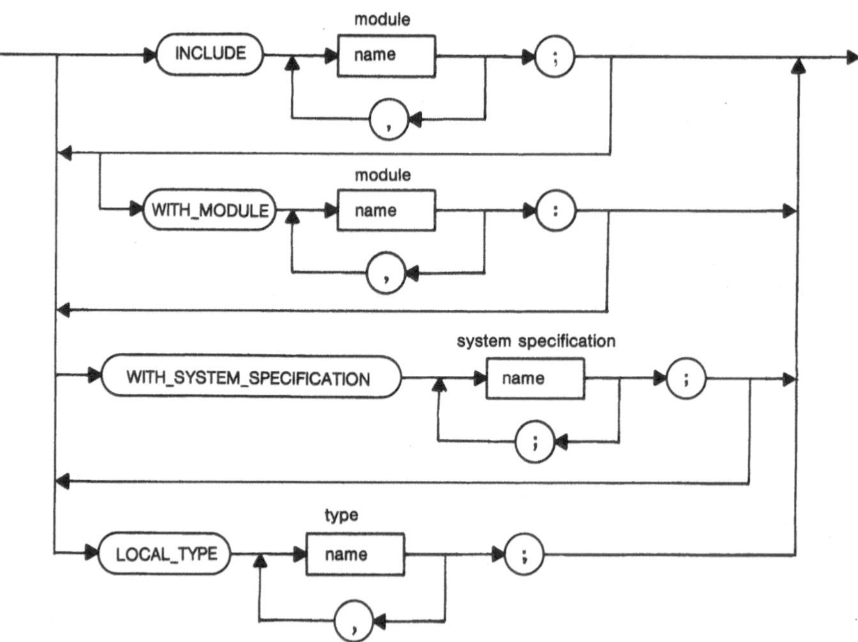

A module can include other modules or system specifications. A local type represents a forward declaration to a data type in the module. Data of this type can be used together with procedures of the module. This implies an implementation possibility of abstract data types.

Global part:

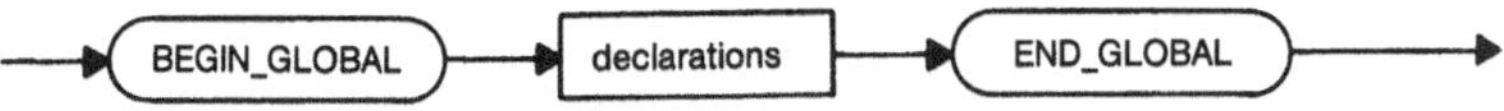

Local part:

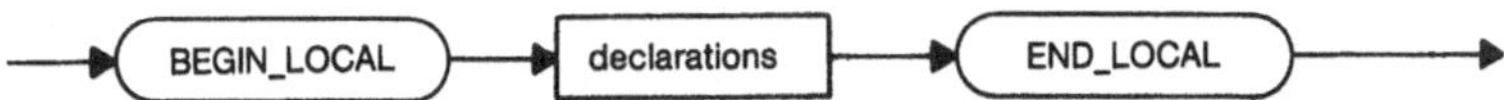

Integer:

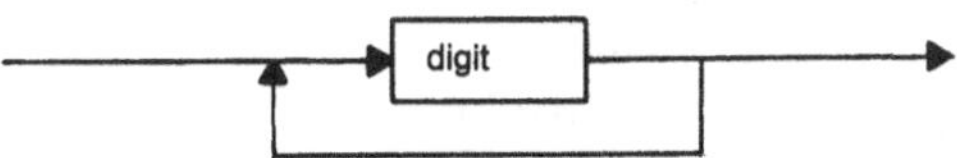

A digit is of 0, 1, …, 9.

Name:

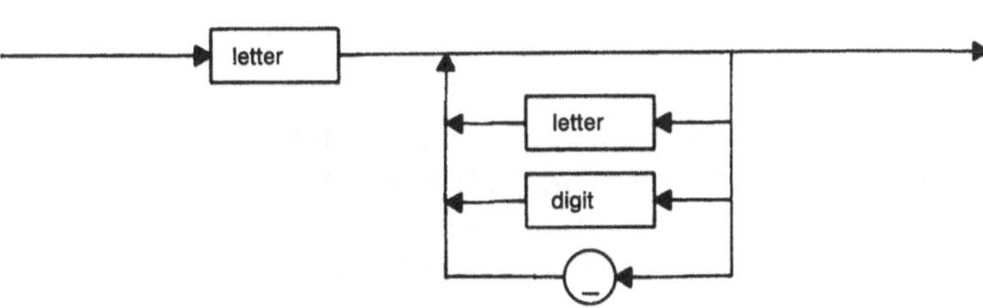

A letter is of a, b, …, z and A, B, …, Z.

Appendix B
PASRO

I. Summary of PASRO Procedures

Vector arithmetic:

makevector	(VAR v: vector; x,y,z: REAL)
vabs	(VAR s: REAL; v: vector)
vadd	(VAR v: vector; v1,v2: vector)
vsub	(VAR v: vector; v1,v2: vector)
vmul	(VAR v: vector; v: vector; s: REAL)
vdiv	(VAR v: vector; v: vector; s: REAL)
vrot	(VAR v: vector; r: rotation; v: vector)
vdot	(VAR s: REAL; v1,v2: vector)
vcross	(VAR v: vector; v1,v2: vector)

Rotations:

makerotation	(VAR r: rotation; v: vector; s: REAL)
rotrot	(VAR r: rotation; r2,r1: rotation)
rotaxis	(VAR v: vector; r: rotation)
rotangle	(VAR s: REAL; r: rotation)

Frame arithmetic:

makeframe	(VAR f: frame; r: rotation; v: vector)
setframe	(VAR f: frame; x,y,z: REAL; v: vector; s: REAL)
frametransl	(VAR f: frame; f: frame; v: vector)
framerot	(VAR f: frame; f: frame; r: rotation)
transframe	(VAR f: frame; f2,f1: frame)
framerel	(VAR f: frame; f1,f2: frame)
frameinv	(VAR f: frame; f: frame)

Robot and gripper control:

drive	(axis: INTEGER; s: REAL)
jdrive	(t: thetai)
pmove	(f: frame)
jmove	(f: frame)

smove	(f: frame)
gripwidth	(s: REAL)
gripopen	
gripclose	
gripforce	(gripf, holdf, time: REAL)
nullpos	
robotoff	
roboton	

Input/Output of the PASRO standard data types:

vread	(VAR v: vector)
vwrite	(v: vector)
rread	(VAR r: rotation)
rwrite	(r: rotation)
rmatrixout	(r: rotation)
fread	(VAR f: frame)
fwrite	(f: frame)
jointout	(f: frame)

Framefile handling:

openframefile	(filename: ffilename; VAR status: INTEGER)
closeframefile	
initialize	(VAR f: frame; VAR gripvalue: REAL; VAR gripstatus: BOOLEAN; fname: framename; VAR status: INTEGER)
createframefile	(filename: ffilename; VAR status: INTEGER)
framewrite	(fname: framename; fin: frame; gvin: REAL; gsin: BOOLEAN; VAR status: INTEGER)

Process I/0:

sigon	(adr: INTEGER)
sigoff	(adr: INTEGER)
sigin	(VAR b: BOOLEAN; adr: INTEGER)
anout	(s1,adr: INTEGER)
anin	(VAR s: INTEGER; adr: INTEGER)

II. Predefined Datatypes and Variables of PASRO

```
TYPE
vector    = RECORD
              x, y, z : REAL
            END;

rotmatrix = RECORD
              t, o, a : vector
            END;

rotation  = RECORD
              axis:   vector;
              angle:  REAL;
              matrix: rotmatrix
            END;

frame     = RECORD
              rot:    rotation;
              transl: vector
            END;

thetai    = ARRAY[0..5] OF REAL;
```

Variables the user may alter:

speedfactor:	REAL;	value between 0.0 (very slow) and 9.0 (very fast)
arobotbase:	REAL;	x-component of the robot base coordinate system
brobotbase:	REAL;	y-component of the robot base coordinate system
crobotbase:	REAL;	z-component of the robot base coordinate system

Variables the user should only ever read and never change:

xaxis:	vector;	unit vector in x-direction
yaxis:	vector;	unit vector in y-direction
zaxis:	vector;	unit vector in z-direction
nilvector:	vector;	all components are zero
nilrot:	rotation;	rotation about the z-axis by zero degrees
nilframe:	frame;	frame with nilrot and nilvector

robotframe:	frame;	describes the actual position and orientation of the robot
errornumber:	INTEGER;	PASRO error number

Systemvariables with which the user is not concerned:

framefile:	FILE OF ffilerecord;	
		file of frames defined by teach-in
robotjoints:	thetai;	describes the current joint angles of the robot
robotsteps:	ARRAY[0..4] OF	INTEGER;
		describes the current number of motor steps for robots with stepper motors (e.g. the MICROROBOT)
sysffopen:	BOOLEAN;	indicates whether the framefile is open or not
sysstatgrip:	gripstatustyp;	represents the gripper status (open or closed)
sysvalgrip:	REAL;	gripwidth of the gripper

Appendix C
PASCAL

Syntax Diagrams

The diagrams for *Letter, Digit, Identifier, Directive, UnsignedInteger, UnsignedNumber,* and *Character String* describe the formation of lexical symbols from characters. The other diagrams describe the formation of syntactic constructs from symbols.

Letter

Digit

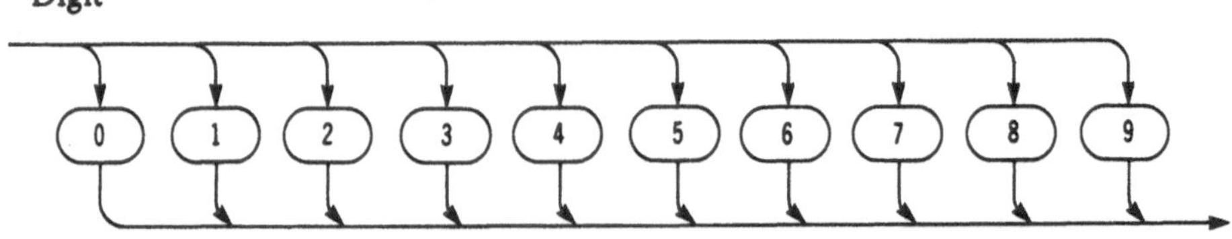

Identifier and Directive

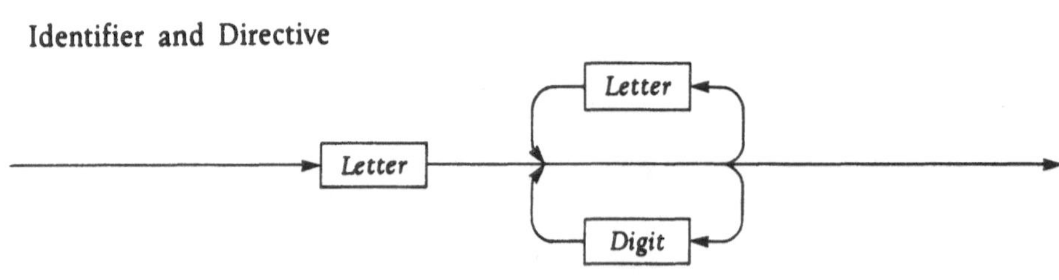

Unsigned Integer

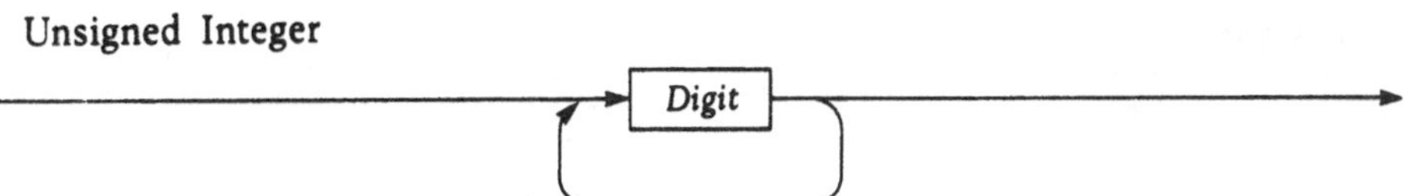

UnsignedNumber

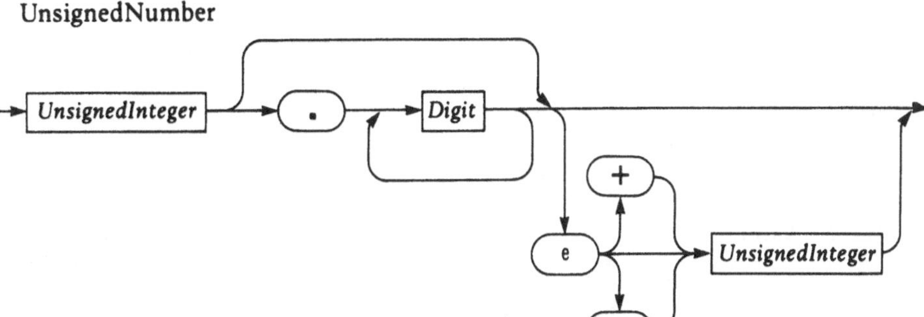

CharacterString

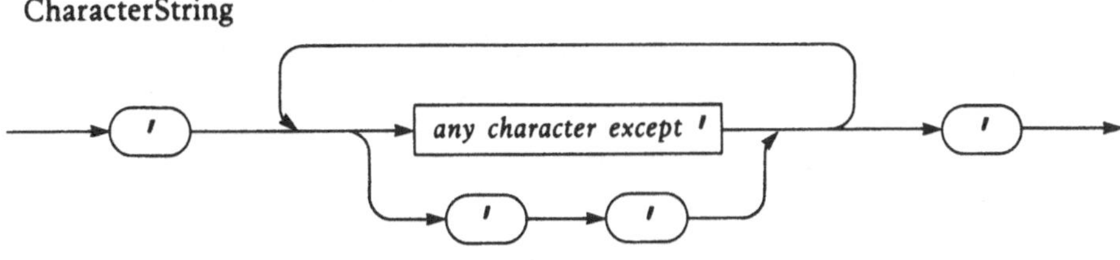

ConstantIdentifier, VariableIdentifier, FieldIdentifier, BoundIdentifier, TypeIdentifier, ProcedureIdentifier and FunctionIdentifier

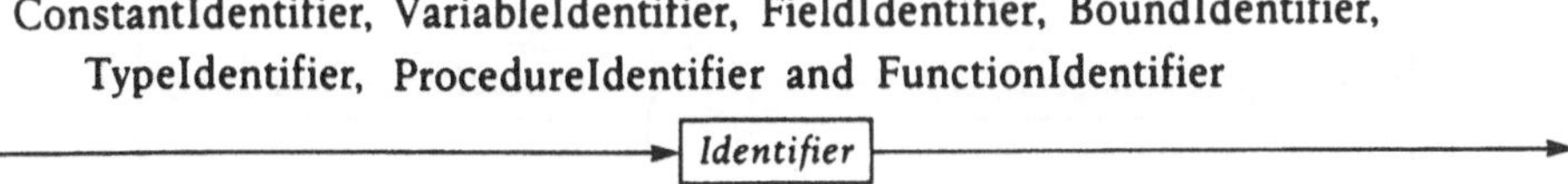

Unsigned Constant

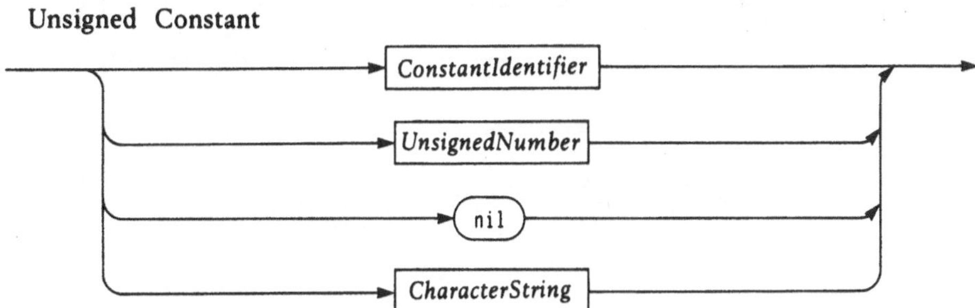

Constant

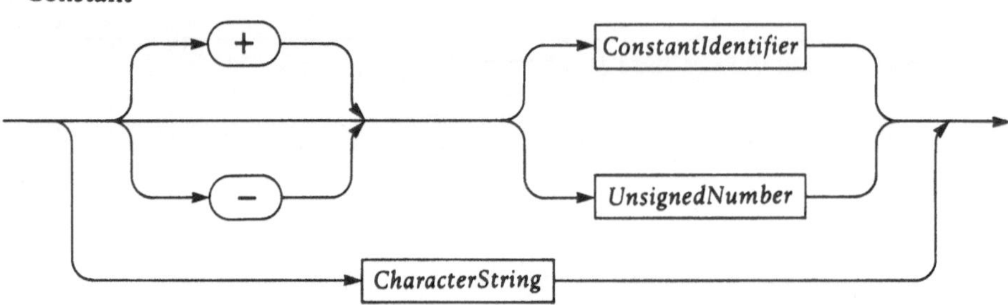

Variable

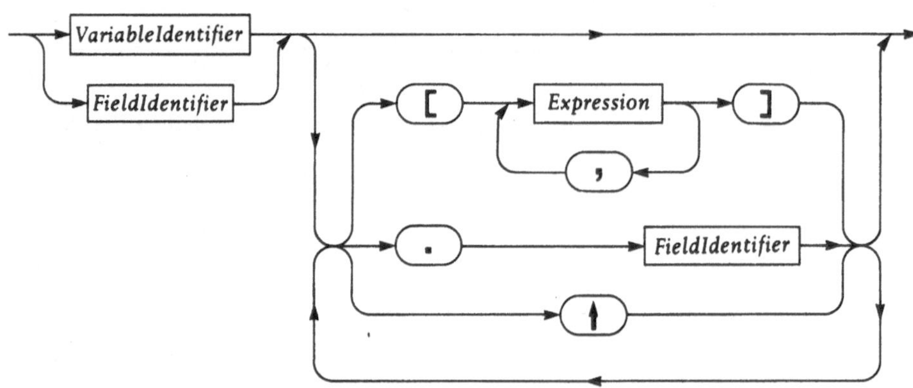

Factor

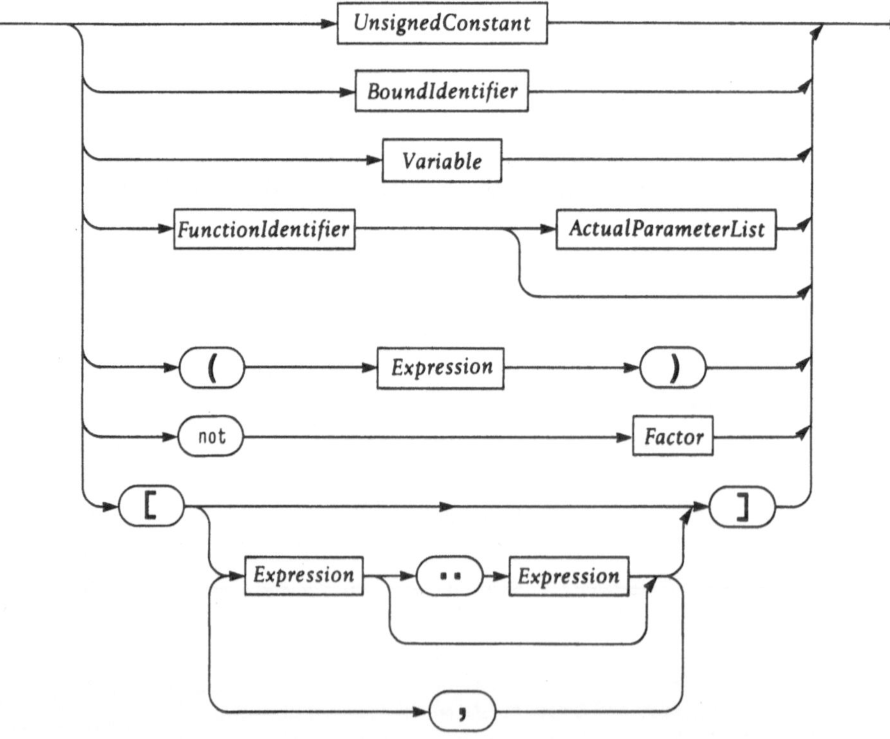

Term

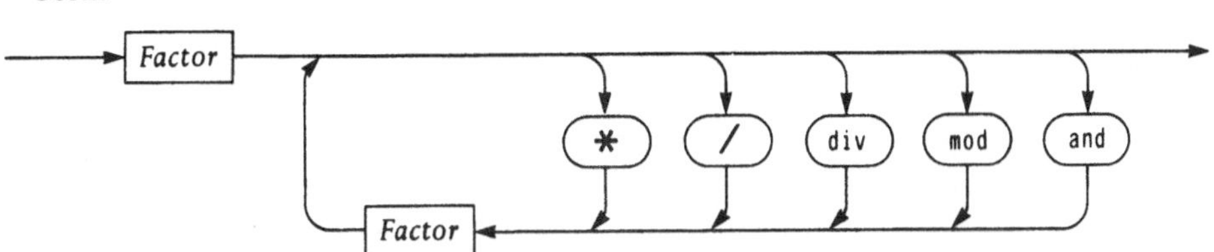

SimpleExpression

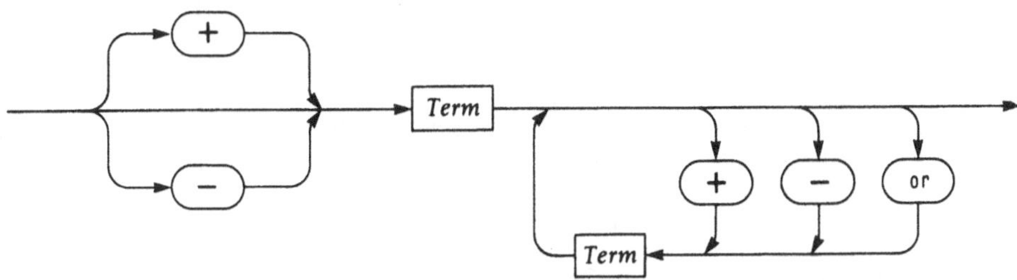

Expression

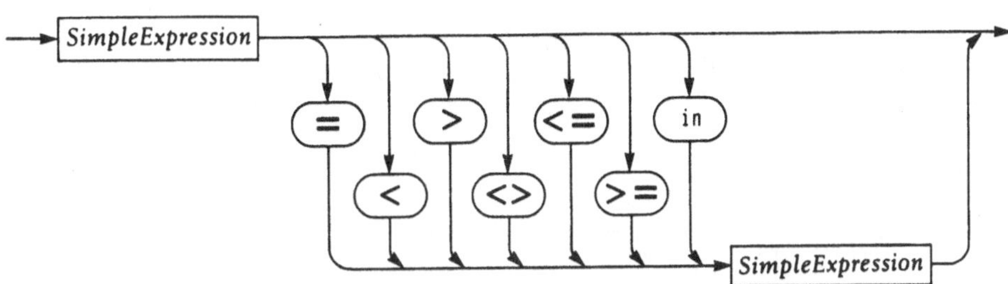

ActualParameterList

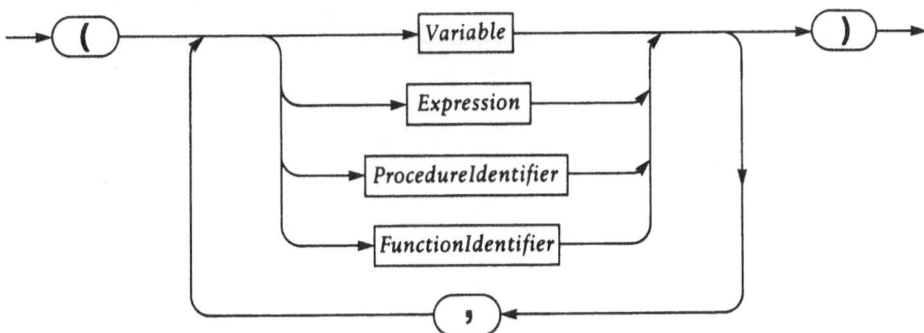

WriteParameterList

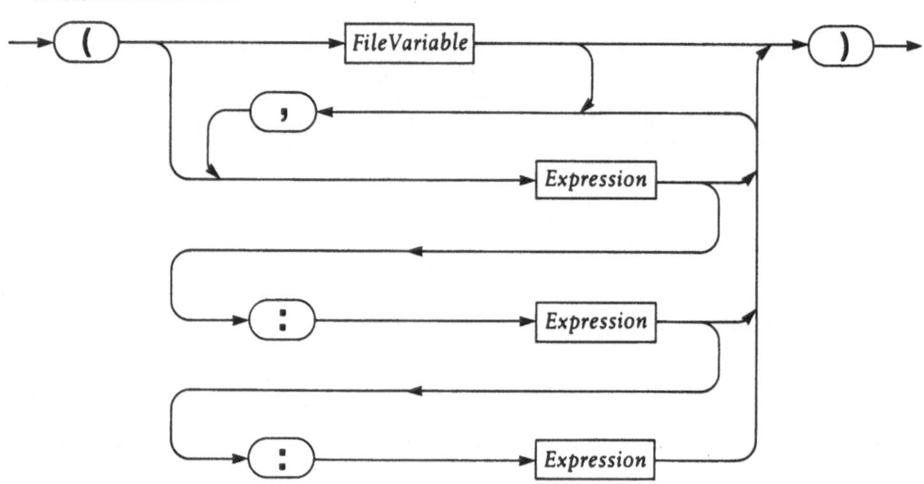

IndexTypeSpecification

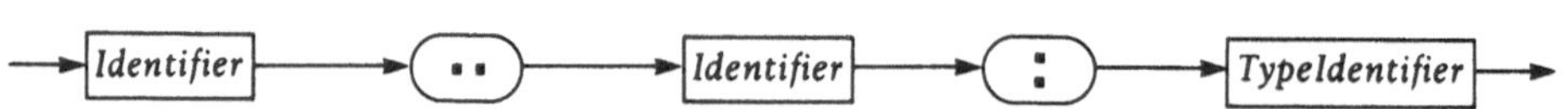

ConformantArraySchema

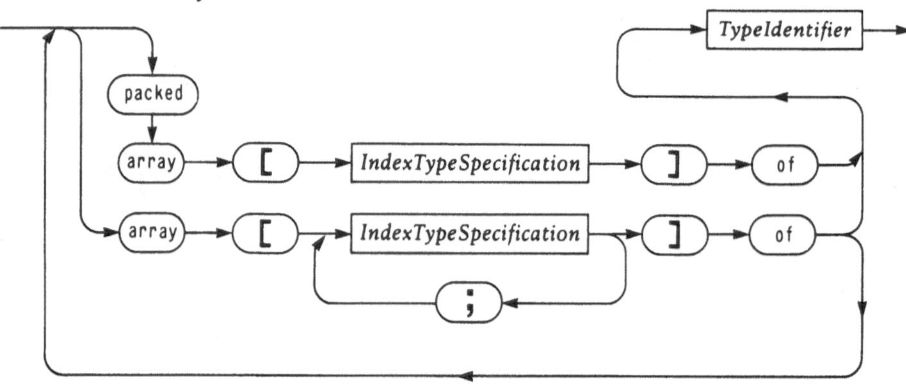

FormalParameterList

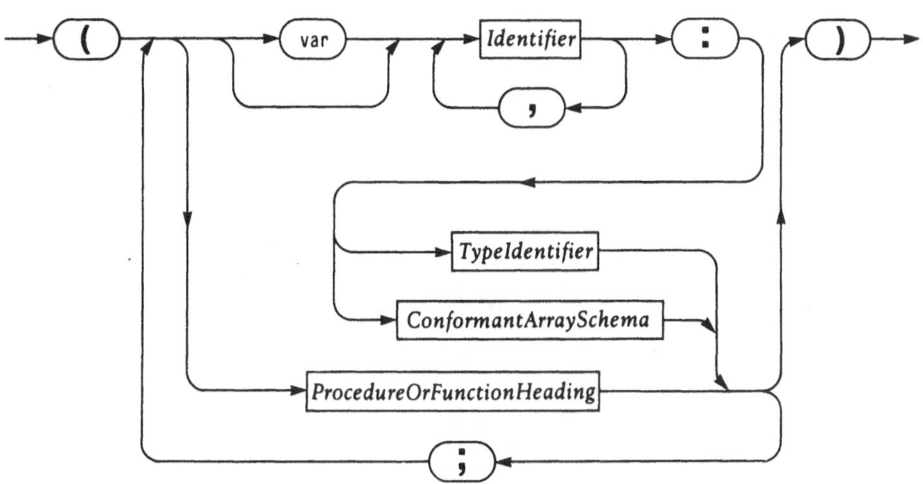

ProcedureOrFunctionHeading

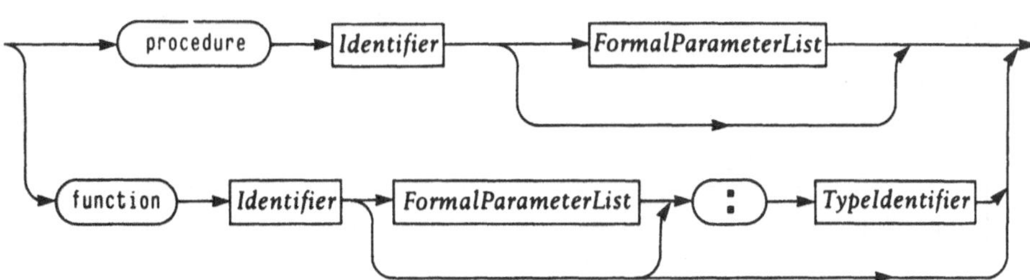

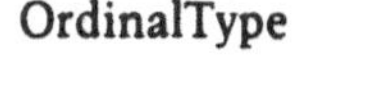

OrdinalType

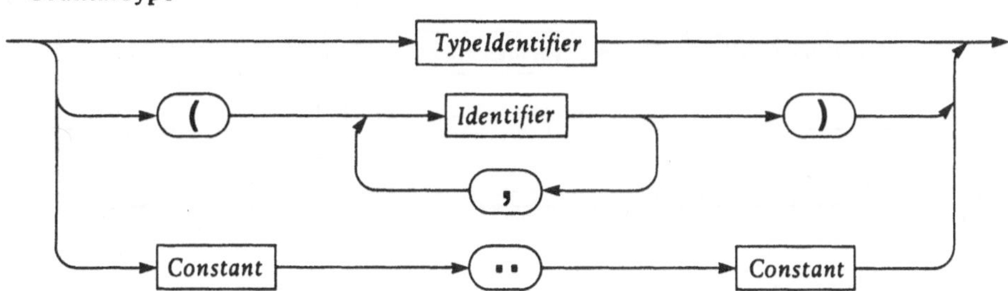

Type

OrdinalType
↑
TypeIdentifier
packed
array [OrdinalType ,] of Type
file of Type
set of OrdinalType
record FieldList end

FieldList

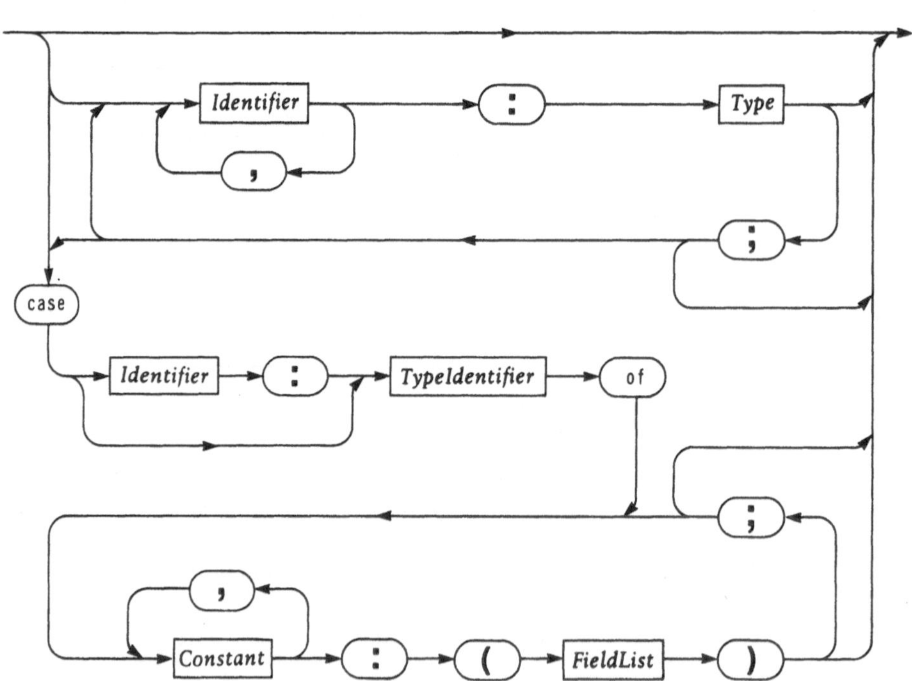

Statement

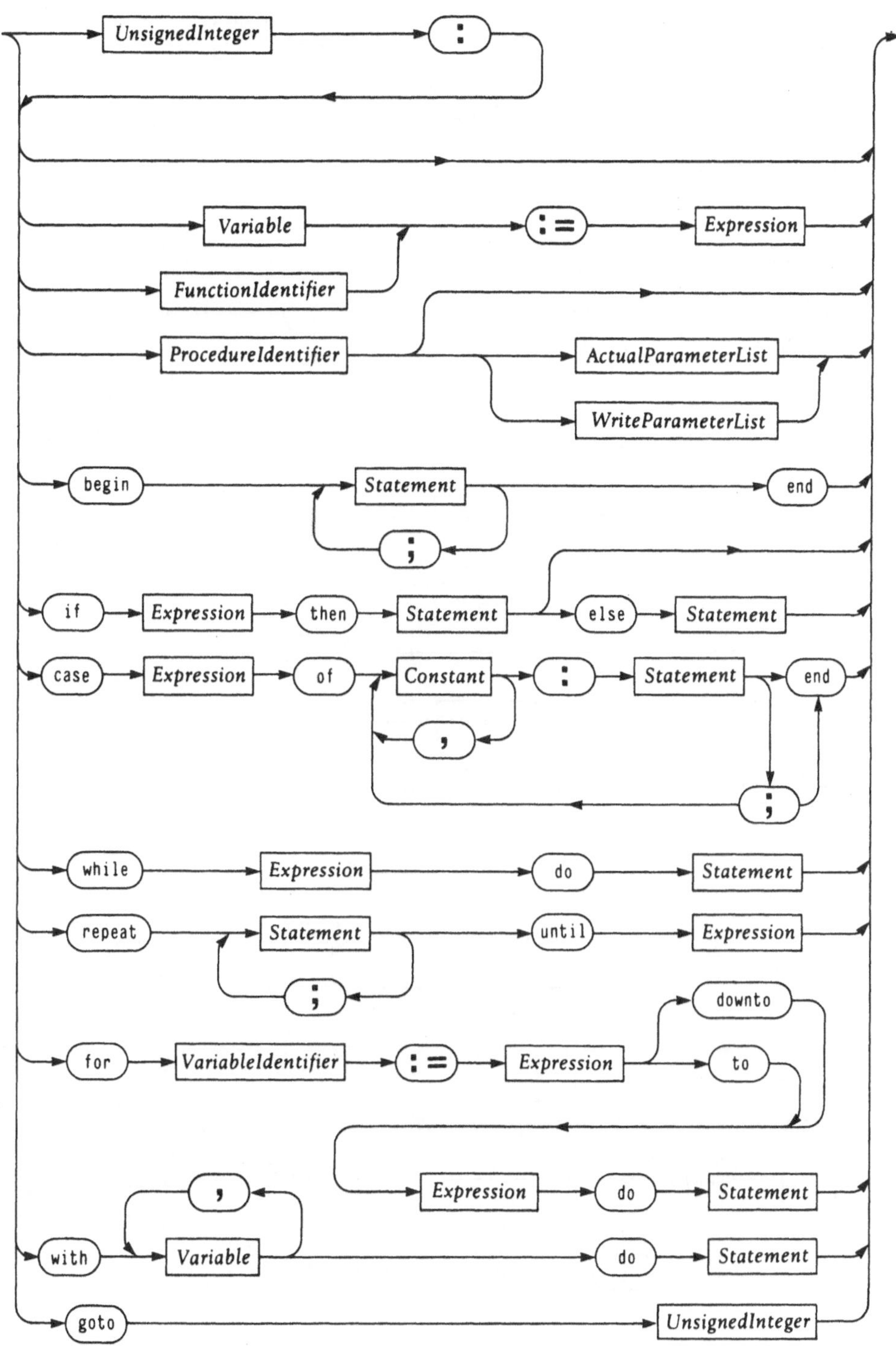
UnsignedInteger
:
Variable
:=
Expression
FunctionIdentifier
ProcedureIdentifier
ActualParameterList
WriteParameterList
begin
Statement
end
;
if
Expression
then
Statement
else
Statement
case
Expression
of
Constant
:
Statement
end
,
;
while
Expression
do
Statement
repeat
Statement
until
Expression
;
downto
for
VariableIdentifier
:=
Expression
to
Expression
do
Statement
,
with
Variable
do
Statement
goto
UnsignedInteger

Block

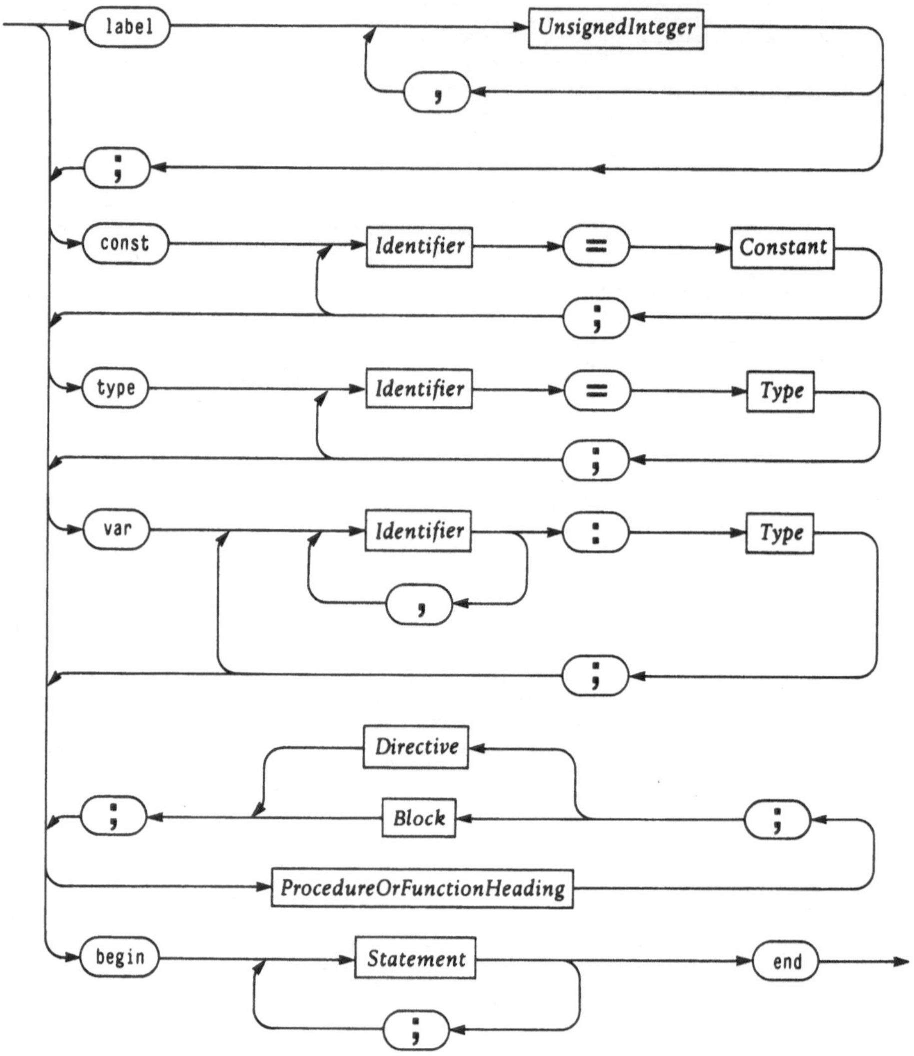

label
UnsignedInteger
,
;
const
Identifier
=
Constant
;
type
Identifier
=
Type
;
var
Identifier
:
Type
,
;
Directive
;
Block
;
ProcedureOrFunctionHeading
begin
Statement
end
;

Program

program
Identifier
;
Block
.
(
Identifier
)
,

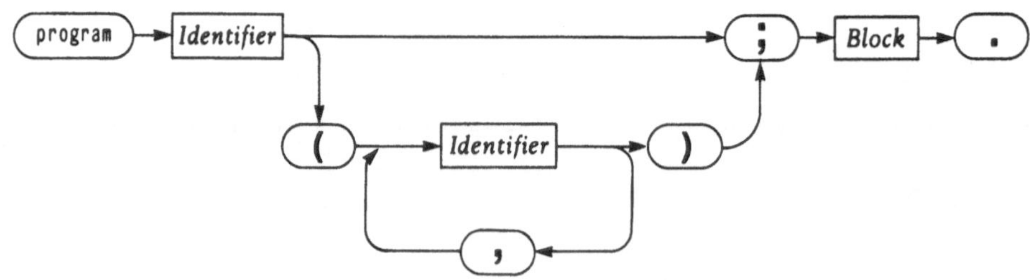

Appendix D
AL

Programm, Block	BEGIN ... END
	COBEGIN ... COEND
Declarations:	
Variable declarations	SCALAR
	VECTOR
	ROT
	FRAME
	TRANS
	EVENT
Array declarations	ARRAY
Procedure declarations	PROCEDURE
Dimension declarations	DIMENSION
Identifier	
Constant	
String	
Instructions:	
Instructions for the world model	AFFIX
	UNFIX
	COPY
Motion instructions	DRIVE
	MOVE
	PARK
	STOP
	DEPROACH
Effector instructions	OPEN
	CLOSE
	CENTER
Sensor instructions	ALWAYS WHEN
Program flow control	IF
	FOR
	WHILE
	UNTIL
	CASE
	(Procedure call)
	RETURN
	PAUSE
	ABORT
	(Tasking)

Synchronization	SIGNAL WAIT
In/output	PRINT PROMPT NOTE (File I/O) DISPLAY
Editing commands	DELETE EDIT REMOVE
Frame constructions	FCONSTRUCT
Assignments	
Arithmetic	

Syntax Diagrams

Program:

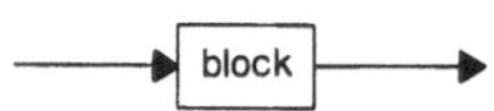

An AL program consists of an enclosing block.

Block:

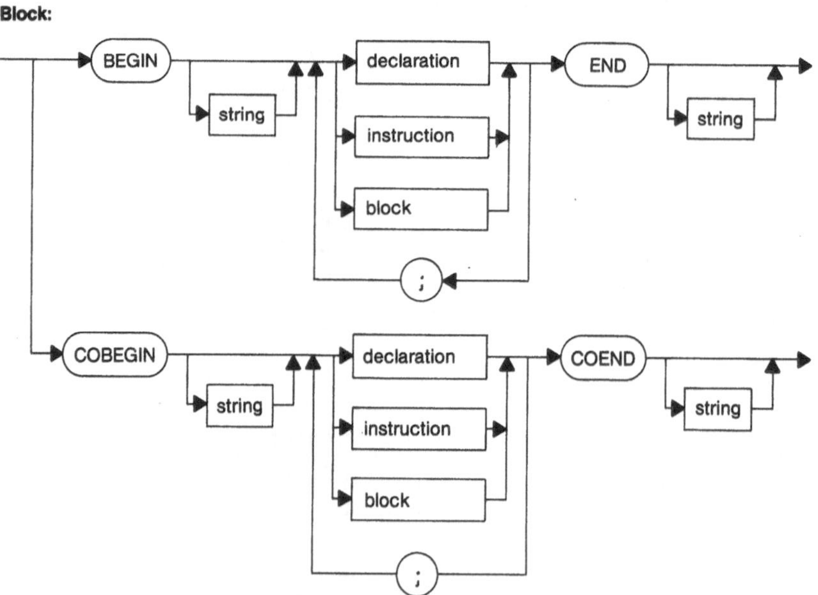

A block contains declarations instructions or further blocks.

Declaration:

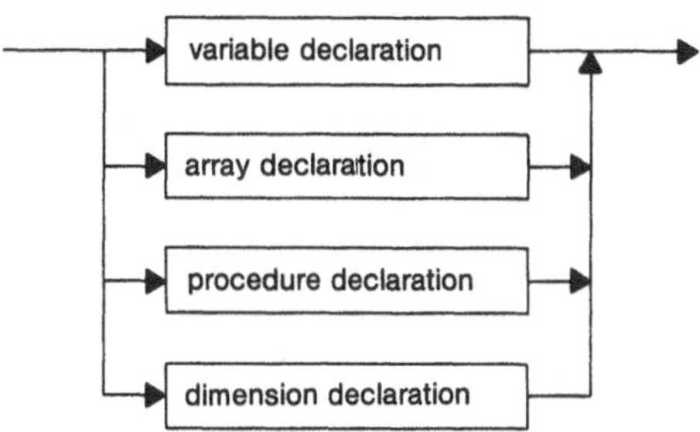

Variable declaration:

dimension
SCALAR
VECTOR
TRANS
ROT
FRAME
EVENT
variable identifier
,

Array declaration:

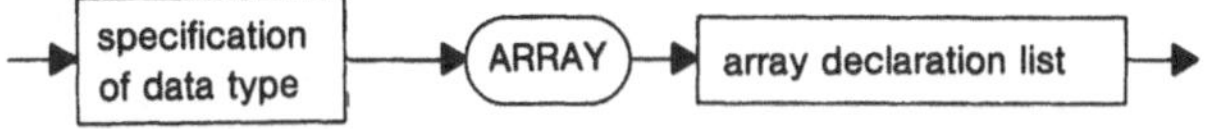

Array declaration list:

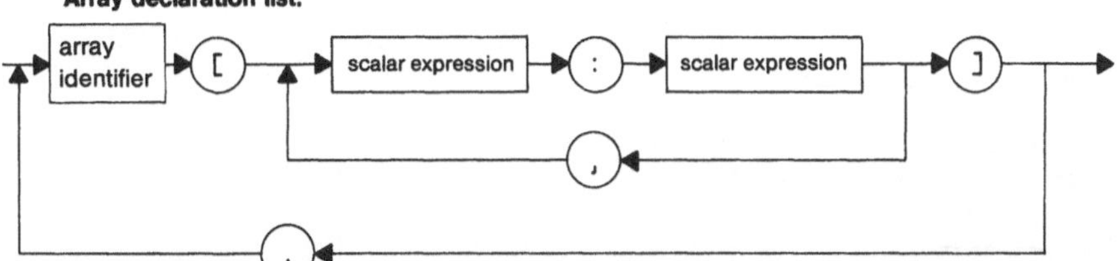

An array consists of components of one data type. Data types may include all those listed in the type declarations, with or without dimensioning, hence especially SCALAR, VECTOR, ROT, FRAME, TRANS and EVENT.

Procedure declaration:

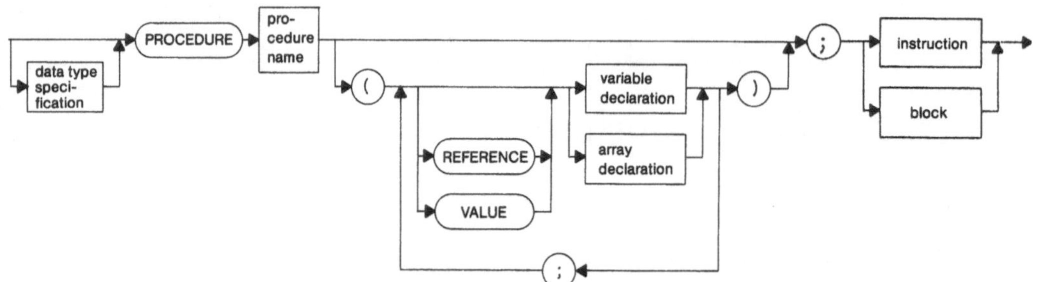

If a procedure has to deliver a result, the data type of the result has to be specified in the procedure declaration.

Dimension declaration:

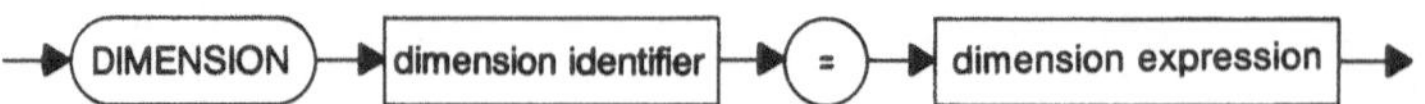

Dimension expression: **Dimensioning factor:**

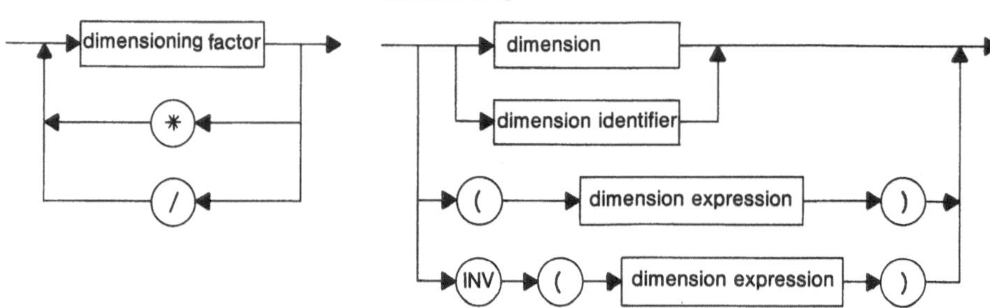

Dimension:

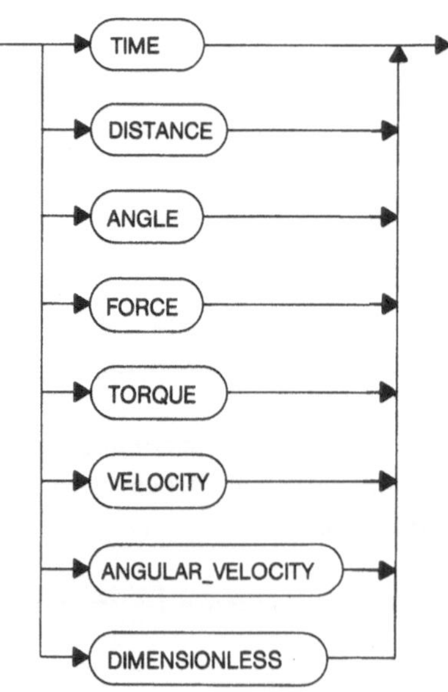

If a variable has been declared with a dimension, the corresponding unit of dimensioning (SEC, CM, DEG, GM, CM/SEC and DEG/SEC or a user-defined unit) has to be specified during an assignment.

Variables, array, procedure and dimension identifiers:

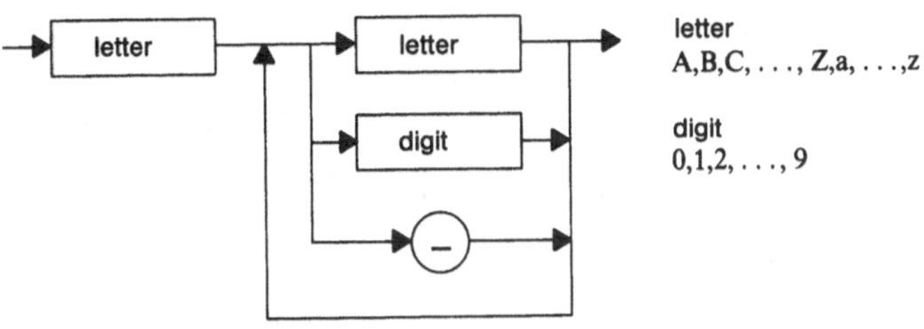

Constant:

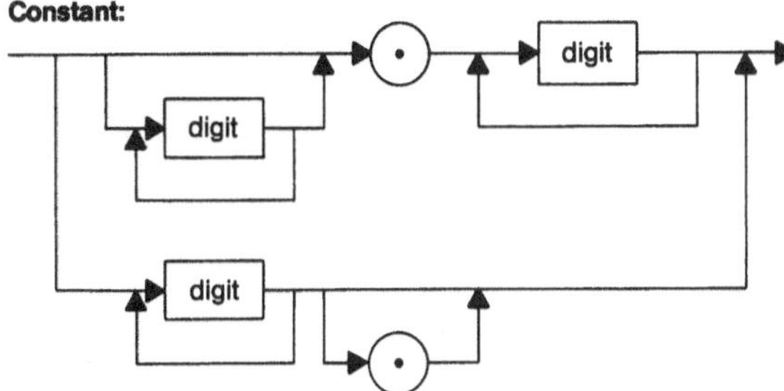

String:

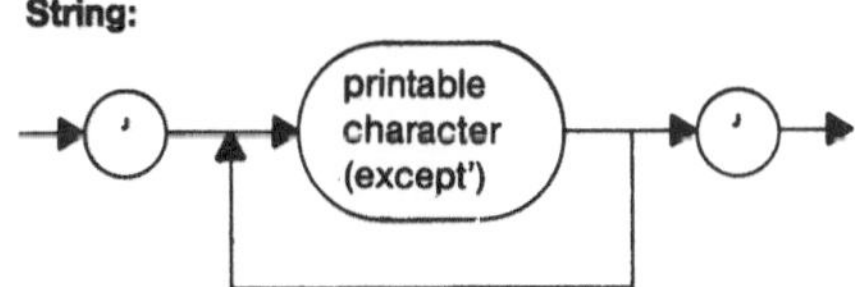

Instructions:

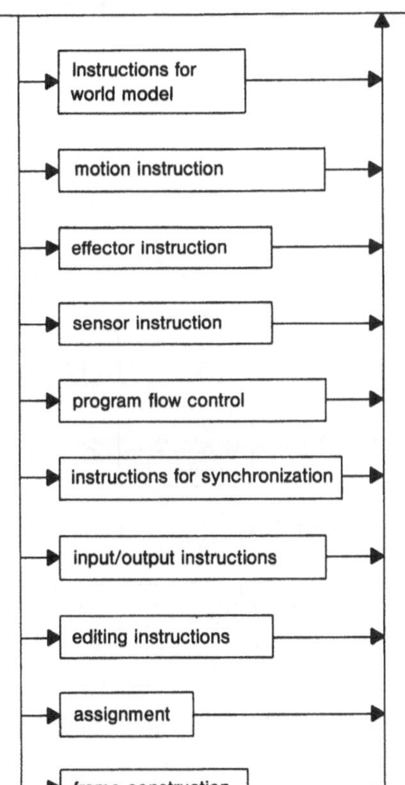

Instructions for world model:

AFFIX instruction:

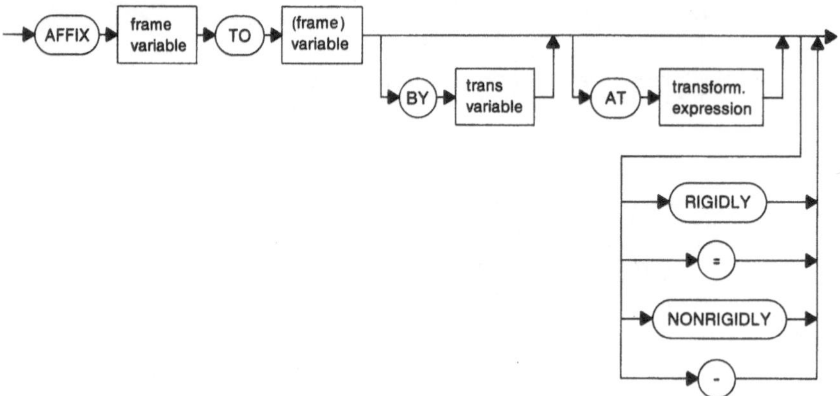

UNFIX instruction:

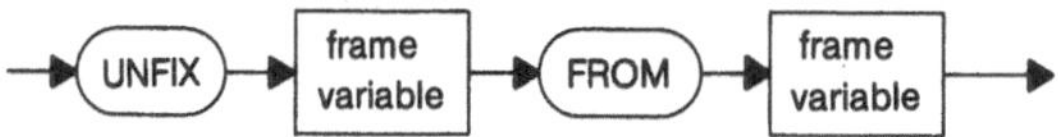

COPY instruction:

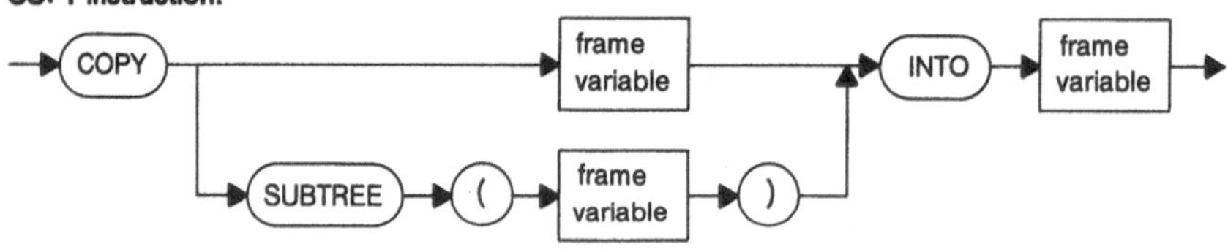

Motion instructions:

DRIVE instruction:

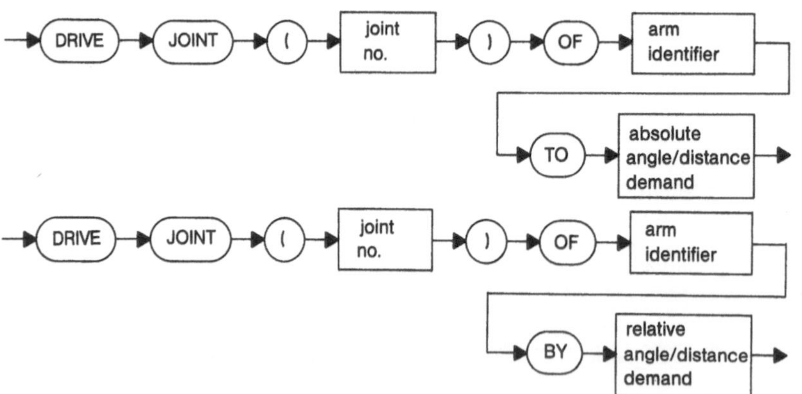

The keywort TO indicates absolut, whereas BY indicates relative angle or distance.

Arm identifier:

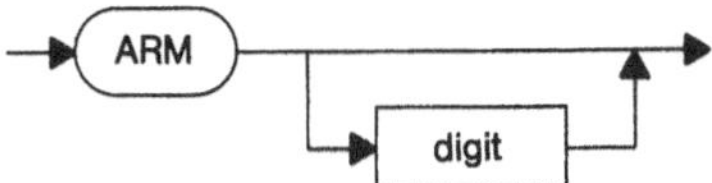

Motion instructions without parameters:

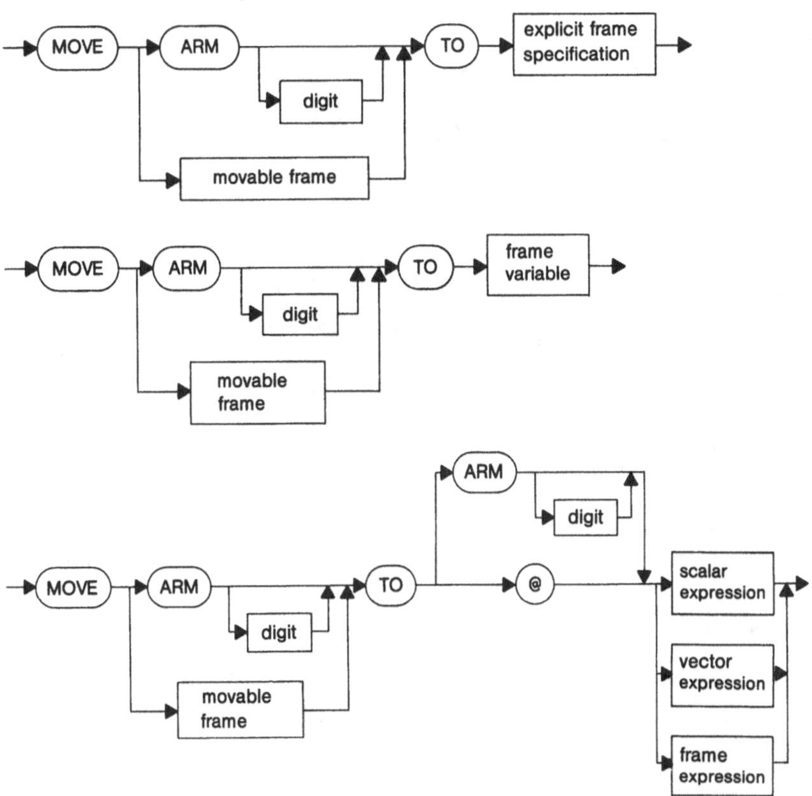

If a frame variable has been specified, it has to have been attached to an arm previously by the instruction AFFIX.

Motion instructions with parameters:

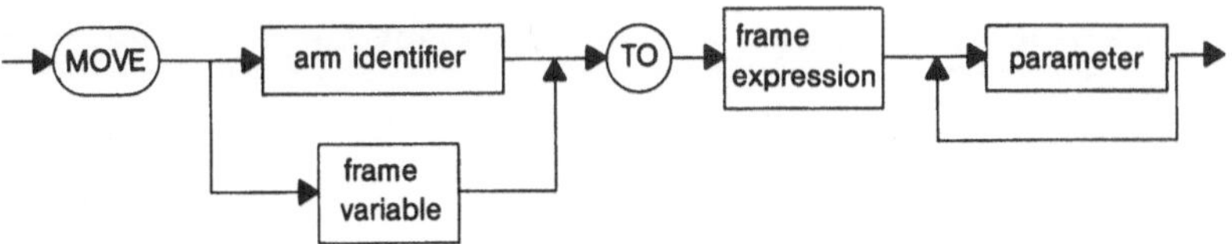

Parameter:

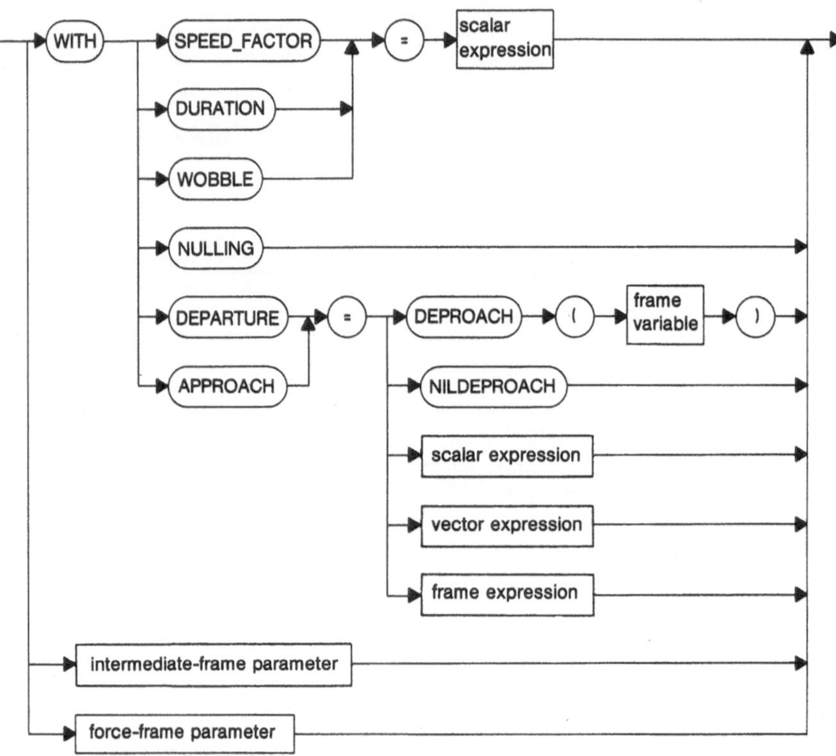

Force-frame parameter:

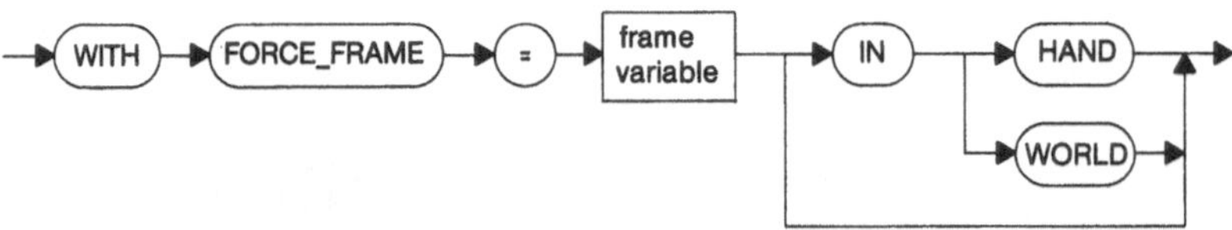

The force-frame parameter refers to movements with force or moment monitoring.

Intermediate-frame parameter:

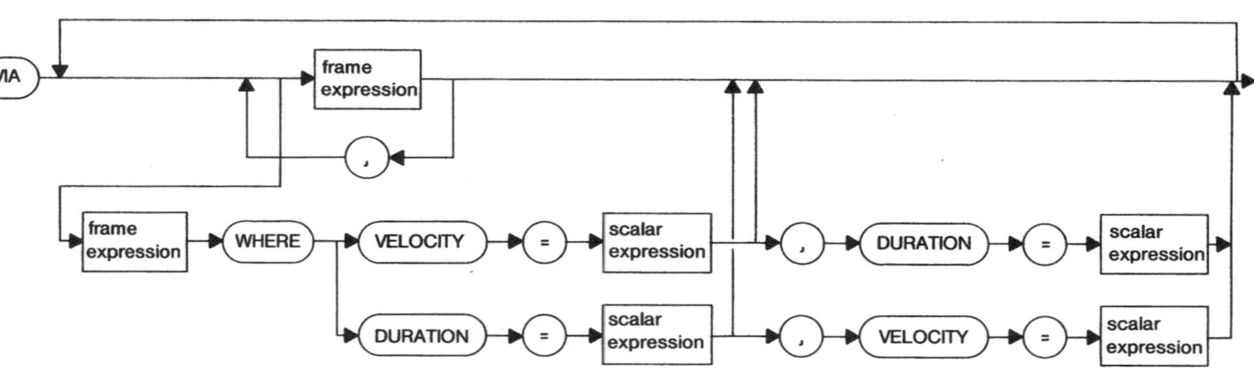

Motion instructions with monitoring of forces, moments, timing and events:

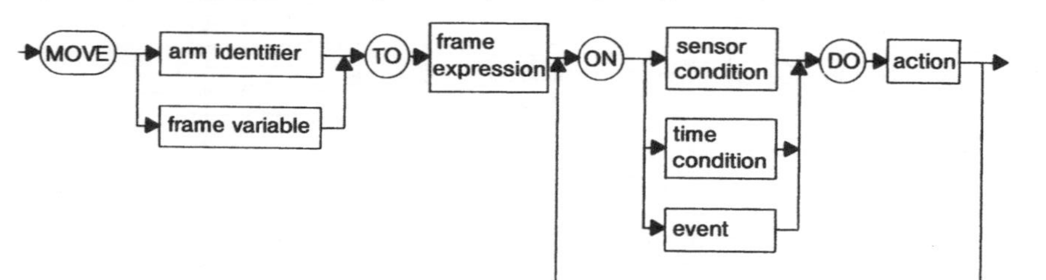

The specification ON <condition> DO <action> may be combined with any parameters.

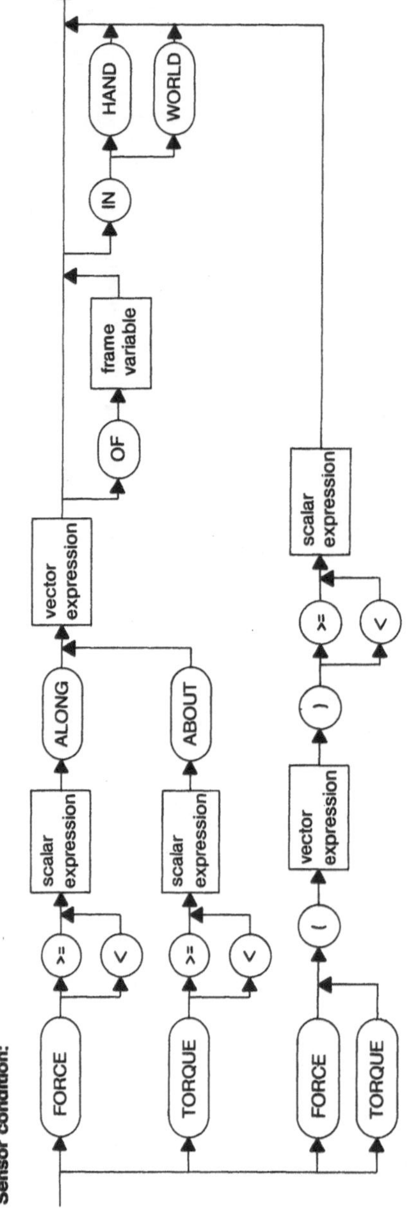

Timing condition:

DURATION >= scalar expression

Action:

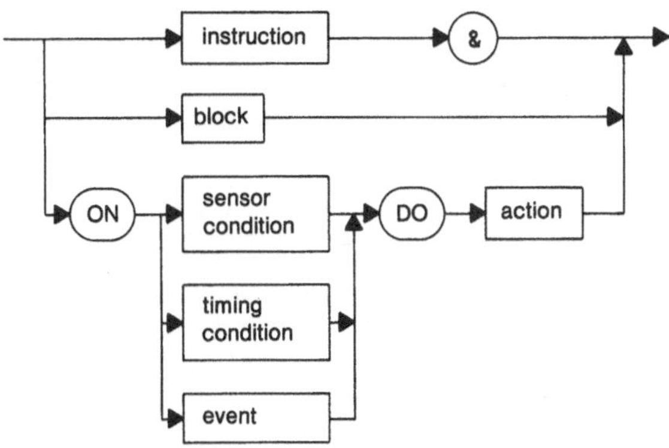

Motion instructions with force or moment parameters:

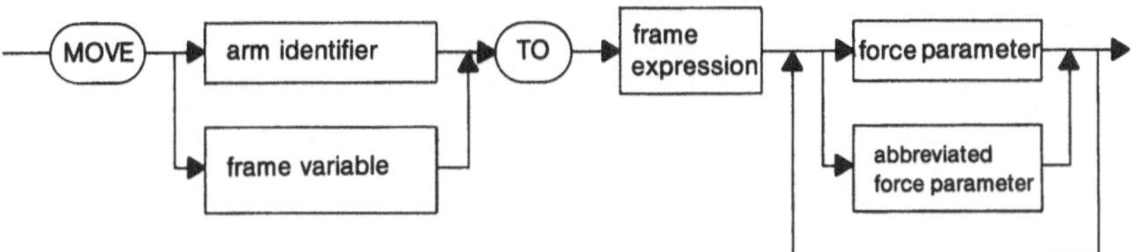

Force parameter:

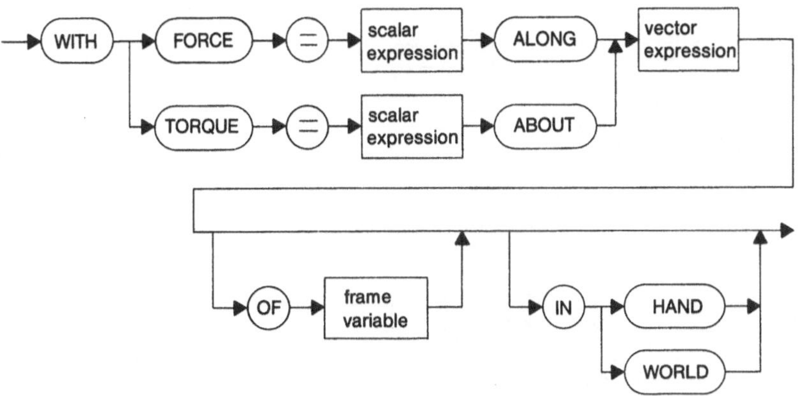

Abbreviated force parameter:

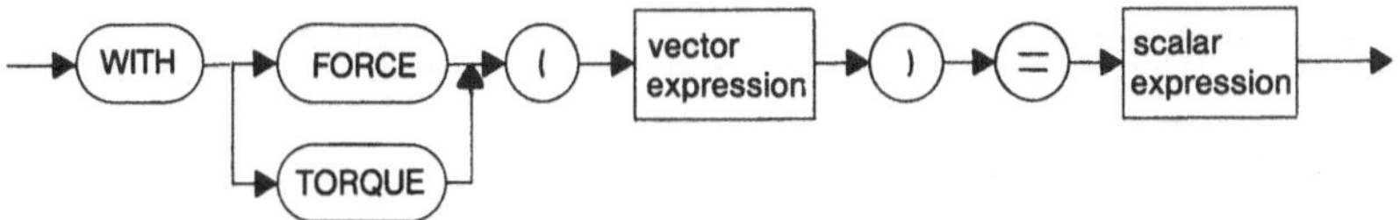

Driving to a home position:

Stopping a robot or effector motion:

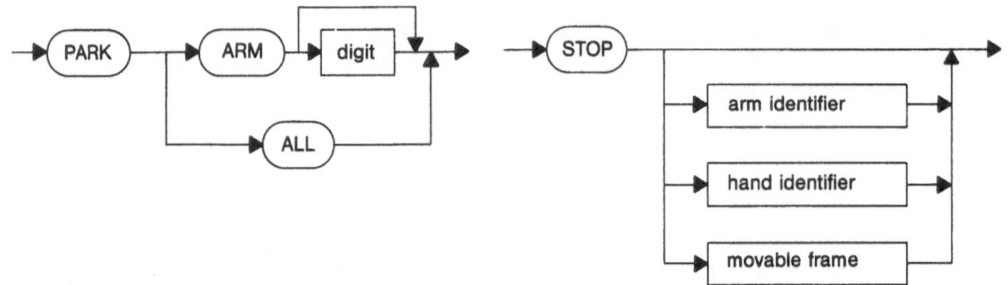

Predefining an approach/departure frame:

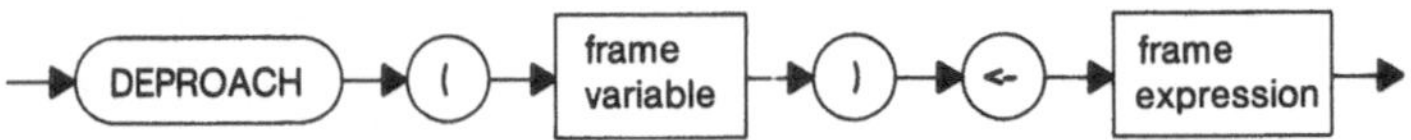

Gripper instruction:

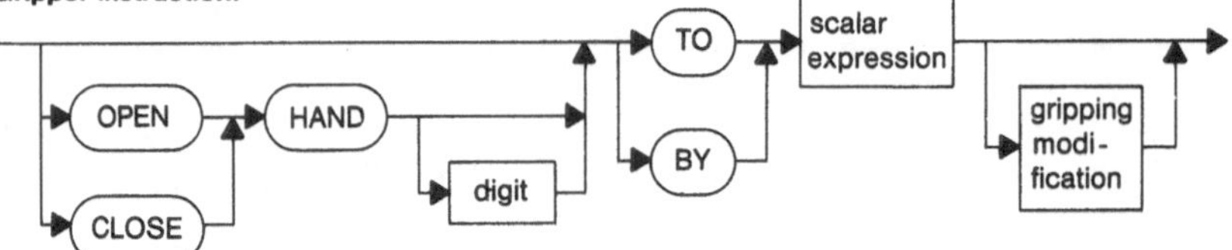

Gripping modification:

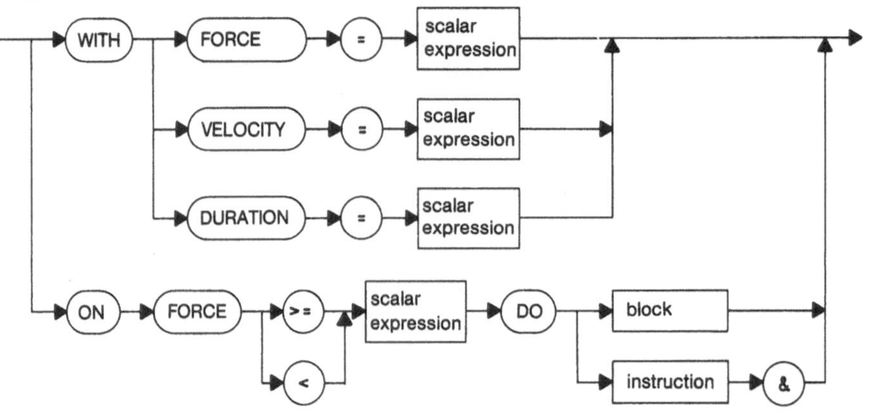

CENTER instruction:

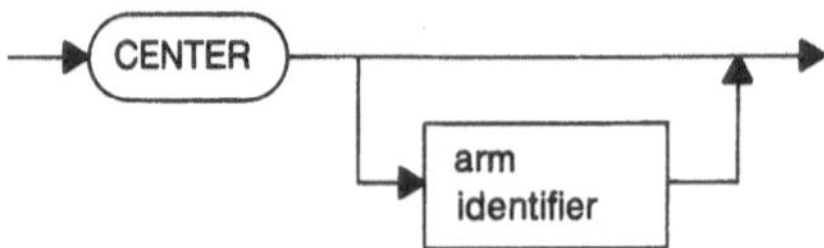

Sensor instruction:

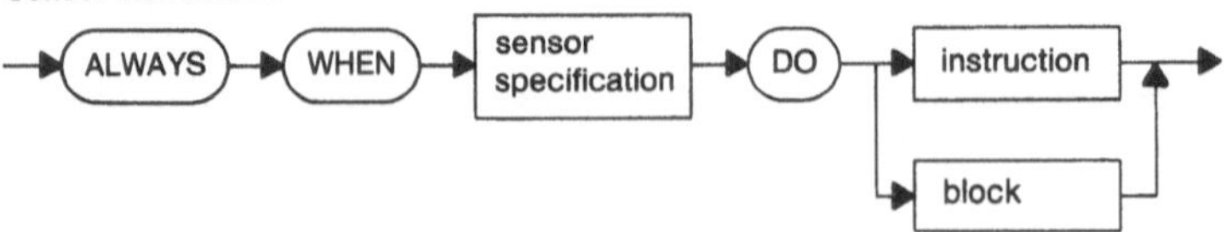

Sensor specification:

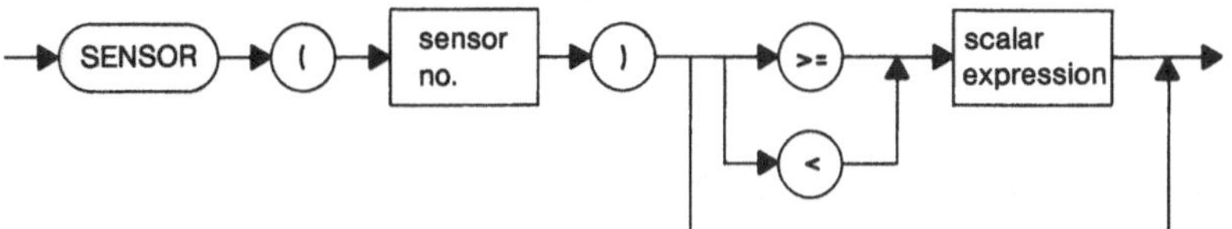

IF instruction:

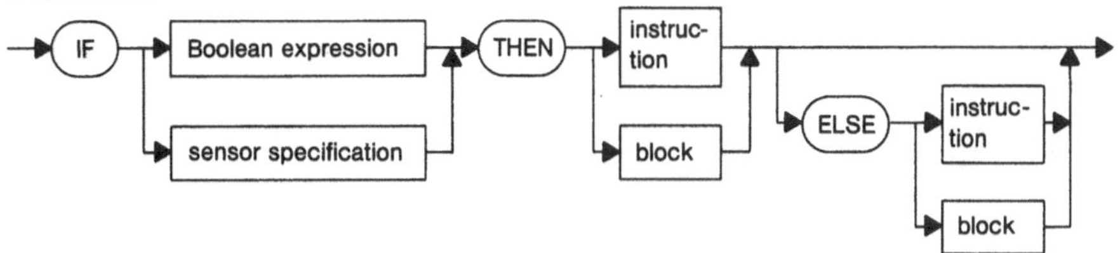

FOR instruction:

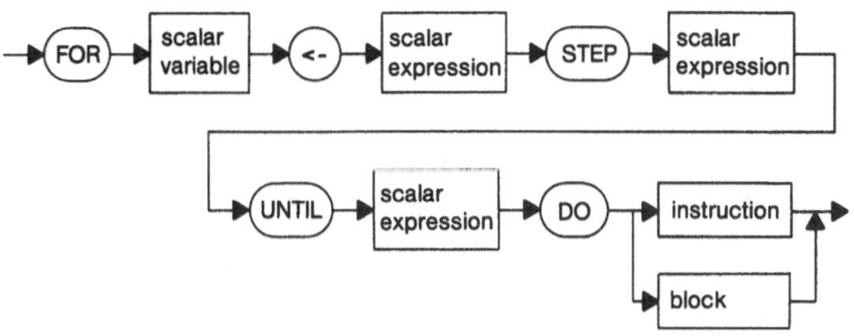

WHILE instruction:

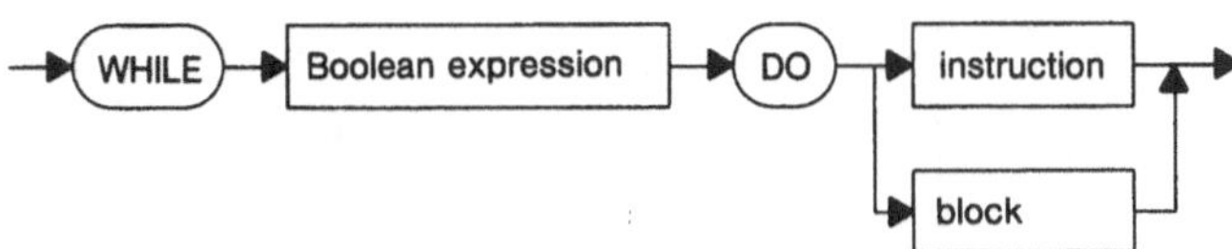

UNTIL instruction:

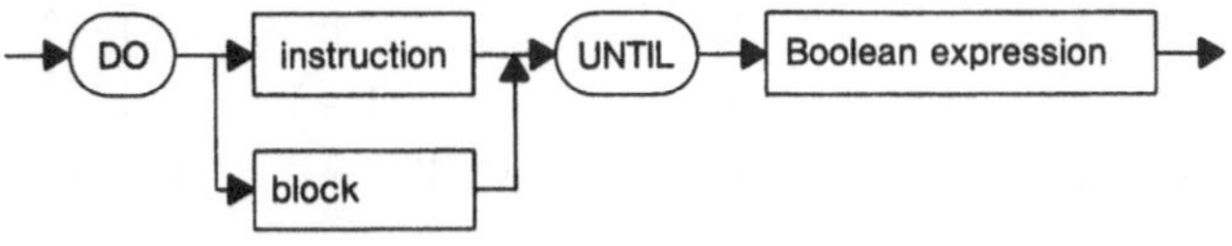

Case instruction:

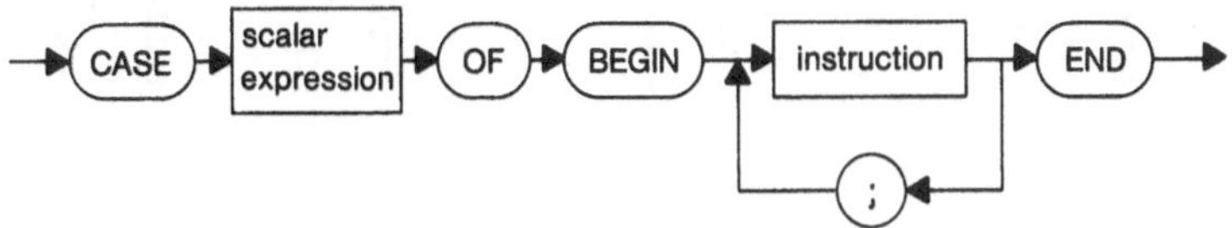

CASE instruction with labels:

CASE
scalar expression
OF
BEGIN
labelled instruction
;
ELSE
instruction
block
;
labelled instruction
END

Labelled instruction:

[
constant
]
instruction
block
;

Procedure call:

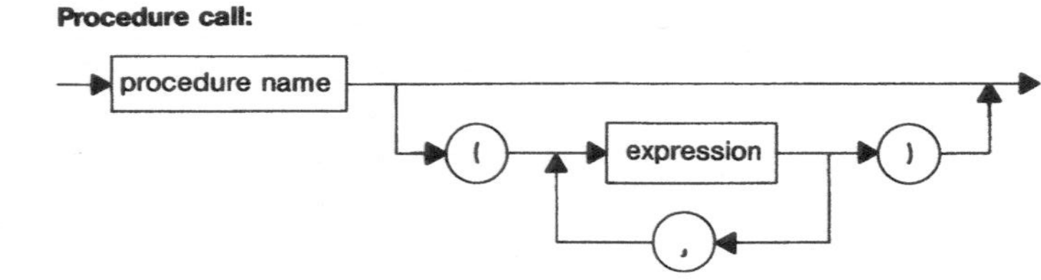

RETURN statement:

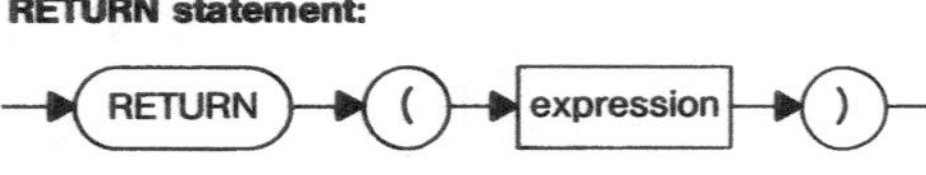

PAUSE instruction:

PAUSE
scalar expression

ABORT instruction:

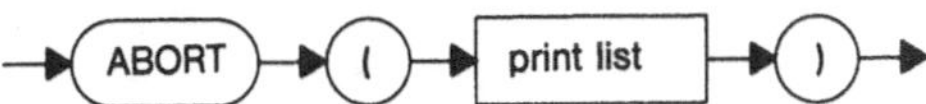

Tasking instruction:

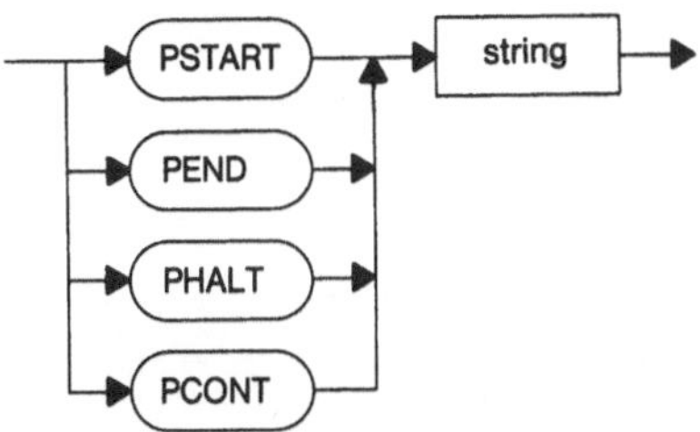

Instructions for synchronization:

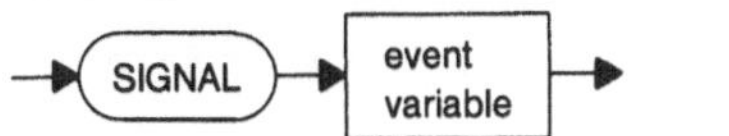

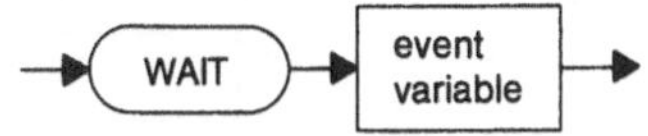

Input/Output:
PRINT instruction:

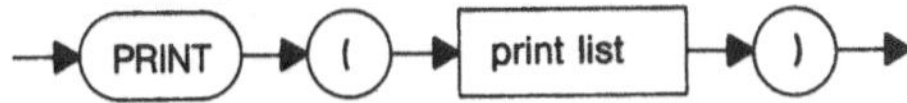

PROMPT instruction:

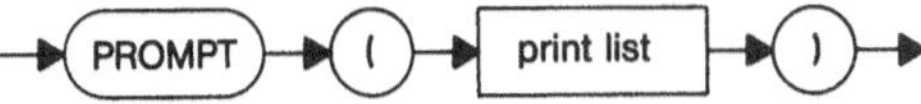

Print list:

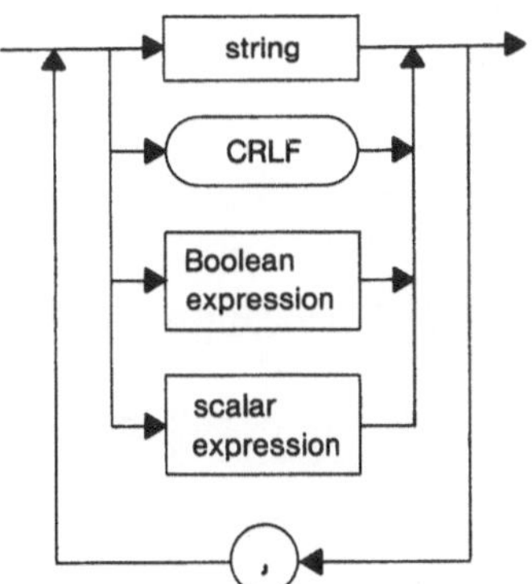

FILE input/output:

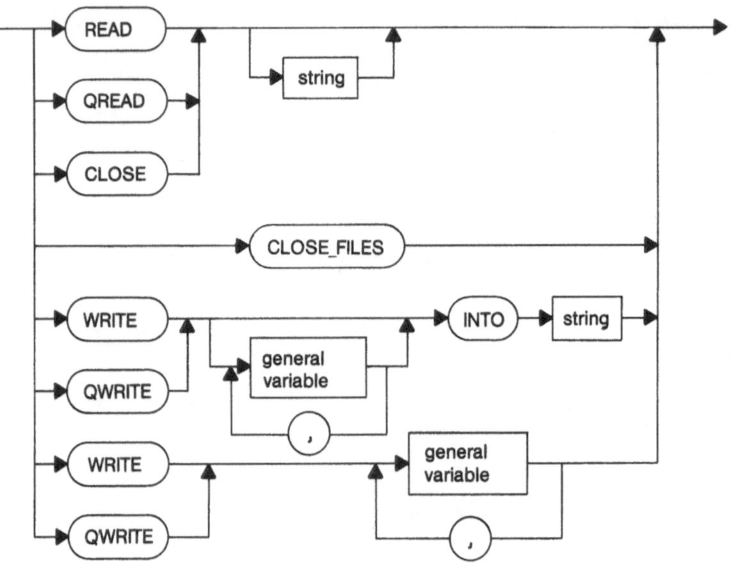

NOTE instruction:

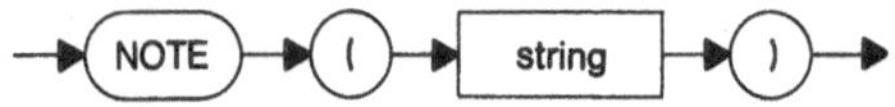

DISPLAY instruction:

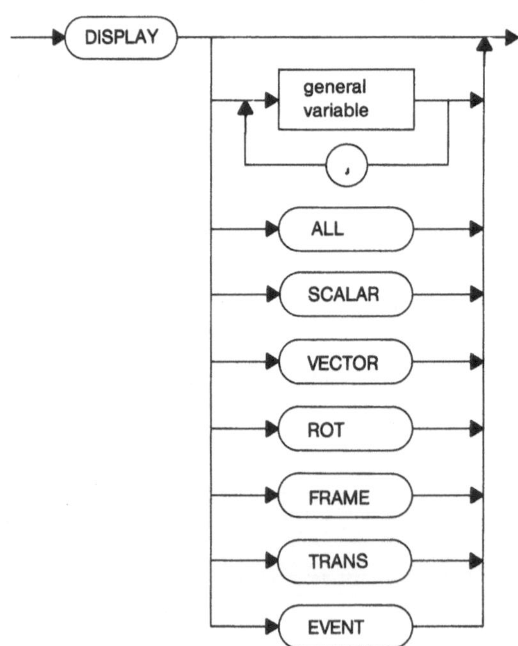

DELETE instruction:

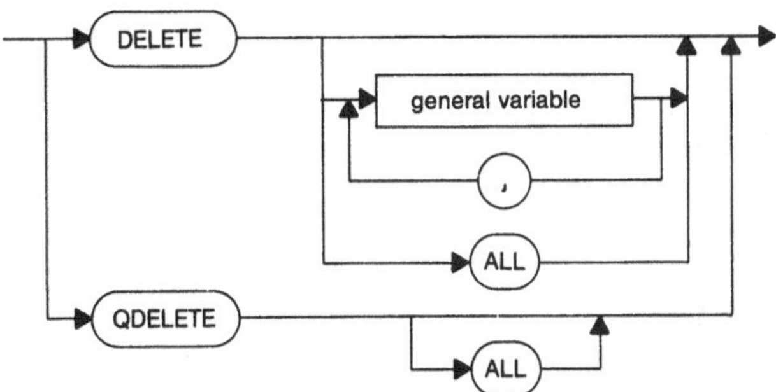

General variable:

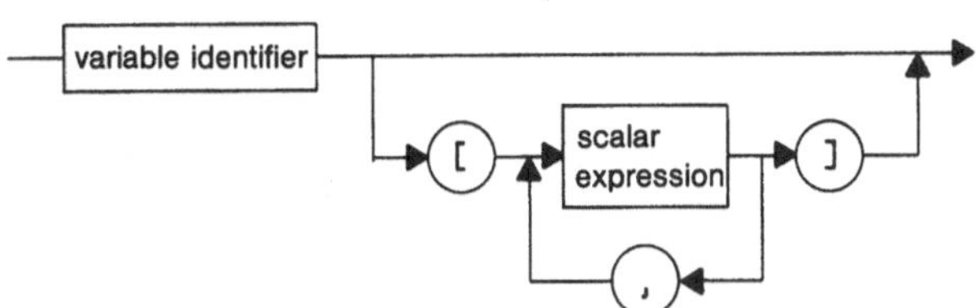

Editing instruction:

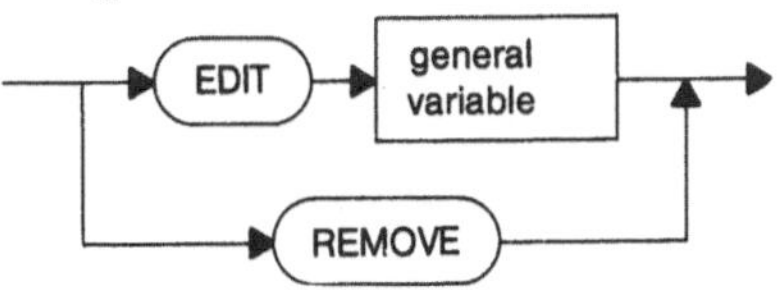

FCONSTRUCT instruction:

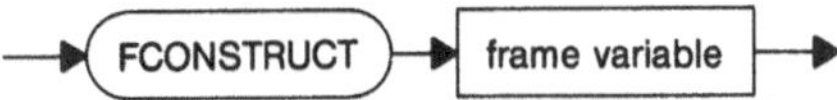

Assignment:

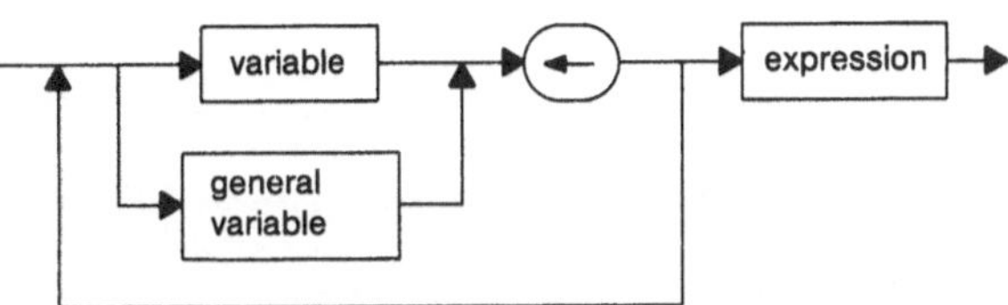

Note: The variable type on the left-hand side of the assignment has to correspond to the one of the expression on the right-hand side (for example: rotationvariable ← rotationexpression)

Expression:

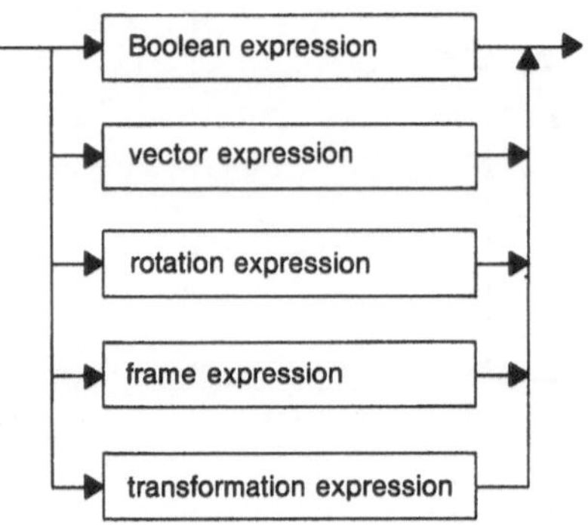

Scalar expression:

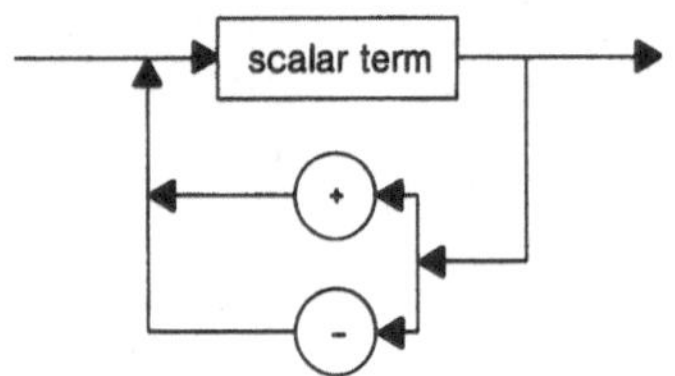

Scalar term:

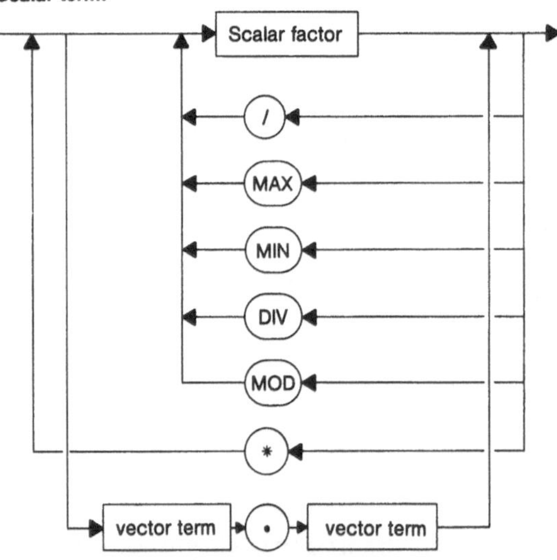

Scalar factor:

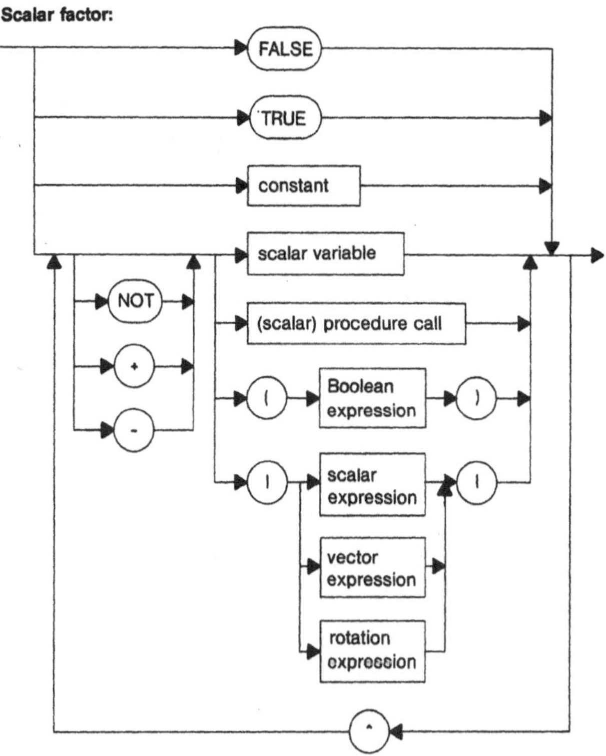

Boolean expression:

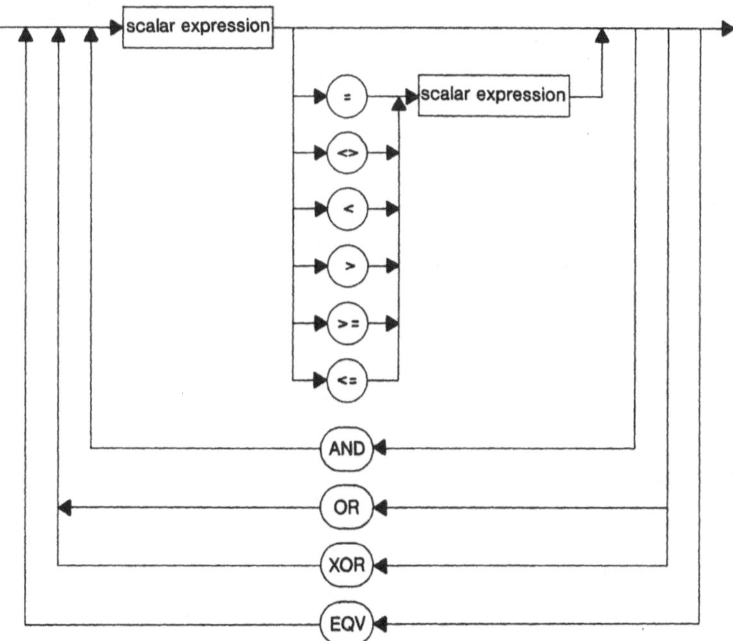

Vector expression:

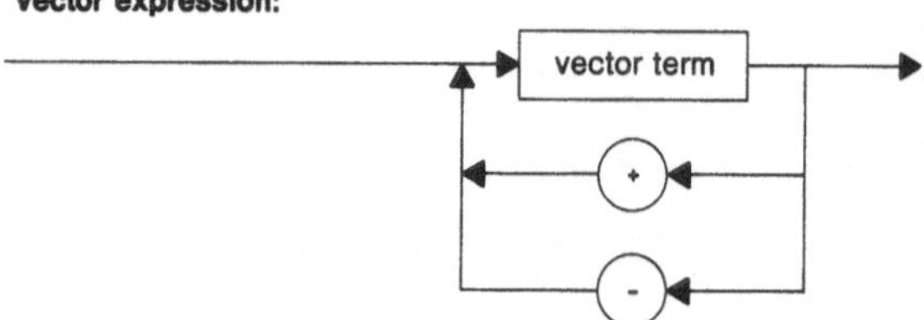

Vector term:

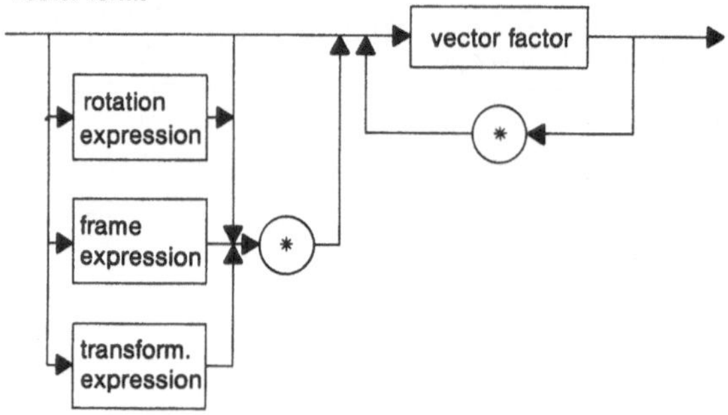

Vector factor:

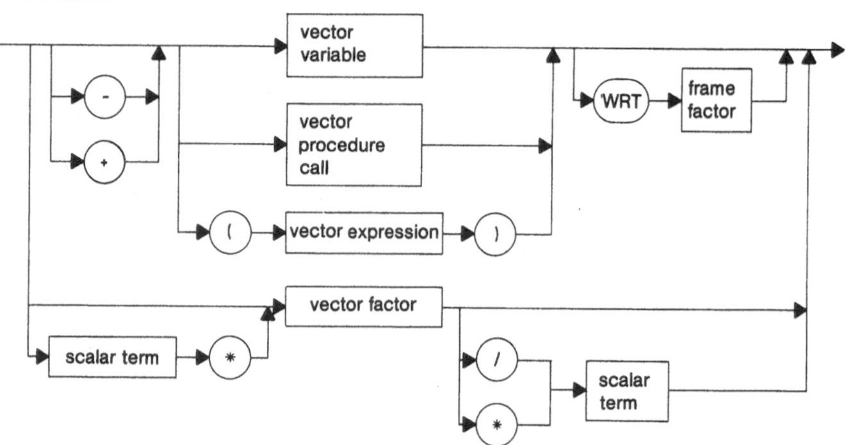

Rotation expression:

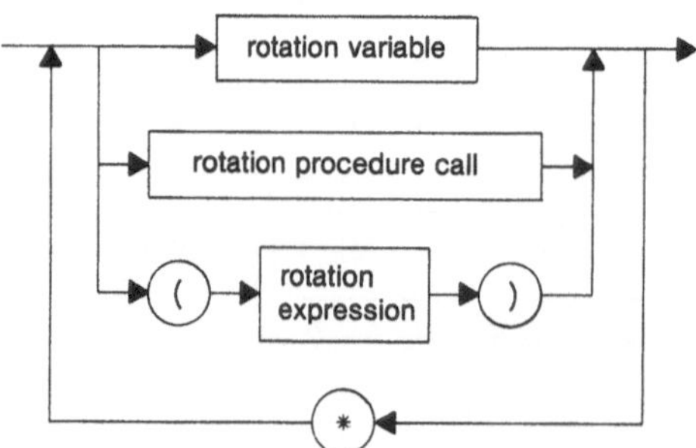

Frame expression:

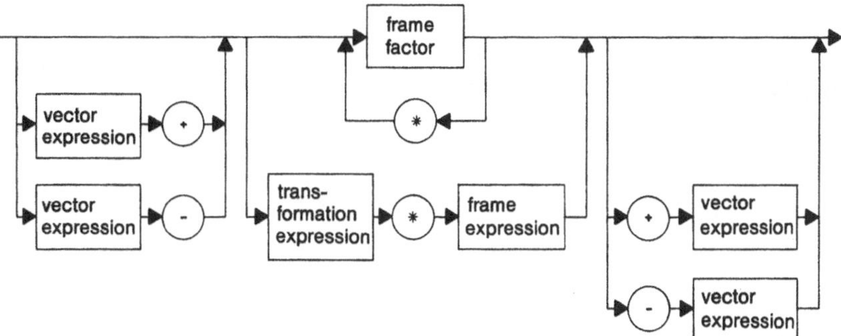

Frame factor:

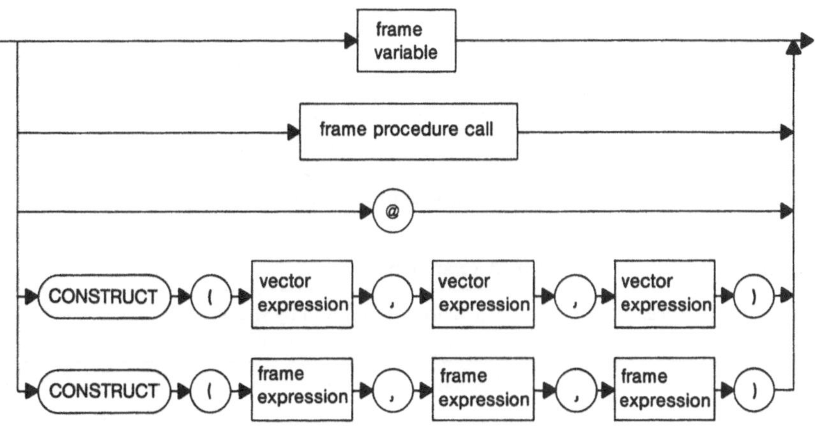

Transformation expression:

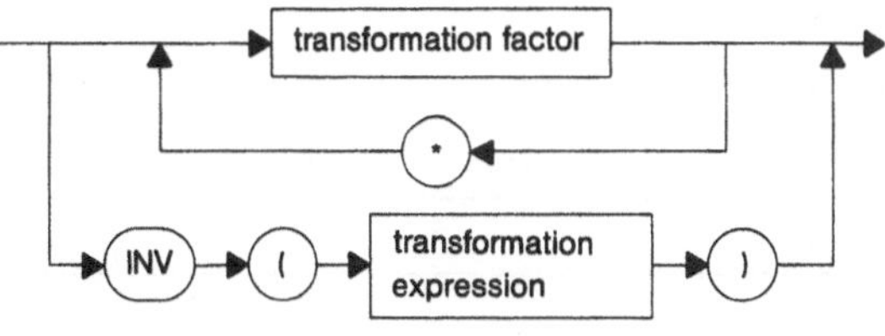

Transformation factor:

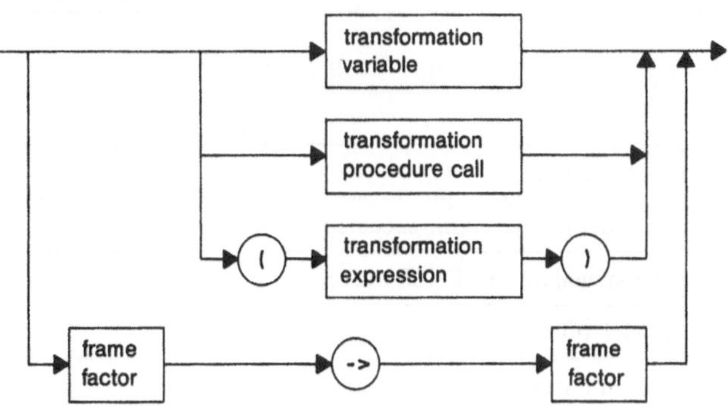

Appendix E
AML

AML subroutines

I. Motion Control

Subroutine	Explanation
ACCEL (factor)	Define acceleration values
AMOVE (axes, target, monitor, parameter)	Execute an asynchronous motion to the specified target
DECEL (factor)	Define decelaration
DMOVE (axes, distance, monitor, parameter)	Execute synchronous motion by specified distance
MOVE (axes, target, monitor, parameter)	Execute synchronous motion to specified target
QGOAL (axes)	Recall target of last motion
QMANIP	Get arm status
QPOSITION (axes)	Get measured position of arm
SETTLE (flag)	Wait (Not wait) for arm to settle
SPEED (factor)	Define speed
STOPMOVE	Abort motion
WAITMOVE	Wait for the completion of motion

II. Safety Instructions

Subroutine	Explanation
ERRTRAP (name)	Activation of a routine executed in case of errors
FREEZE	Put system into FREEZE mode
SHUTDOWN	Switch off hydraulics
STARTUP	Activate system

III. Sensor Instructions

DEFIO (....)	Define the characteristics of a user provided sensor
DELIO (....)	Erase the definition of a sensor
ENDMONITOR (monitorset)	Terminate sensor monitoring
MONITOR (....)	Determine the parameters for monitoring the sensors and activate it
QIODEF (number)	Interrogate the characteristics of a sensor
QMONDEF (number, label)	Interrogate the monitor parameters
QMONITOR (monitorset)	Interrogate the monitor status
SENSIO (number, template)	Execute sensor input/ouput
REMONITOR (monitorset)	Reactivate a group of monitors

IV. Arithmetic

Subroutine	Explanation
ABS (number)	Absolute value of the number
ACOS (number)	Function: arc cosine
AGGSIZE (aggregate)	Number of elements in an aggregate
APPLY (number,parameter)	Dynamic execution of a subroutine or computing instruction
ASIN (number)	Function: arc sine
ATAN (number1, number2)	Function: arc tangent
BOXTRANS (....)	Determines rotation matrix and translation vector for given coordinate transformation
COS (degree)	Function: cosine
CROSS (u,v)	Cross product of two vectors
CVTBN (aggregate)	Convert an aggregate of sixteen binary elements into an integer
CVTNB (number)	Convert an integer into an aggregate of sixteen binary numbers
CVTNS (number)	Convert a number into a string
CVTSN (string, errorlabel)	Convert a string to a number
DOT (u,v)	Scalar product of u and v
EULERTRANS (....)	Evaluate rotation matrix and position vector for given Eulertransformation
EVAL (string)	Evaluation of a string, which may contain an AML expression, definition of a variable or AML instructions
INVERSE (m)	Inversion of a 3 x 3 matrix
IOTA (n)	Produces the aggregate of integers from 1 to n

Subroutine	Explanation
LENGTH (string)	Number of characters in a string
MAG (u)	Magnitude of vector u
MAP (routine, aggregate)	Execution of a subroutine for each element in the aggregate
SELECT (mask, aggregate)	Select elements from an aggregate
SIN (degree)	Function: sine
SQRT (number)	Function: square root
STRING (n)	Produces a string of n spaces
SUBSTR (string, begin, count)	Selection of a substring
TAN (degree)	Function: tangent
TRANSBOX (....)	Transformation of the coordinates of all axes for a given rotation matrix and position vector
TRANSEULER (....)	Evaluates the Euler transformation for a given rotation matrix and position vector
TRANSPOSE (m)	Transpose the matrix m

V. Data Manipulation

Subroutine	Explanation
ALLOCATE (name, length, number)	Space allocation for a disk file
BRANCH (label)	Branching to a label
CLEANUP (name)	Activation of a routine to be executed before the program is terminated
CLOCK (startvalue)	Set the clock
CLOSE (channel)	Close output channel
COLUMN (channel)	Evaluates the position to which the next character has to be written
COMPRESS (disk)	Make contiguous space on the disk
COPYFILE (file1, file2, copy)	Copy files
COPYVOL (disk1, disk2, copy)	Copy a complete disk
DELAY (n)	Delay program execution by n seconds
DISPLAY (data,..,..)	Display data on terminal
EOD	Set end of data for a file
ERASE (file)	Delete file from disk
FMTNUM (....)	Format a number for display

Subroutine	Explanation
INITDISK (name, type, disk)	Initialize disk
KEY (key, function)	Assign a function to a key
LISTFILE (name, channel, begin,end)	Recording a file
MESSAGE (number)	Make text available for a message
OPEN (file, errorlabel)	Open a file on disk
POINT (channel, recordnum, errorlabel)	Position file pointer
PREFILL (data)	Fill the input array from terminal with data
PRINT (channel, ‹data,..›)	Display data as strings
QCHAN (channel, errorlabel)	Ask for channel status
QDISK (disk)	Returns the free space on the disk
QFILES (disk, qualifier, channel)	Displays the files on the disk and their attributes
QUIT (level)	Return to a higher programming level
QTERM	Make the last terminal entry available
QVOLS	Make the name of the last used volume available
READ (channel, template, errorlabel)	Read a record from a file or a terminal
RENAMEFILE (name1, name2)	Rename a file
RENAMEVOL (name1, name2)	Rename a volume
RETURN (value)	Return to the calling program
SCROLL (flag)	(No) Scroll of terminal
SETCOL (position, channel)	Defines the first position on the screen
SHOW (menue, parameter)	Call a general, menu-driven dialogue for the terminal
TIME (hour, min, sec, msec, month,day,year)	Set/recall time and date
TP3780 (....)	Host computer communication
USERVOL (name)	Define default disk
WRITE (channel, data..., errorlabel)	Output a set of data

VI. Calibration

Subroutine	Explanation
CALLROLL (angle, errorlabel)	Calibration of rotation axis
CALLYP (errorlabel)	Calibration of pitch axis
CGRASP (width, force)	Close gripper while centering on the object
FINDCUP	Align with the bottom of a cylinder
FINDEDGE (axes, target, errorlabel)	Find the edge of an object by using the cross fire sensor
FINDPOST (errorlabel)	Find coordinates of a vertical post
LEDSEARCH (axes, target, parameter,errorlab)	Move arm until cross fire sensor triggers
SEARCH (....)	Move arm until a sensor triggers

Appendix F
VAL-II

I. Monitor Commands

In the following commands, “location” refers to a “precesion point” or a “compound” unless otherwise noted. A “compound” can be a transformation, a transformation function, or a compound transformation.
The minimum abbreviation currently accepted for each command is shown along with the command. The length of the minimum abbreviation may differ if your VAL-II system includes optional features such as vision.

```
Defining Locations
        B         BASE {<dX>}, {<DY>}, {<dZ>}, {<Z rotation>}
        H         HERE <location variable>
        PO        POINT <location variable> {= <locationvalue>}
        TEA       TEACH <location variable>
        TO        TOOL {<compound>}
        W         WHERE {<integer>}

Program Editing
        ED        EDIT {<program>}, {<step>}
                  <any program instruction>
                  <carriage return>
                  C <program>, {<step>}
                  D {<steps>}
                  E
                  I
                  L
                  P {<steps>}
                  R <character string>
                  R, <character string>
                  S {<step number>}
                  T <location variable>
                  TS <location variable>

Program and Data Listing
        DIR       DIRECTORY
        LISTL     LISTL {<location>} {, ..., <location>}
        LISTP     LISTP {<program>} {, ..., <program>}
        LISTR     LISTR {<expression>} {, ..., <expression>}
```

```
Program and Data Storage
        COMP        COMPRESS
        DELETEF     DELETEF <file specification>
        FO          FORMAT
        LISTF       LISTF
        LOA         LOAD <file specification>
        STORE       STORE <file spec> {=<program>} {, ..., <program>}
        STOREL      STOREL <file spec> {= <program>} {, ..., <program>}
        STOREP      STOREP <file spec> {= <program>} {, ..., <program>}
        STORER      STORER <file spec> {= <program>} {, ..., <program>}

Program Manipulation
        COP         COPY <new program name> = <old program name>
        REN         RENAME <new program name> = <old program name>

Program and Data Deletion
        DELETE      DELETE <program> {, ..., <program>}
        DELETEL     DELETEL <location variable>{, ..., <location variable>}
        DELETEP     DELETEP <program>{, ..., <program>}
        DELETER     DELETER <real variable>{, ..., <real variable>}

Program Control
        A           ABORT
        DO          DO {<program instruction>}
        EX          EXECUTE {<program>}, {<loops>}, {<step>}
        PR          PROCEED
        RET         RETRY
        SP          SPEED <value>
        X           XSTEP {<program>}, {<loops>}, {<step>}

Control of Process Control Programs
        PCA         PCABORT
        PCEN        PCEND
        PCEX        PCEXECUTE {<program>}, {<loops>}, {<step>}
        PCP         PCPROCEED
        PCR         PCRETRY

System Status and Control
        CA          CALIBRATE
        DON         DONE
        FR          FREE
        LIM         LIMP
        PCS         PCSTATUS
        STA         STATUS
        ZE          ZERO
```

```
System Parameters and Switches
        DIS       DISABLE <switch> {, ..., <switch>}
        EN        ENABLE <switch>{, ..., <switch>}
        PAR       PARAMETER <parameter name> {= <value>}
        SW        SWITCH {<switch>} {, ..., <switch>}

Binary Signals
        IO        IO
        PC        PC <signal> {,<bits>} = <expression>
        RES       RESET
        SI        SIGNAL <signal> {, ..., <signal>}

Miscellaneous
        IOP       IOPUT <address> = <value>

Supervisory Communication
        ID        ID
        LOO       LOOPBACK {<mode>}
        N         NET
        PAN       PANIC
        TES       TESTP <program>

Command-Program Control
        COMM      COMMANDS <program>

System Diagnostics and Modification
        DIA       DIAGNOSTIC <file specification>
        O         OVERLAY <file specification>
```

II. Program Instructions

In the following commands, "location" refers to a "precision point" or a "compound" unless otherwise noted. A "compound" can be a transformation, a transformation function, or a compound transformation.
A numeric constant, real-variable name, or arithmetic expression can be provided for all arguments which require a numeric value.
The minimum abbreviation currently accepted for each instruction is shown along with the instruction. The length of the minimum abbreviation may differ if your VAL-II system includes optional features such as vision.

```
Real Variable Assignment
                  <variable> = <value>
        DEC       DECOMPOSE <arrayname>[] = <location>
```

```
Location Variable Assignment
        HE          HERE <location variable>
        SE          SET <location variable> = <location value>
        TO          TOOL {<compound>}

Motion
        ALI         ALIGN
        APPRO       APPRO <location>{!}, <distance>
        APPROS      APPROS <location>{!}, <distance>
        BRA         BRAKE
        BRE         BREAK
        DEL         DELAY <time>
        DEPART      DEPART <distance>
        DEPARTS     DEPARTS <distance>
        DR          DRIVE <joint>, <change>, <speed>
        MOVE        MOVE <location> {!}
        MOVES       MOVES <location> {!}
        MOVEST      MOVEST <location> {!}, <hand opening>
        MOVET       MOVET <location>{!}, <hand opening>
        NE          NEST
        READ        READY

Hand Control
        CLOSE       CLOSE {<hand opening>}
        CLOSEI      CLOSEI {<hand opening>}
        GR          GRASP <hand opening>, {<label>}
        OPEN        OPEN {<hand opening>}
        OPENI       OPENI {<hand opening>}
        RELAX       RELAX
        RELAXI      RELAXI

Program Control
        CAL         CALL <program>
        GO          GOTO <label>
        HA          HALT
        IF          IF <logical expression> GOTO <label>
        PAU         PAUSE
        RET         RETURN
        ST          STOP
        WA          WAIT {<expression>}
```

Structured Constructs

```
CAS       CASE <expression> OF
V           VALUE <expression> {, ..., <expression>}:
              <group of steps>
            VALUE <expression> {, ...<expression>}:
              <group of steps>
            :
            VALUE <expression> {, ...<expression>}:
              <group of steps>
AN          {ANY
              <group of steps>}
END       END

DO        DO
            {<group of steps>}
U         UNTIL <logical expression>

FO        FOR <variable> = <expression> TO <expr> {STEP <expr>}
            {<group of steps>}
END       END

IF        IF <logical expression> THEN
            <group of steps>
EL        {ELSE
            <group of steps>}
END       END

WH        WHILE <expression> DO
            {<group of steps>}
END       END
```

Control of Process Control Programs

```
PCEN      PCEND
PCEX      PCEXECUTE {<program>}, {<loops>}, {<step>}
```

Binary Signals

```
PC        PC <signal>, {<bits>} = <value>
RES       RESET
RU        RUNSIG <signal>
SI        SIGNAL <signal>{, ..., <signal>}
```

Asynchronous Processing

```
IG        IGNORE <signal>
LO        LOCK <priority>
REACT     REACT <signal>, <program>, {<priority>}
REACTE    REACTE {<program>}
REACTI    REACTI <signal>, {<program>, {<priority>}}
```

```
Configuration Control
        AB          ABOVE
        BE          BELOW
        FL          FLIP
        LE          LEFTY
        NOF         NOFLIP
        RI          RIGHTY

Trajectory Control
        CO          COARSE {ALWAYS}
        FI          FINE {ALWAYS}
        INTOF       INTOFF {ALWAYS}
        INTON       INTON {ALWAYS}
        NON         NONULL {ALWAYS}
        NU          NULL {ALWAYS}
        SP          SPEED <value> {<units>} {ALWAYS}

Input and Output
        DA          DAC <channel> = <value>
        PR          PROMPT "{<string>}" {, <variable>} {, ..., <variable>}
        TY          TYPE {<output spec.>} {, ..., <output spec.>}
        IO          IOPUT <address> = <value>

Miscellaneous
        AT          ATTACH
        BA          BASE {<dX>}, {<dY>}, {<dZ>}, {<Z rotation>}
        DET         DETACH
        DI          DISABLE <switch> {, ..., <switch>}
        ENA         ENABLE <switch> {, ..., <switch>}
        PAR         PARAMETER <parameter> = <value>
        TI          TIMER <number> = <value>

Path-Control Communication
        ALTE        ALTER (<channel>,<mode>, {<program>,{<priority>}})
                                                         {<data>}
        ALTO        ALTOUT {<value>}, ..., {<value>}
        NOA         NOALTER
```

III. Real-Value Functions

The following functions all return real values. The names of real-valued functions cannot be abbreviated, unlike command and instruction names.
In all cases where a numeric value is required, the argument can be a numeric constant, a real-variable name, or an arithmetic expression.

Mathematical

ABS(<expression>)
ATAN2(<expression>, <expression>)
COS(<expression>)
FRACT(<expression>)
INT(<expression>)
SIGN(<expression>)
SIN(<expression>)
SQR(<expression>)
SQRT(<expression>)

Robot Control

HAND
PENDANT(<expression>)
PRIORITY
SIG(<signal> {, ..., <signal>})
SPEED(<expression>)
STATE(<expression>)

Robot Location

DISTANCE(<compound>, <compound>)
DX(<compound>)
DY(<compound>)
DZ(<compound>)
INRANGE(<compound>)

Constant Value

FALSE
ID(<expression>)
PI
TOANG
TODIS
TPS
TRUE

Miscellaneous

ADC(<channel>)
BCD(<expression>)
BITS(<signal>,{<bits>}
DCB(<expression>)
ERROR(<expression>)
IOGET(<expression>)
LLAST(<location array name>[])
PARAMETER(<parameter>)
RANDOM
RLAST(<real array name>[])
TIMER(<expression>)

IV. Location Functions

The following functions all return location values. Unlike command and instruction names, function names cannot be abbreviated.

```
Transformation
        BASE
        DEST
        FRAME(<compound>, <compound>, <compound>, <compound>)
        HERE
        INVERSE(<compound>)
        NORMAL(<compound>)
        NULL
        SCALE(<compound> BY <expression>)
        SHIFT(<compound> BY {<expr>}, {<expr>}, {<expr>})
        TOOL
        TRANS({<expr>}, {<expr>}, {<expr>}, {<expr>}, {<expr>},
                                                      {<expr>})

Precision Point
        PPOINT({<expr>}, {<expr>}, {<expr>}, {<expr>}, {<expr>},
                                                      {<expr>})
```

Appendix G
HELP

HELP Assembly Instructions

HELP language is adapted for assembly with the assistance of a set of subroutines and a group of logic arithmetic functions. These functions make it easy for the programmer to use external devices avoiding the need each time of a complicated command sequence. Assembly cycle programming instructions are thus grouped:

1. **System control instructions**

a) HOLD subroutine	it has the same function as the HOLD pushbutton on the control panel
b) ZERO subroutine	it has the same function as the ZERO pushbutton on the control panel
c) MAYRUN subroutine	it waits the pressing of the RUN pushbutton
d) ENDPNT, NOENDP subroutines	they permit the management of the END pushbutton pressing effect
e) RT 11 subroutine:	it permits changeover from program execution (HELP) to program development (RT 11) environment
f) TIME ( ) arithmetic function	it permits cycle execution time measurement

2. **Movement Instructions**

a) HALT subroutine	it permits to stop one or more arms
b) MANUAL subroutine	it enables arms control via joy-stick
c) SPEED subroutine	it sets up arms movement speed
d) MOVE subroutine	it moves axes of one arm
e) SMOVE subroutine	it moves axes of one arm
f) EOM subroutine	it waits the end of arm movement controlled by SMOVE
g) FORCE subroutine	it enables a force sensor

3. **Input/Output Instructions**

s) SET subroutine	it energizes one or more outputs
b) RESET subroutine	it de-energizes one or more outputs
c) PULSE subroutine	it pulse energizes one or more outputs
d) HIGH subroutine	it waits an input modification
e) STROKE subroutine	it generates an output command for a fixture and checks correct execution
f) TESTB logic function	it checks an input state
g) BIN special function (optional)	it reads and interprets a string of inputs as binary data
h) BCD special function (optional)	it reads and interprets a string of inputs as BCD data

4. **Position Measurement Instructions**

a) COORD subroutine	it obtains an arm position information
b) AX arithmetic function	it permits X axis coordinate reading
c) AY arithmetic function	it permits Y axis coordinate reading
d) AZ arithmetic function	it permits Z axis coordinate reading
e) AR arithmetic function	it permits ROLL axis coordinate reading
f) AP arithmetic function	it permits PITCH axis coordinate reading
g) AYW arithmetic function	it permits YAW axis coordinate reading
h) IVALUE and VALUE arithmetic functions	they permit measurement of a sensor output value

Program Structure

SENTENCE ● Each sentence is closed by «;» ● In this manual sentences are identified by symbol f_i	● This is a single instruction building up an ELEMENT ● It is immediately compiled but not performed
ELEMENT ● A sentence may account for an element ● In this manual elements are identified by symbol e_i	● It is an instruction sequence performed immediately after compiling
BLOCK f_1; : f_n ENDBLOCK;	DYNAMIC BLOCK ● It includes a group of sentences inside an element ● The block content is: 1. integrally analyzed 2. integrally executed 3. cancelled
STATIC BLOCK f_1; : f_n ENDBLOCK; ● After translation it is not executed ● It is the most suitable location for a subroutine	STATIC BLOCK ● It includes a group of sentences in an element ● The block content is: 1. integrally analyzed 2. retained
INIT BLOCK f_1; : f_n ENDBLOCK; ● It has execution priority over static and dynamic block	INIT BLOCK ● It includes a group of sentences in an element 1. the block content is completely analyzed and performed with priority 2. its text is cancelled

STATEMENTS DEFINE name ($n_1, n_2, \ldots n_i$); ● This is a direct order to the machine ● The HELP recognize the first 6 characters of the name only ● The first character must be an alphabetic one, the others must be alphanumeric ● n_i must be a number and indicates a dimension of the matrix ● Several matrixes may be defined at a time: DEFINE A (1,5), POI (2, 3, 4);	DIMENSION STATEMENT ● It assigns matrix name a suitable memory space
LET SYM name = /string/; ● This is a direct order to the machine ● The HELP recognizes the first 6 characters of the name only ● The first character must be alphabetic, the others must be alphanumeric ● Several symbols may be stated at a time: LET SYM name1 = /string1/,SYM name2 = /string2/;	SYMBOL STATEMENT ● It allows the HELP to interprete the name assigned by the user if it were the HELP procedure indicated in the string
LET VAR name AT address; ● The use is recommended only to users with a sound knowledge of the HELP translator structures	VARIABLE ADDRESS STATEMENT ● It assigns the indicated variable a memory area stated by the user
LET STRUCT name AT address; ● The use is recommended only to users with a good knowledge of the HELP translator structure	MATRIX ADDRESS STATEMENT ● It assigns the indicated matrix a memory area stated by the user
SUBROUTINE name: f_1; ⋮ f_n; RETURN; ● It has the same use as in other languages (FORTRAN, BASIC etc.) ● There is no need to redefine variables ● The HELP recognizes the first 6 alphanumeric characters of the name only ● It is advisable to include the subroutine within a static block	SUBROUTINE ● This part of the program may be temporarily referred to during program execution; after it, system management goes back to the main program

CLOSURES

SENTENCE CLOSURE ;	It closes a sentence or an element.
ENDBLOCK; ● It should not be preceded by «;» ● If followed by a «STOP», it loses the «;»	BLOCK CLOSURE ● It closes: BLOCK STATIC BLOCK INIT BLOCK
RETURN ● It must be followed by a «;», STOP or ENDBLOCK ● The sentence preceding it should be closed by a «;»	SUBROUTINE CLOSURE ● It transfers the program management to the instruction following the GO-SUB
STOP	END-OF-PROGRAM
● It should not be preceded nor followed by «;»	● It indicates the end of a program

LANGUAGE

NUMERIC CONSTANT ● It must lie between 10^{-39} and 10^{+38}	● It is expressed with floating point
name ● The HELP recognizes the first 6 alpha-numeric characters of the name only ● The first character must be alphabetic ● It is allocated in the symbol table ● Scalar and vectorial variables are not allowed to have the same name	SCALAR VARIABLE ● It indicates the name assigned to a memory where numeric values may be allocated LEGAL EXAMPLES: A ALT B5A0 B$POI ILLEGAL EXAMPLES: 4A A.1 P.0L
name (i,j, . . .) ● The HELP recognizes the first 6 alpha-numeric characters of the name only ● The first character must be alphanumeric ● It is allocated in the compiling area ● Indexes i, j, must be integer numbers; eventual fractional indexes are rounded off to the nearest integer ● Scalar and vectorial variables are not allowed to have the same name	VECTORIAL VARIABLE ● It indicates element (i,j, . . .) of matrix called name EXAMPLES: FINE(2,5) FINE(A,B) FINE(A+B,B) FINE(SQRT(4),5) FINE(V(I),V(J))

variable: = arithmetic expr

ASSIGNMENT
It assigns the result of the arithmetic expression to the variable indicated on the left of symbol «: =»

ARITHMETIC OPERATORS

Operator		Execution priority
*	multiplication	1
/	division	1
−	subtraction	2
+	addition	2
( )	parentheses	0

---

function (subject)
- It results in a numeric value
- The only function of the standard HELP is
  ABS(subject)
  extracting the absolute value of subject

FUNCTION
- It processes the subject according to the indicated function

---

LOGIC EXPRESSIONS

```
IF logic expr THEN sentence;
                   sentence
                   END;
```

FLOW CHART

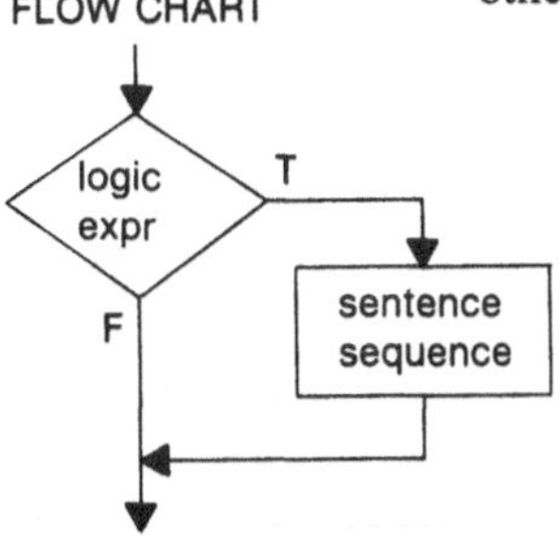

HALF-SELECTION
- If the expression is true, the program performs the sequence of sentences, otherwise it jumps beyond the END

```
IF logic expr THEN sentence 1;
                   sentence 1
         ELSE
                   sentence 2;
                   sentence 2
         END;
```

COMPLETE SELECTION
- If the logic expression is true, it performs the sequence of sentences 1, if it is false it performs the sequence of sentences 2

FLOW CHART

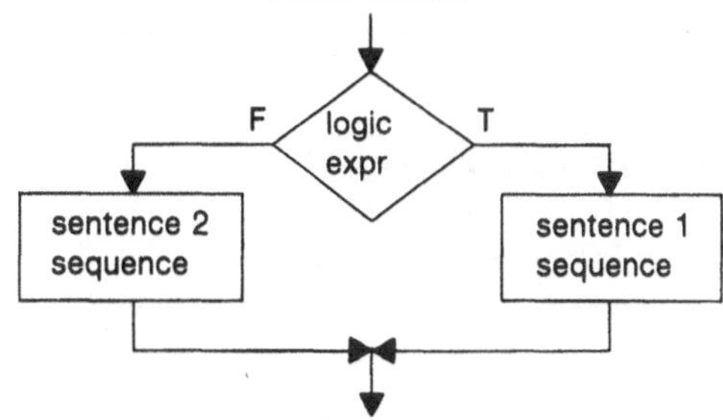

SIMPLE LOGIC EXPRESSION
> higher
< lower
= equal

COMPOSED LOGIC EXPRESSION

	evaluation priority
logic expr1 AND expr	1
logic expr1 OR expr	2
NOT logic expr	0

EXAMPLES
IF A>B THEN f_1; f_2; f_3 END;
IF A>B THEN f_1; f_2 ELSE f_3; f_4 END;
IF A>B AND c<D THEN f_1; f_2 ELSE f_3; f_4 END;
IF NOT A>B THEN f_1; f_2 END;

JUMP

GOTO label;
- No return is allowed

GOTO
- It allows jumping to perform what follows the indicated label

---

GOTO [val.] lab1,lab2, . . .,labn;
- The value must be an integer; if not, it is rounded off
- No jump is made, if the value exceeds the number of labels
- The value may be the result of a numeric expression

COMPUTED GOTO
- It allows to jump to label 1, 2 or n, if the value is 1, 2 or n

---

GOSUB label;
- After performing the subroutine, the program goes back to the sentence following the GOSUB

GOTO SUBROUTINE
- It allows to jump to the subroutine called label

GOSUB [val.] lab1, lab2, . . ., labn;
- The value must be an integer, otherwise it is rounded off
- No jump is made, if the value exceeds the number of labels
- The value may be the result of an arithmetic expression

COMPUTED GOSUB
- It allows to jump to label 1, 2 or n, if the value is 1, 2, or n

---

EXAMPLES
GOTO label;
GOTO [I] PR,MAT,ZERO;
GOTO [A+2*B]PR,MAT,ZERO;
GOTO [V(I)] PR,MAT,ZERO;

- They are valid for GOSUB as well

ITERATION

FOR variab.sc.: = e.a.1,e.a.2, . . ., e.a.i DO f_1; f_2; REP;	FOR OF TYPE 1 ● For scalar variable equal to e.a.1, e.a.2, . . . e.a.i, the program performs the sequence of sentences
EXAMPLES FOR I: = 1,3,5.5 DO sentenceREP; FOR VAR: = A,B,5 DO sentenceREP; FOR I: = A,E(1),(A + E(3)) DO sentence REP;	
FOR variab.sc.: = e.a.1 TO e.a.2 BY e.a.3 DO f_1; f_2 REP; ● The step is assumed equal to 1, if «BY e.a.3» is missing ● The variable must not be a vectorial one	FOR OF TYPE 2 ● For scalar variables assuming values between e.a.1 and e.a.2 with step equal to e.a.3, the program performs the sequence of sentences
EXAMPLES FOR I: = 1 TO 10 BY 2 DO f_1; f_2 REP; FOR I: = 1 TO 5 DO f_1; f_2 REP;	
WHILE logic expr DO f_1; f_2 REP; ● Make sure the logic function may result FALSE in the sequence of sentences	WHILE ● As long as the logic expr results true, the program repeats the sequence of sentences

PROGRAM MANAGEMENT

PERIPHERAL CODES KB : typewriter or alphanumeric video DX : single density floppy disk DY : dual density floppy disk DK : hard disk PR : tape punch LP : high speed printer CT : cartridge unit SY : any specific peripheral unit stated as SY during system generation	These codes may be combined with a number in order to distinguish different peripherals of the same type.
STORE KB PR CT DXO:NAME.EST DX1:NAME.EST DY0:NAME.EST DY1:NAME.EST DK:NAME.EST	STORE It loads into the central memory the program coming from the indicated peripheral

RUN KB
PR
CT
DX0:NAME.EXT
DX1:NAME.EXT
DY0:NAME.EXT
DY1:NAME.EXT
DK:NAME.EXT

RUN
It starts execution of the program coming from the indicated peripheral

---

LANGUAGE OPTIONS

FUNCTIONS

SIN (e.a.)	sine in degrees
SINR (e.a.)	sine in radian
COS (e.a.)	cosine in degrees
COSR (e.a.)	cosine in radian
ARCTG (e.a.)	arctang in degrees
ARCTGR (e.a.)	arctang in radian
SQRT (e.a.)	square root
ABS (e.a.)	absolute value

DIALOG

ASK (‹text›)

- It is a logic relation in every respect, and should be used as such
- Its text does in no way affect the program
- The text is a memo for the operator
- The program stops until an answer is received

LOGIC REQUEST

- The program asks the operator for a true or false evaluation by printing out: text [Y/N]?:
- The operator answers by entering from keyboard Y or N and pressing the carriage return key

---

ASKN (‹text›)

- The text does in no way affect the program
- The text is a memo for the operator
- The program stops until an answer is received

NUMERIC REQUEST

- The program asks the operator for a numeric value by printing out the text
- The operator answers by entering from keyboard a numeric value and pressing the carriage return key

---

PRINTOUT

PRINT (list of parameters)

- Commands for the printer may be included in the list of parameters
  // 0 Carriage return and line feed
  // 1 Carriage return
  // 2 Line feed
  // 3 Two-line feeds
  // 4 Five acoustic signals

PRINTOUT FROM TYPEWRITER

// 5 Takes the carriage on tabulator position
// 6 Causes printout of ‹HELP> ›text

PUNCH (list of parameters)	Tape PUNCHING ● It causes the listed parameters to be punched on tape
BINOCT (value)	Writes the words expressing the value in brackets
● Does not require any statement	STANDARD FORMAT ● The printout has 7 integers and 4 decimals
FORMAT (integers, decimals, filling ch.) ● It is modal	FORMAT ● The printout has the format stated in brackets
FILE MANAGEMENT	
OPEN (‹NAME›, # n)	OPENS FILE ● On RT-11 it opens file number n, called NAME, on floppy disk 1 ● n is set by the user
CLOSE (# n)	CLOSES FILE
WRITE (# n,expression)	WRITES IN FILE It writes the quoted expression in file n
WRITE (# n,‹text›)	WRITING EXPRESSIONS ● It writes the text in file n ● The text may be the instructions of a program to be started by command: RUN DX1:FILE NAME
WRITE (# n,data)	● It writes one or more data in file n
WRITE (# n,‹[›)	● It writes the character identifying the opening of a record, in file n
WRITE (# n,‹]›)	● It writes the character identifying the closing of a record, in file n
LOAD (# n)	FILE LOADING INTO BUFFER ● It loads the file content into buffer, one data at a time
TOPREC (# n)	BEGINNING OF RECORD ● It takes the file pointer to the beginning of record

TOPFILE (# n)	BEGINNING OF FILE ● It takes the file pointer to the beginning of file
DXLIST (# n)	FILE LISTING ● It lists the file content
EOF (# n) ● The pointer reads the file through and stops after the first encountered ‹[›. If no ‹[› is encountered, the logic function is true ● The pointer is shifted	CHECKS END OF FILE ● The logic function is true, if the record has ended
EOR (# n) ● The pointer checks whether it is on character ‹]›. In the positive case, the logic function is true ● The pointer is not shifted	CHECKS END OF RECORD ● The logic function is true, if the record has ended
SEARCH (# n,‹text›) ● The pointer searches the ‹text› in file n and stops after the last character of the text ● If it does not find the ‹text›, it reads the whole file through ● The pointer is shifted	TEXT SEARCH ● The logic function is true, if the text is found
PSEUDO MANAGEMENT FUNCTIONS	
● They may be included in a HELP program ● They should not be considered when defining the HELP syntax	● These are utility functions
$CORE	prints out the memory map
ECHO	enables printout of the main program contemporarily to execution
NOECHO	disables the ECHO
BREAK	it holds over the actual input and enables the typewriter
NOBREA	starts again the input from where it was left
GOSIZE	enables counting the memory areas occupied by each sentence and the number of sentences

ONSIZE	prints out the memory areas occupied by each sentence
EXSIZE	prints out what computed by GOSIZE
PSRT11	bootstraps the RT-11 system
RT-11	stops the HELP process and recalls the operator system
PROCESS SYNCHRONIZING	
	● Process management sentences to be included in the HELP texts
ASSIGN(n,REF label)	assigns a number to the process identified by the label
ACTIVE(n_1, n_2, . . ., n_n)	enables the process activities identified by the numbers with priority n_1 on n_2 etc.
ERASE	disables all activities
SIGNAL(n)	activates flag n
WAIT(n)	stops the process until flag n is activated. After process activating, it deactivates the flag
DELAY(time)	stops the process for the time indicated
TEST(n)	logic function. It returns true, if flag n is active

Appendix H
SIGLA

Keyword index of SIGLA

If the question OPER? is answered by IN, the system is prepared to receive the instructions below. They are grouped as follows:

A Commands which are not stored:

A1 Control commands which are employed by the user during the input phase
- CO: console = input
- NM: no motion
- RM: retry motion

A2 Editing commands for correcting already stored programming commands
- DE: delete
- RE: replace
- PL: place, insert

B Commands which are stored:

B1 General working commands, available during all execution and input phases
- MO: motor
- OM: null points
- AX: auxiliary axis

B2 Interrupt instructions
- HL: hold
- WA: wait

B3 Interrogation of external conditions
- PP: part present

B4 Geometric relations
- II: Start of incremental programming
- IF: End of incremental programming
- QA: distance for collisions
- SI: set bias of origin
- CI: call bias of origin

B5 Arithmetic instructions
- SE: set
- IC: add an increment
- NE: negate
- DM: define memory values
- AM: add memory values

B6 Output commands for the terminal
NT: note on the terminal

B7 Commands for logic
they enable comparisons, take logical decisions, control the execution sequence
NU: label for jump command
JU: jump command
ES: set event
EW: wait for an event
BE: branch if equal
BG: branch if greater
BL: branch if less
EX: execute subroutines

B8 Typical machine commands
RP: search for surface
MT: transfer measurement

Appendix I
ROBEX

I. SHORT REFERENCE MANUAL OF THE ROBEX LANGUAGE A LANGUAGE FOR PROGRAMMING OF NC-ROBOTS

* * * * * * * * * * * * * ABBREVIATIONS * * * * * * * * * * * * * *

a = X COMPONENT OF THE NORMAL VECTOR OR LOGICAL CHANNEL NUMBER
b = Y-COMPONENT OF THE NORMAL VECTOR OR LOGICAL CHANNEL NUMBER
c = Z-COMPONENT OF THE NORMAL VECTOR
e = LOGICAL CHANNEL NUMBER
n = LABEL
s = FEEDRATE VALUE
w = VALUE (Generally)
A, B, C, D = COEFFICIENTS OF THE PLANE EQUATION

nx = NUMBER OF POINTS IN X-DIRECTION
ny = NUMBER OF POINTS IN Y-DIRECTION
nz = NUMBER OF POINTS IN Z-DIRECTION
az = NUMBER OF POINTS

sc = SYMBOL OF A CIRCLE
sl = SYMBOL OF A LINE
sp = SYMBOL OF A POINT
sm = SYMBOL OF A MATRIX
spl = SYMBOL OF A PLANE
smf = SYMBOL OF A SURFACE
sb = SYMBOL OF A BEAM
scy = SYMBOL OF A CYLINDER
sph = SYMBOL OF A SPHERE
sco = SYMBOL OF A CONE
sbo = SYMBOL OF A BODY
spa = SYMBOL OF A PART
spn = SYMBOL OF A PATTERN
spp = SYMBOL OF A ROBOT-INTERNAL PALETTIZING PATTERN
sv = SYMBOL OF A VECTOR

pi = POINT-INDEX

wa = VALUE OF A DISTANCE
wh = VALUE OF THE HALF-ANGLE
wr = VALUE OF A RADIUS
ww = VALUE OF AN ANGLE
wx = VALUE OF A X-COORDINATE
wy = VALUE OF A Y-COORDINATE
wz = VALUE OF A Z-COORDINATE

cl = VALUE OF A LINEAR INCREMENT
dw = VALUE OF AN ANGULAR INCREMENT

```
dx=VALUE OF A X-DISTANCE
cy=VALUE OF A Y-DISTANCE
cz=VALUE OF A Z-DISTANCE

* * * * * * * * * * * * * BASIC DEFINITIONS * * * * * * * * * * * * *
CHARACTERS:
A B C D E F G H I J K L M N O P Q R S T U V W X Y Z
0 1 2 3 4 5 6 7 8 9
LANGUAGE WORDS OR VARIABLES HAVE 1 TO 6 CHARACTERS STARTING WITH A
LETTER
BASIC OPERATORS: + - * /
STANDARD FUNCTIONS: SIN COS ATAN SQRT ABS LOG NLOG IFIX INT DIST EXP
MEASURING UNITS: LENGTH IN MM OR INCH, ANGLES IN DEGREES

** * * * * * COORDINATE TRANSFORMATION COMMANDS * * * * * * *
sm=MATRIX/TRANSL,wx,wy,wz

sm=MATRIX/XYROT,ww
          YZROT
          ZXROT

sm=MATRIX/XYROT,ww1,XYROT,ww2
          YZROT      YZROT
          ZXROT      ZXROT

sm=MATRIX/XYROT,ww1,XYROT,ww2,XYROT,ww3
          YZROT      YZROT      YZROT
          ZXROT      ZXROT      ZXROT

sm=MATRIX/XYROT,ww,TRANSL,wx,wy,wz
          YZROT
          ZXROT
sm=MATRIX/XYROT,ww1,XYROT,ww2,XYROT,ww3,TRANSL,wx,wy,wz
          YZROT      YZROT      YZROT
          ZXROT      ZXROT      ZXROT
ENABLE/DISABLE COORDINATE TRANSFORMATION COMMANDS
TRASYS/sm
TRASYS/NOMORE
FIXED Z PLANE COMMAND
ZSURF/wz

* * * * * * * * * * * * GEOMETRIC ELEMENTS * * * * * * * * * * * *
sp=POINT/wx,wy,wz
sp=POINT/wx,wy                    USING FIXED Z-PLANE
sp=POINT/sp,DELTA,dx,dy,dz
sp=POINT/sp,DELTA,dx,dy
sp=POINT/INTOF,sl1,sl2

sp=POINT/XSMALL,INTOF,sc1,sc2
         XLARGE
         YSMALL
         YLARGE

sp=POINT/XSMALL,INTOF,sl,sc
         XLARGE
         YSMALL
         YLARGE

sp=POINT/RTHETA,XYFLAN,wr,ww
                YZFLAN
                ZXFLAN
```

```
sl = LINE/sp1,sp2
sl = LINE/wx1,wy1,wx2,wy2
sl = LINE/sp,RIGHT,TANTO,sc
             LEFT
sl = LINE/RIGHT,TANTO,sc1,RIGHT,TANTO,sc2
          LEFT              LEFT
sl = LINE/sp,PERPTO,sl
sl = LINE/sp,PARLEL,sl
sl = LINE/PARLEL,sl,XSMALL,wa
                    XLARGE
                    YSMALL
                    XLARGE
sl = LINE/sp,ATANGL,ww
sl = LINE/XPAR,wa
          YPAR
sl = LINE/INTOF,spl1,spl2
sc = CIRCLE/wx,wy,wr
sc = CIRCLE/CENTER,sp,RADIUS,wr
sc = CIRCLE/TANTO,sl,XSMALL,sp,RADIUS,wr
                     XLARGE
                     YSMALL
                     YLARGE
sc = CIRCLE/XSMALL,sl1,XSMALL,sl2,RADIUS,wr
            XLARGE     XLARGE
            YSMALL     YSMALL
            YLARGE     YLARGE

sc = CIRCLE/XSMALL,sl,XSMALL,IN,sc,RADIUS,wr
            XLARGE    XLARGE OUT
            YSMALL    YSMALL
            YLARGE    YLARGE
spm = PATTERN/LINEAR,sp1,sp2,az
spm = PATTERN/LINEAR,sp,ATANGL,ww,INCR,az,AT,dl
spm = PATTERN/ARC,sc,ww,CLW,az
                        CCLW
spm = PATTERN/ARC,sc,ww,CLW,INCR,az,AT,dw
                        CCLW
spm = PATTERN/ARC,sc,ww1,ww2,CLW,az
                             CCLW
spm = PATTERN/TRAFO,spm1,spm2
spm = PATTERN/TRAFO,spm1,spm2,XYROT,cw
spm = PATTERN/MIRROR,sl,spm
spm = PATTERN/RANDOM,spm1,spm2,sp1,spm3,spr
sp = POINT/spm,pi
PATTERN MODIFICATION PARAMETERS
INVERS, RETAIN, OMIT
sv = VECTOR/wx,wy,wz
sv = VECTOR/wx1,wy1,wz1,wx2,wy2,wz2
sv = VECTOR/sp1,sp2
sv = VECTOR/LENGTH,wa,ATANGL,ww,XYPLAN
                                YZPLAN
                                ZXPLAN
```

```
spl = PLANE/sp1,sp2,sp3
spl = PLANE/sp,PARLEL,spl1

spl = PLANE/PARLEL,spl1,XLARGE,wa
                        XSMALL
                        YLARGE
                        YSMALL
                        ZLARGE
                        ZSMALL
spl = PLANE/A,B,C,D
spl = PLANE/sp,PERPTO,sv
spl = PLANE/sp1,sp2,PERPTO,spl1
sb = BEAM/sp1,sp2,sp3
sb = BEAM/wx1,wy1,wz1,wx2,wy2,wz2
scy = CYLNDR/wx,wy,wz,a,b,c,wr
scy = CYLNDR/sp,sv,wr
sph = SPHERE/wx,wy,wz,wr
sph = SPHERE/CENTER,sp,RADIUS,wr
sco = CONE/sp,sv,wh
sco = CONE/wx,wy,wz,a,b,c,wh

sbo = BODY/IN  ,BEA,smf,BASE,wz,LENGTH,wa
           OUT CIR      TOP  spl
               CYL
               CON
               SPH
               CUR

sbo = BODY/IN  ,BEA,smf,BASE,wz,TOP,wz
           OUT CIR           spl    spl
               CYL
               CON
               SPH
               CUR
spa = PART/[sbo,] 1 to n
```

* * * * * * * * * * EXPLICIT MOTION COMMANDS * * * * * * * * *

```
- - - Definition of a safe position or surface - -
SAFPOS/wx,wy,wz
SAFPOS/wz
SAFPOS/NOMORE
- - - Go to a position - - -
GOTO/wx,wy,wz
GOTO/SAFP
GOTO/sp
GOTO/spm
Unchanged wx,wy,wz may be replaced by CONST
- - - Go a specified distance - - -
GODLTA/dx,dy,dz [,EVENT,a [,ELSE,m]]
GODLTA/cz [,EVENT,a [,ELSE,m]]
- - - Turn to an orientation - - -
TURN/XYROT,ww1,[XYROT,ww2,[XYROT,ww3]] [,EVENT,a[,ELSE,m]]
     YZROT      YZROT      YZROT
     ZXROT      ZXROT      ZXROT
```

* * * * * * COMPLEX HANDLING COMMANDS (PRELIMINARY) * * * * * *

```
MOVE/spa,spl1,AGAINST,spa2,spl2[,EVENT,a [,ELSE,m]]
              COPLANAR
```

* * * * * * * * * * * * * ADDITIONAL PARAMETERS * * * * * * * * * * *

.[,TEACH]. . . = online teach-in required
.[,EX] = Go to exact point coordinates

✻ ✻ ✻ ✻ ✻ ✻ COMMANDS FOR ROBOT-INTERNAL PALETTIZING ✻ ✻ ✻ ✻ ✻ ✻

– – – Load or unload a palette – – –

GET/spp
PUT/spp

✻ ✻ ✻ ✻ ✻ ✻ ✻ ✻ ✻ ✻ TOOL OR GRIPPER COMMANDS ✻ ✻ ✻ ✻ ✻ ✻ ✻ ✻ ✻ ✻

GRIPNO/e[,f] or TOOLNO/e[,f]	Select gripper or tool
OPENGR [/a] [,EVENT,b [,ELSE,m]]	Open gripper
CLOSGR [/a] [,EVENT,b [,ELSE,m]]	Close gripper

✻ ✻ ✻ ✻ ✻ ✻ ✻ ✻ ✻ ✻ ✻ MACHINE COMMANDS ✻ ✻ ✻ ✻ ✻ ✻ ✻ ✻ ✻ ✻ ✻

STOP	Stop program execution
CPSTOP	Stop on operator signal
FEDRAT/s	Velocity select
RAPID	Rapid traverse

✻ ✻ ✻ ✻ SYNCHRONIZATION AND PROGRAM CONTROL COMMANDS ✻ ✻ ✻ ✻
(PERFORMED BY THE ROBOT-NC)

DELAY/a	Wait for a seconds
WAIT/EVENT,a	Wait for signal a
ONSIG/EVENT,a,JMP,m	On signal a jump to m
SWITCH/a,ON OFF	Switch logic channel a
JUMP/m	Unconditional jump

✻ ✻ ✻ ✻ ✻ ✻ ✻ ✻ ✻ ✻ ✻ PALETTIZING COMMANDS ✻ ✻ ✻ ✻ ✻ ✻ ✻ ✻ ✻ ✻ ✻
(FOR ROBOT INTERNAL PALETTIZING)

spp = PALTZ/LINEAR,sp,dx,dy,dz,nx,ny,nz	Create palettize pattern
RESET/spp	Reset palettize pointer

✻ ✻ ✻ ✻ ✻ ✻ ✻ ✻ ✻ ✻ ✻ ✻ ✻ PARTS HANDLING ✻ ✻ ✻ ✻ ✻ ✻ ✻ ✻ ✻ ✻ ✻ ✻ ✻

TIED [spa] 2 to n	Define parts as tied together
UNTIED/ [spa] 2 to r	Define parts as untied

✻ ✻ ✻ ✻ ✻ ✻ ✻ ✻ ✻ PROGRAM CONTROL STATEMENTS ✻ ✻ ✻ ✻ ✻ ✻ ✻ ✻ ✻
(only for internal processor control)

LOOPST	Loop control start
LOOPND	Loop control end
JUMPTO/m	Unconditional jump
IF(w1,EQ,w2) m NE LT GT LE GE	Conditional jumps
IF (value) m1,m2,m3	
JUMPTO/(m1,m2,m3,. . .,mn),w	
name = MACRO[parameters]	Macro definition
name = EXMAC/IN[,parameters],OUT[,parameters]	
TERMAC	Macro end
CALL name [,parameters]	Call macro

✻ ✻ ✻ ✻ ✻ ✻ ✻ ✻ ✻ PROCESSOR CONTROL STATEMENTS ✻ ✻ ✻ ✻ ✻ ✻ ✻ ✻ ✻

PARTNO/text
REMARK/text
$$text
PAGEFM/number of characters, lines on page
PRINT/text
INSERT/direct robot command
MACHIN/symbol of post-processor,parameters
NOPOST
PPFUN/list

```
CLPRNT[/ON]
        OFF
PPWORD/[ new pp-language word,new code-integer] 1 to n
EXWORD/[ new language word,new code-integer] 1 to n
UNITS/MM[,a]
       INCH
SYN/syn1,word,syn2,word,. . .,syn n,word
LIST/ON
     OFF
PRINT/r,ON [,parameters]
        OFF
FINI
```

II. ADDENDUM: PLANNED OR REALIZED EXTENSIONS TO ROBEX

During the completion of the manuscript, the authors were informed about intended or realized extensions to ROBEX. The following listing attempts to give an overview:

1. **Variable**
   One of the most important extensions is the admissions of variables. The data types FRAME, REAL, BOOLEAN, as well as possibly INTEGER and STRING are under consideration.

2. **Reading and writing of data**
   It has not yet been decided which data types will be admitted to this.

3. **Procedures**
   It is intended to declare procedures with the introduction of
   SUBRTN/⟨symbol⟩, ⟨parameters⟩
   and terminate it with the RETURN statement. Execution of the procedure is intended to be invoked via
   CALL/⟨symbol⟩, ⟨parameters⟩.
   The question about parameter passing is still open, as is the permission of recursion, neither is it clear whether functions will be featured.

4. **Parallel processes, tasks**
   It is intended to execute parallel processes and tasks, but the final form has not yet been decided upon. Possibilities for synchronization and any commands associated with a task concept are still open, such as activation, termination, suspension and continuation of a task.

5. **Interrupt programming**
   It is intended to define program sections in order to serve interrupts. Additionally, it is possible to enable and disable interrupts.

6. **Motion commands**
   It is intended to combine the GOTO and TURN instructions into the DRIVE command for explicit control of the axes. Both commands can be equipped with sensor commands or can be used for teaching.

The first five points now break with the NC concept of exclusively executing the control. So far, NC control could have been fixed wired electronics, but now it has to be implemented on a microcomputer on which an interpreter can be executed. Furthermore, the integration of modern concepts such as variables, procedures and tasks do not allow the current NC programmers to use the language without any further training. However, this was one of the main arguments to hold on to ROBEX, because the NC tradition could be continued with relatively few changes for robots.

The ROBEX variant ROBEX-M is interesting in this context and is listed with a quick reference guide. ROBEX-M does contain variables, arithmetic and logic operators as well as elements for program flow control. However, the implicit geometric parameters typical for NC languages have been omitted, such as lines, areas and body models and subsequent trajectory calculations and collision checking.

III. ROBEX-M (QUICK REFERENCE)

VARIABLES

Any length, without blanks. Beginning with a character. Integer, Real, Matrix, Point, Frame, Boolean, String

ARRAYS

1, 2, or 3-dimensional of any variable. Index in brackets

ARITHMETIC OPERATORS

=	Assigns value or variable to variable
+	Addition
−	Subtraction
*	Multiplication
/	Division
**	Raising to a power

LOGICAL OPERATORS

AND	Logical "And"
OR	Logical Inclusive "Or"
EXOR	Logical Exclusive "Or"
NOT	Logical "Not"

RELATIONSHIP OPERATORS

=	Equal
< >	Not Equal To
<	Less Than
>	Greater Than
< =	Less Than or Equal To
> =	Greater Than or Equal To

SYSTEM COMMANDS

CRDOFF	Suppress source code numeration
PRINT	Compiler debug

GEOMETRIC DEFINITIONS

POINT	Points in carthes. coordinates (x,y,z)
FRAME	Position (with orientation)
MATRIX	Coordinate transformation
TRASYS	Enable/disable transformation
ORIGIN	New reference coordinate system

MOTION COMMANDS

MOVE	PTP motion
CPMOVE	CP motion
COMOVE	CP motion, constant orientation
DRIVE	Drive joint
HOME	Reference position

TECHNOLOGIC COMMANDS

FEDRAT	Set speed
RAPID	Maximum speed
MOVTIM	Motion time between positions
ACCEL	Acceleration
DECEL	Deceleration
MACHIN	Select robot
ARMNO	Select arm
TOOLNO	Select tool
CLOSGR	Close gripper
OPENGR	Open gripper
STOP	Stop program execution
OPSTOP	Stop on request
DELAY	Delays milliseconds
WAIT	Wait for signal
INSERT	Insert robot command
SMOOTH	Smooth motion
TEXT	Display text

SENSOR COMMANDS

EVENT	Binary input
SWITCH	Binary output
RECEIV	Read external data
SEND	Send external data

PROGRAM FLOW

PARTNO	Start of program
FINI	End of program
IF	Conditional branch
DO	Do loop
DOEND	End of do loop
REPEAT	Repeat the following
UNTIL	Until condition is true
CASE	Selective branch
SUBRTN	Subroutine
RETURN	End of subroutine
CYCLE	Repetitive commands
JUMP	Unconditional jump
NOP	No operation
INTRPT	Interrupt definition
INTRET	Interrupt return
PARBEG	Parallel execution
PAREND	End of parallel execution
TASK	Task definition
TSKEND	Task end
MODUL	Modul definition
EXTERN	External modul declaration

GENERAL

$$	Comment follows
$	Statement continues next line
UNIT	Measuring unit

Appendix J
RAIL

VISION FEATURE NUMBERS

The built-in RAIL function OBJ_FEAT can be used to obtain the value of any of the vision features for the current blob. OBJ_FEAT requires as an input parameter the number of the vision feature to be calculated. There are currently 45 vision features. The table below shows the correspondence between feature numbers and vision feature names:

1	OBJ_COLOR	16	OBJ_AREA	31	OBJ_HOLEAREA
2	OBJ_NCELLS	17	OBJ_XCENT	32	OBJ_HOLERATIO
3	OBJ_TOTALCELLS	18	OBJ_YCENT	33	OBJ_AXRATIO
4	OBJ_NHOLES	19	OBJ_MAJOR	34	OBJ_PPDA
5	OBJ_XMIN	20	OBJ_MINOR	35	OBJ_RADRATIO
6	OBJ_XMAX	21	OBJ_ANGLE	36	OBJ_LENRATIO
7	OBJ_YMIN	22	OBJ_PERIMETER	37	OBJ_YDIFF
8	OBJ_YMAX	23	OBJ_TOTALAREA	38	OBJ_XDIFF
9	OBJ_XPERIM	24	OBJ_RMIN	39	OBJ_BOXAREA
10	OBJ_YPERIM	25	OBJ_RMAX	40	OBJ_BXARATIO
11	OBJ_SIGX	26	OBJ_RMINANG	41	OBJ_CGDIST
12	OBJ_SIGY	27	OBJ_RMAXANG	42	OBJ_PEROUND
13	OBJ_SIGXX	28	OBJ_AVRAD	43	OBJ_MAXMINRAD
14	OBJ_SIGXY	29	OBJ_LENGTH	44	OBJ_X3SIGN
15	OBJ_SIGYY	30	OBJ_WIDTH	45	OBJ_Y3SIGN

VISION OPTION SWITCHES

The table below lists all of the option switches in the Autovision II vision system. The initial setting of each switch (ON or OFF) is shown after the switch name. The valid assginments in RAIL are ON or OFF; for example, to turn off the training mode, type in
VIS_TRAINING=OFF

VISION SWITCHES:

VIS_TRAINING	=ON Perform training with prototype objects.
VIS_NNRECOG	=ON Use nearest-neighbor recognition.
VIS_CHISTYP	=OFF Type out contributions to nearest neighbor recognition chisquare by each feature value for each prototype.
VIS_DISPLAY	=ON Draw a picture of object on display.
VIS_OUTLINE	=ON Draw outline around selected blob if DISPLAY and DOPERIM switches are ON.
VIS_WINDOWING	=ON This switch is inoperative in current versions; the windowing is now performed by the function SETWINDOW.
VIS_NEGATIVE	=ON Select picture polarity.

VIS_STROBEON	=OFF Enable strobe flashing. The strobe will actual flash when the PICTURE command is given.
VIS_DO1MOM	=ON Calculate first moments of area (needed for centroid position).
VIS_DO2MOM	=ON Calculate second moments of area (needed for orientation angle of object).
VIS_DOPERIM	=ON Set up perimeter array and calculate perimeter information.
VIS_DORADII	=OFF Accumulate information to enable calculation of features RMIN, RMAX, AVRAD, RADRATIO using the perimeter (DOPERIM must be ON).
VIS_DOLENWID	=ON Accumulate information to enable calculation of features LENGTH, WIDTH, LENRATIO, X3SIGN, Y3SIGN using the perimeter (DOPERIM must be ON).
VIS_DOCONNECT	=ON Perform connectivity analysis. If VIS_DOCONNECT=OFF, then the vision system will only calculate the area of the object within the current window; no connectivity analysis is performed. The result of the area calculation is stored in the RAIL variable VIS_PIXELS.
VIS_KEEPBLOBS	=OFF If OFF, keep only the largest of the blobs within the background, otherwise keep all.
VIS_HOLMOM	=OFF If ON, first and second moments of parts are not calculated, but first and second moments of holes within those parts are. Used to save processing time.

VISION PARAMETERS

The table below lists all of the global parameters in the Autovision II vision system. The initial setting of each parameter is shown after the parameter name. The valid assignments in RAIL are integer or real numbers, with conversion done if necessary. As an example, to change the lower threshold value to 7, type in

VIS_LOTHRESH=7

VISION PARAMETERS:

VIS_MINBLOB	=10 Minimum blob area in pixels for a blob to be kept in linked blob list.
VIS_XBASE	=0 X origin of pixel-based coordinate system.
VIS_CAMERANO	=1 Camera number in range 1 to 8.
VIS_BUFFERNO	=0 hardware buffer number: 0 or 1.
VIS_LOTHRESH	=8 Cutoff threshold between light and dark. Pixels with threshold values less than VIS_LOTHRESH are considered to be dark, all others are considered to be light. Valid threshold values are between 0 and 14.
VIS_WINXLEFT	=1
VIS_WINXRIGHT	=236
VIS_WINYTOP	=1
VIS_WINYBOTTOM	=243 Pixel numbers for window limits in X and Y for the camera.

VIS_XSIZE	=1.0
VIS_YSIZE	=1.0
	X and Y coordinate scale factor or pixel size: 1.0 for pixel-based, other after calibration.
VIS_XZERO	=0.0
VIS_YZERO	=0.0
	X and Y coordinate system origin. The four parameters XSIZE, YSIZE, XZERO and YZERO are calculated internally during execution of the CALIBRATE command.
VIS_LDIFFMIN	=1.3333
VIS_WDIFFMIN	=1.3333
	See description of Features #44, #15 in "Autovision (TM) II Vision Program Commands."
VIS_REJTHRESH	=4.0
	Rejection threshold of normalized chi-square used in recognition. In nearest-neighbor recognition, there are two rejection thresholds to enable separation of "good", "marginal", and "rejected" parts. Initially both these thresholds are set to REJTHRESH.

RECOGNITION FEATURES

The table below lists all of the recognition features in the Autovision II vision system. The initial setting of each feature switch (ON or OFF) is shown after the feature name. Features that are turned OFF will not be calculated during object recognition. The valid assignments in RAIL are ON or OFF, for example, to turn off the perimeter feature, type in

REC_PERIMETER=OFF

RECOGNITION FEATURES:

REC_AREA	=ON	REC_CGDIST	=OFF
REC_HOLEAREA	=ON	REC_COLOR	=OFF
REC_LENGTH	=ON	REC_HOLERATIO	=OFF
REC_MAJOR	=ON	REC_LENRATIO	=OFF
REC_MINOR	=ON	REC_NHOLES	=OFF
REC_PERIMETER	=ON	REC_PEROUND	=OFF
REC_TOTALAREA	=ON	REC_PPDA	=OFF
REC_WIDTH	=ON	REC_RADRATIO	=OFF
		REC_RMAX	=OFF
REC_AVRAD	=OFF	REC_RMIN	=OFF
REC_AXRATIO	=OFF	REC_XDIFF	=OFF
REC_BOXAREA	=OFF	REC_YDIFF	=OFF
REC_BXARATIO	=OFF		

The RAIL variable REC_NFEATS contains the number of features currently being used during recognition. This parameter is for display only and cannot be directly changed by the user.

Appendix K*
IRDATA

Introduction to IRDATA

Work on standardization of the interface for offline-programming of industrial robots in Germany began approximately four years ago. The work is based on the existing standards in the field of NC-technology.

These are:	ISO 840/	DIN 66024	NC-Code
	ISO 1056–1058/	DIN 66025	
	ISO 4343/	DIN 66215	CLDATA
		DIN 66267	DNC Protocol

An interface similar to the NC-CLDATA-code has been defined in the draft standard VDI 2863.

The IRDATA-interface (*I*ndustrial *R*obot-*DATA*) is defined open. Extensions to the existing instruction set for user specific requirements are possible.

In VDI 2863 part 1 the instruction and program structure had been defined, as well as the representation of IRDATA-code and data transfer between programming system and robot control based on the DIN 66267 DNC-proposal.

An alphanumeric representation has been choosen.

These standard should enable the robot user to easily combine his specific robot systems with available programming systems.

4 General Structure of IRDATA-Code

This standard does not focus to a particular type of controller (e.g. 16 bit). If there are any restrictions related to the possible range of values (Sects. 4.1.3, 4.2.2 and 4.2.3), they apply only to the implementation purposes. This increases the portability of IRDATA-code but should not imply a particular data format.

4.1 Structure of IRDATA-Code

4.1.1 IRDATA-Code

Each IRDATA-coded program consists of a sequence of records, terminated by semicolon. The records are specified in Chap. 6.

4.1.2 Record

Each record consists of a sequence of words. These words are seperated by commas.

* Extract from "VDI-Standard 2863, Part 1 (1963)" reproduced by kind permission of the publisher, VDI-Verlag, Düsseldorf. For applications, the latest version of the standard is essential and obtainable from Beuth-Verlag, Burggrafen-Strasse 4–10, 1000 Berlin 30.

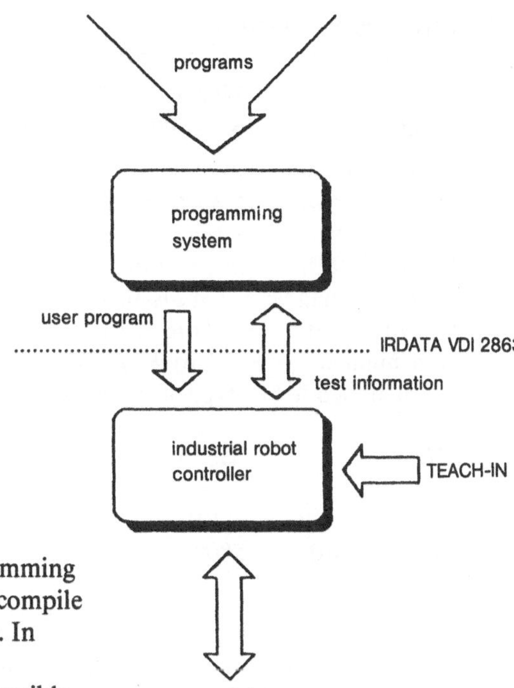

The controller independent programming system as shown in Fig. 1 is able to compile user programs into IRDATA-code. In addition, the interchange of test information in IRDATA-code is possible.

4.1.3 Word

Each word consists of a sequence of characters. It can be of the types

- INTEGER
- UNSIGNED INTEGER
- REAL
- STRING

The range of integer values according to 32 bits is:
- 2147483647 ... + 2147483647

The range of unsigned integer is:
0 ... 4294967295

The range of real values according to 32 bits is:
1.7 E 38 ... + 1.7 E 38

These ranges are only minimum values for each IRDATA implementation but there are no reglementations according to the precision of real values.

4.2 Composition of Records

4.2.1 Representation of Words

The representation of the words W1 to Wn of a specific record is as follows:

W1	serial record number (unsigned integer)
W2	instruction code, composed of type and code number (integer)
W3 to Wn	arguments according to the instruction code

4.2.2 Serial Record Number

The word W1 of a record is representing the serial record number. Data type of W1 is unsigned integer ranging from 1 to 65535. Within IRDATA-coded programs records are numbered consecutively.

4.2.3 Instruction Code

The instruction code W2 of each record is composed of the type of record and a code number. The type of record is representing a specific task oriented group of IRDATA-instructions. The code number classifies the particular instruction. Data type of W2 is unsigned integer ranging from 1000 to 32767.

type of record	=((Word W2)DIV 1000) × 1000
code number	=(Word W2)MOD 1000
instructioncode	=(type of record)+(code number)
DIV	= Integer devision
MOD	= Modulo division

Examples:
W2=5022

type of record	=5000
code number	= 22

4.2.4 Arguments

The words W 3 to Wn are representing arguments according to the instruction code W2. If not specified, the data type of W3 to Wn is always real.

4.3 Structure of Programs in IRDATA-Code

The structure of an executable IRDATA-code program is as follows:

1) Instruction code for the begin of program (PBEG, type of record = 22000).
2) Declaration part.
3) Executable part of the main program. It is included within a block.
4) Instruction code for the end of program (PEND, type of record = 22000)

4.3.1 Declaration Part

The declaration part consists of procedure- and task-declarations (see also type of record = 22000) as well as lists of data (type of record = 14000). This part could be empty. The sequence of declarations is not specified.

4.3.2 Executable Part

The executable part consists of a set of IRDATA instructions without declarations or lists of data. This part could be empty.

4.3.3 Declaration of Procedures and Tasks

A declaration of a procedure or task is composed of

1) the instruction code for the begin of a procedure or task (PRBEG, TSKBEG, type of record = 22000)
2) the executable part, which is part of a block, if the procedure or task has local variables or parameters
3) the instruction code for the end of a procedure or task (RETURN, TSKEND, type of record = 22000).

4.4 Type of Records

Record 1000

This record contains the linenumber and the designation of the user program statements together with additional user information.

Record 2000

This record contains technical specifications for the robot motion, e.g. speed, accelaration etc.

Record 5000

This record contains motion statements.

Record 9000

This record contains definitions of units of measurements, length and angles as well as scaling factors.

Record 14000

This record contains statements to transfer lists of data.

Record 17000

This record contains definitions of tool dimensions.

Record 19000

This record contains definitions of robot kinematic and its workspace.

Record 21000

This record contains statements for mathematical operations with reference to the data types described in Chap. 5.1.

Record 22000

This record contains statements for program structure, program flow and memory organization.

Record 23000

This record contains statements for communication with peripheral devices, operator panels and robot controllers.

Record 24000

This record contains definitions and instructions for tracking (see part 2).

Record 25000–27000

Reserved for future extensions.

Record 28000–32000

Reserved for user requirements and application specific statements.

5 Data Types and Adressing Methods

5.1 Data Types

According to this standard, the following data types as shown in Tab. 1 can be used in IRDATA-code. In addition, any new data types can be specified by user specific commands. The representation of these data types depends on the implementation and the control architecture and is not part of this standard. The specific codes to represent data types are unsigned integer.

Table 1. Data types

Data type	Abbreviation	Code	Description
BOOLEAN	B	129	A data type covering the logical values "TRUE" and "FALSE". These logic values are defined as follows: TRUE = 1, FALSE = 0
INTEGER	I	130	A data type covering a range of nonfractional numbers as defined in Chap. 4.1.3.
REAL	R	132	A data type covering a range of fractional numbers as defined in Chap. 4.1.3.
VECTOR	VEC	4	As data type composed of 3 real numbers describing a position by 3 cathesian coordinates as defined in Chap. 5.2, using the current unit of measurement.
	VECX VECY VECZ	5 6 7	These sub data types are defined to handle with particular vector components.
ORIENTATION	ORI	8	A data type composed of 3 real numbers describing the orientation of end effectors by 3 revolving angles relative to a cartesian coordinate system as defined in Chap. 5.2, using the current unit of measurement.
	ORIO ORIA ORIT	9 10 11	These sub data types are defined to handle with the orientation components.
ADDITIONAL AXIS	ADX	32	A data type composed of up to 31 real numbers describing the position of any additional axis by joint coordinates.
	ADX1 ⋮ ADX31	33 63	These sub data types are defined to handle with the particular ADX components.
WORLD	WLD	16	A data type composed of VECTOR and ORIENTATION. They define the position and orientation of end deffectors with reference to a cartesian coordinate system of Chap. 5.2, using the current unit of measurement.
	WLD POS WLD POSX WLD POSY WLD POSZ WLD ORI WLD ORIO WLD ORIA WLD ORIT	20 21 22 23 24 25 26 27	These data types are defined to handle with the WORLD components.
JOINT	JOINT	64	A data type of up to 31 real numbers describing a position with reference to a joint coordinate system. The kinematic structure is defined by a record with the instruction code 19010.
	JOINT1 ⋮ JOINT31	65 95	These sub data types are defined to handle with the JOINT components.

Table 1. (continued)

Data type	Abbreviation	Code	Description
CHARACTER	C	136	A data type ranging from 0...255. The first 128 values are representing the ASCII code (Chap. 7.1).
STRING	STR	144	A data type composed of an array of character. The index starts with 1, the possible length should be not less than 255.
POINTER	P	160	A data type for addressing memory locations. Only unsigned integers are allowed.
	up to	4095	Reserved for extensions
	from up to	4096 65535	User specified data types.

5.3 Adressing Methods

The following adressing methods are defined to handle with any kind of data in IRDATA records.

Argu-ment Wi	Argument Wi+1	Addressing method
1	INTEGER REAL STRING	Boolean or integer constant real-constant character- or stringconstant
2	bst*s+index	block relative addressing (unsigned integer)
4	adress	absolute addressing (unsigned integer)
9 10 12	0 0 0	constant on stack block relative address on stack absolute address on stack (POINTER)
16	STRING	symbol (name)

Explanations

a) Blockrelative addressing method:
bst = block indenting, unsigned integer 0...65535
index = index with reference to the address table, unsigned integer 0...65535
s = 65536 (2^{16})

b) Absolute addresses are calculated by means of instruction ABSADR (type of record 22000), which enables a general indirect addressing method.

c) The elements of the symbol character set is limited to ASCII-code greater than 31. The first element of a symbol is always a letter.

d) The string constant can be empty. The character constant is represented by a string with one element.

e) The organization of blockrelative addresses is performed by using a memory space independent address table as shown in Fig. 6. Several instruction codes are responsible for the reservation of memory space (BLBEG, DEFVAR and GENARR).

f) Addressing variables by symbolic names are considered as an extension to all other addressing methods and therefore not a minimum requirement.

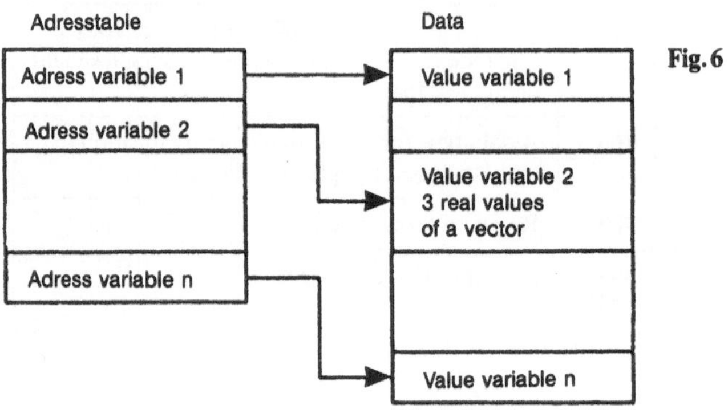

Fig. 6

Appendix L

Table of Comparison

Table of Comparison between AL, AML, PASRO, ROBEX, SIGLA, SRL and VAL

Language	AL	AML	PASRO	ROBEX	SIGLA	SRL	VAL
Simple motion instructions:							
– Specification of motor steps	–	–	–	–	MO/	–	–
– Specification of axis joints/distances	DRIVE	–	DRIVE DRIVE	DRIVE	–	DRIVE	DRIVE precise points (FRAME)
– Explicit Cartesian coordinates	FRAME	REAL	FRAME	FRAME	–	FRAME	
– Frame variables	yes	–	yes	only at compile time	–	yes	yes
– Relative motion	FRAME+ FRAME FRAME+ VECTOR	DMOVE	via computation	GODLTA	–	FRAME expression	DRAW
Motion instructions with parameters:							
– Velocity	SPEED-FACTOR	SPEED	SPEED-FACTOR	RAPID FEDRAT	–	VELOCITY	SPEED
– Duration	DURATION	–	*	–	–	DURATION	–

Table of Comparison between AL, AML, PASRO, ROBEX, SIGLA, SRL and VAL

Language	AL	AML	PASRO	ROBEX	SIGLA	SRL	VAL
– Robot configuration	–	–	–	–	–	POSTURE	RIGHTY, LEFTY, ABOVE, BELOW
– Control accuracy	NULLING NONULLING	SETTLE	–	EX	–	EXACT	COARSE, FINE, NONULL, NULL, INTOFF, INTON
– Parallel task execution	yes, generally	AMOVE WAITMOVE	*	**	yes, via tasks	yes, via SECTION	MOVET, MOVEST (gripping only)
– Superimposed weaving motion	WOBBLE	–	–	–	–	WOBBLE	WEAVE
– Intermediate frames	VIA	–	*	–	–	VIAFRAMES	CP
a. With change in speed	VELOCITY	–	–	–	–	VELOCITY	SPEED
b. With duration	DURATION	–	–	–	–	DURATION	–
– Approach frames	APPROACH	–	calculated	simulated	–	APPRO	APPRO
– Departure frames	DEPARTURE	–	calculated	simulated	–	DEP	DEPART
Motion instructions with sensory monitoring:							
– Force at gripper	FORCE	in MOVE	*	–	RP	ALWAYS WHEN	–
– Torque at gripper	TORQUE	in MOVE	*	–	–	ALWAYS WHEN	–
– Any other sensors	–	–	–	–	–	ALWAYS WHEN	–
Motion instructions with event monitoring:							
– Interrupts	EVENT-variable	–	*	EVENT-specifi-cation	–	ALWAYS WHEN interrupt	REACT (IGNORE)

Motion instruction with time monitoring	DURATION	–	*	–	–	ALWAYS WHEN	–
End effector control							
– Gripping	OPEN, CLOSE	in MOVE	GRIPOPEN, GRIPCLOSE, GRIPWIDTH	OPEN, CLOSE	–	OPEN, CLOSE, GRIPWIDTH	OPEN, CLOSE
– Execution of operations	–	–	–	–	–	OPERATE	–
Sensor instructions	–	yes	yes	limited	–	yes	–
Standard data types							
– Integer	SCALAR	INT	INTEGER	–	counter	INTEGER	integer
– Real	SCALAR	REAL	REAL	–	–	REAL	****
– Boolean	SCALAR	–	BOOLEAN	–	–	BOOLEAN	–
– Character	–	–	CHAR	–	–	CHAR	–
– String	–	STRING	***	–	–	STRING	–
– Event	EVENT	–	–	–	–	SEMAPHORE, INTERRUPT	–
Geometric data types	VECTOR, ROT, FRAME, TRANS	–	VECTOR, ROTATION, FRAME, THETAI	POINT, (VECTOR), frame	–	VECTOR, ROTATION, FRAME, ROBOTJOINTS	(POINTS)
Structured data types	ARRAY	aggregate	ARRAY, RECORD	–	–	ARRAY, RECORD	–
Arithmetic operations							
– Basic operations: +, −, *, /	yes	yes	yes	**	+, − only	yes	yes
– Relations	yes	yes	yes	**	limited	yes	yes
– Higher maths functions	yes	yes	yes	**	–	yes	–

Table of Comparison between AL, AML, PASRO, ROBEX, SIGLA, SRL and VAL

Language	AL	AML	PASRO	ROBEX	SIGLA	SRL	VAL
Flow control							
- Jumps	-	yes	restricted	yes	yes	-	yes
- Branching	yes	yes	yes	limited	limited	yes	limited
- Loops	yes	yes	yes	-	-	yes	-
Subroutines	yes	yes	yes	yes	yes	yes	yes
Procedures	yes	yes	yes	-	-	yes	-
Multi tasking	limited	-	*	limited	yes	yes	-

Type identifiers of the relevant languages are given here in capital letters.

* Implementation-dependent, e.g., in CONCURRENT PASCAL, MicroPower/Pascal or Modula-2

** Planned

*** Many PASCAL implementations include strings, even though they are not standard

**** VAL II allows scalar values, which cover integer and reals

Subject Index

Springer